AutoCAD® 2002
3D Modeling, A Visual Approach

AutoCAD® 2002
3D Modeling, A Visual Approach

JOHN WILSON

ALAN J. KALAMEJA

autodesk
press

Australia • Canada • Mexico • Singapore • Spain • United Kingdom • United States

autodesk Press

Autocad® 2002 3D Modeling, A Visual Approach
by John Wilson and Alan J. Kalameja

Autodesk Press Staff

Business Unit Director:
Alar Elken

Editorial Assistant:
Jasmine Hartman

Production Manager:
Larry Main

Executive Editor:
Sandy Clark

Executive Marketing Manager:
Mary Johnson

Production Editor:
Stacy Masucci

Acquisitions Editor:
James DeVoe

Marketing Coordinator:
Karen Smith

Art/Design Coordinator:
Mary Beth Vought

Development Editor:
John Fisher

Executive Production Manager:
Mary Ellen Black

Library of Congress Cataloging-in-Publication Data:

Kalameja, Alan J.
   AutoCAD 2002 : 3D modeling, a visual approach / Alan Kalameja, John Wilson.
      p.cm.
      ISBN 0-7668-3851-X
   1. Computer graphics. 2. AutoCAD. 3. Three-dimensional display systems. I. Wilson, John, 1933- II. Title.

T385 .K34382002
620'.0042'02855369--dc21

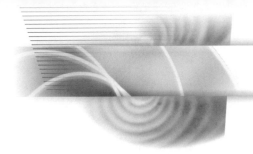

CONTENTS

INTRODUCTION

HOW TO USE THIS BOOK

AutoCAD 2002: 3D Modeling, A Visual Approach is a comprehensive, extensively illustrated guide to all of the 3D features of AutoCAD 2002. It uses numerous examples, demonstrations, and exercises in explaining how to manage 3D space; how to make 3D wireframe, surface, and solid models; how to modify them; and how to display and use them.

When you have finished this book, you will be able to construct any surface or solid model that AutoCAD is capable of making, create production drawings from the model, and make renderings of it. The book is logically divided into chapters covering 3D AutoCAD (introduction), Working in 3D Space, Wireframe Models, Surface Models, Solid Models, Creating and Editing Solid Models, Analyzing Solid Models, Paper Space and 2D Output, and Rendering. Each chapter builds on information in the previous chapters.

Each chapter is further divided into subjects that are based on 3D concepts. Within each subject:

- The 3D concept is explained.

- Every AutoCAD command for implementing that concept is fully described.

- Major commands are listed and reviewed.

- Related system variables are listed and described.

- Tips give you practical information for using the command – including suggestions, shortcuts, and warnings.

- Demonstrations and examples of how to apply the command are often given.

- Step by step instructions lead you through one or more practical exercises involving 3D models of real-world objects to help you understand the concept and give you experience in using the command.

Each chapter begins with a list of learning objectives and concludes with review questions. These review questions will reinforce and test your knowledge of the concepts and facts in the chapter.

THE BOOK'S CD-ROM

The accompanying CD-ROM contains AutoCAD 2002 drawing files for the exercises in this book, as well as examples of 3D objects. The name of the applicable file will be listed in the description for each exercise, and in the text that shows an example or describes the application of a command.

Use the Windows Explorer to copy these files to a folder of your choice on your computer. All of the copied files will have their read-only attribute set, because they are from a CD-ROM. You must clear the read-only attribute of each file, if you intend to work with and modify the file in AutoCAD. To clear the read-only attribute of a file, right-click the file's name in the Windows Explorer. In the menu that will appear, select Properties. A dialog box titles Properties will appear. In the General tab of that dialog box, clear the check box labeled Read-only, and then click the OK button.

The names of all of the files on the CD-ROM are given in the following list:

File Name	Description
3d_ch2_01	Simple wireframe with dimensions
3d_ch2_02	Bracket wireframe with dimensions
3d_ch2_03	Supplemental wireframe exercises 1 through 5
3d_ch3_01	Helix – rough and smoothed using the Helix.lsp routine
3d_ch3_02	Example of a closed spline
3d_ch3_03	Electronic display case wireframe
3d_ch3_04	Closed spline with two additional points
3d_ch3_05	Wireframe exercise of a sheet metal part
3d_ch3_06	Wireframe of a boat hull
3d_ch4_01	Table made with extruded objects
3d_ch4_02	Simple wireframe covered with 3D faces
3d_ch4_03	Display enclosure with 3D faces added

STYLE CONVENTIONS

To assist you in understanding the AutoCAD command line syntax, and related descriptions and comments, the following format is used in this book:

Convention: Command Names are in small caps

Example: The MOVE command

Convention: Menu names appear with the first letter capitalized

Example: Draw pull down menu

Convention: Toolbar menu names appear with the first letter capitalized

Example: Standard toolbar

Convention: Command sequences are indented. User inputs are indicated by bold-face. Instructions are indicated by italics and are enclosed in parentheses.

Example:

> **Command: MOVE**
>
> **Enter variable name or [?]: SNAPMODE**
>
> **Enter group name: (*Enter group name*)**

3D AutoCAD

LEARNING OBJECTIVES

This chapter will introduce the basic concepts of 3D and the 3D capabilities of AutoCAD. When you have completed Chapter 1, you will:

- Understand the differences between 2D drafting and 3D modeling.
- Understand the differences between wireframe, surface, and solid models, and know what rendering means.
- Know some of the advantages that 3D models have over 2D drawings, and some practical uses for 3D models.
- Be acquainted with some of the 3D capabilities and limitations of AutoCAD.

DIFFERENCES BETWEEN 3D AND 2D

To AutoCAD, there is no difference between 3D and 2D. AutoCAD is always fully 3D. For most users, however, there are significant differences between working in 3D and 2D. The fundamental difference is that 3D has another direction to work with—in addition to height and width, the objects you create have depth. One result of this extra dimension will be a change in your input methods. You will probably use object snaps and typed-in coordinates when specifying points and displacements more often in 3D work, even though AutoCAD has an assortment of tools to assist you in using your pointing device in 3D.

This extra dimension will also affect the way you look at the object you create. For 2D work, you invariably look straight down on the drawing plane, whereas in 3D you usually look at your object from an angle because it will often have objects directly over other objects. Furthermore, you will use a variety of viewing directions as you construct your model, and you are likely to have several viewports on your computer screen that simultaneously show your model from different viewpoints.

Another difference is that entities are more concentrated in 3D than they are in 2D.

Even though a 3D wireframe model may not have any more entities—lines, circles, and so on—than a 2D drawing of the same object, the entities will all be clustered in one location, rather than spread out in several different views. Surface models, with their surface mesh lines, can be especially congested.

To control these densely packed entities you will use more layers in 3D than in 2D, and you will be freezing and thawing layers more frequently. You will need all of the layers in 3D that you used in 2D, plus extra ones to help get entities out of the way as well as to selectively isolate pieces of the model for easier viewing and working and to control the visibility of the model's components.

An obvious difference between 3D and 2D is that there are over 80 AutoCAD commands related primarily to 3D for you to learn and use. Although a few of these commands—such as UCS, VPORTS, and those related to paper space—are also useful in 2D work, most are specifically for 3D objects.

These extra commands are used in addition to—not in place of—the 2D commands. In 3D, you will regularly use most of the commands you have been using for 2D work, plus the 3D commands. Moreover, some familiar commands, such as FILLET and CHAMFER, work in a different way when applied to 3D solids.

Last, your approach and your thinking as you build 3D models will be subtly different than when making 2D drawings. In 3D you are actually constructing an object, not just drawing views of it as if seen from different viewpoints. In some respects, 3D construction is more exacting than 2D drawing. You must be very precise in locating and positioning objects. This exactness can be an advantage, as it forces designs to be precise and accurate.

SOME FREQUENTLY USED 3D TERMS

Even though we will use a minimum of jargon as we explore the 3D features of AutoCAD, there are a few specialized terms that are unavoidable. In fact, we have already used some of these terms. Therefore, we need to define a few basic terms before going any further. Others will be defined as we cover their subject area.

MODEL

3D objects made in AutoCAD are generally called *models*—the same term used for clay, plaster, or cardboard representations of some real or planned object. Like those physical models, AutoCAD models are fully 3D and relatively easy to build and modify, but unlike the physical models, you cannot directly touch them.

The process of making a model is called *modeling*. Although the terms *model* and *modeling* can be applied to 2D AutoCAD objects, they are usually reserved for 3D models—with 2D work being referred to as *drawing* or *drafting*.

WIREFRAME MODEL

Wireframe models represent an object by its edges only. Nothing is between the edges. Therefore, wireframes cannot hide objects that are behind them. Figure 1.1 shows a simple wireframe model. It is made of 15 lines, with two circles representing the edges of a round hole. In actuality, a hole has no meaning in a wireframe model because there is nothing in which to make a hole.

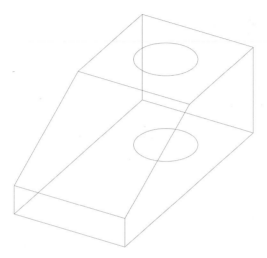

Figure 1.1

SURFACE MODEL

Surface models have an infinitely thin computer-calculated surface between their edges. Although they appear to be solid, they are an empty shell. The model on the left in Figure 1.2 shows the previous object as a surface model. There appears to be a round hole through the model. The hole is actually a tube simulating the surface of a hole, as can be seen in the center model, in which a surface panel has been removed.

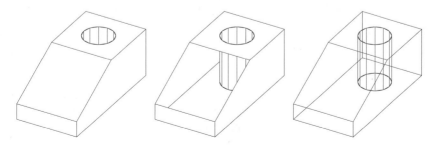

Figure 1.2

Surface models often use wireframe models as a frame for their surfaces; it is not unusual for models to be part wireframe and part surface. Because surfaces are transparent unless a command for hidden line removal has been invoked, surface models often look like wireframe models (as shown on the right in Figure 1.2). Furthermore, surface models are sometimes even referred to as wireframes. Although this is an incorrect designation, it is not uncommon, even in AutoCAD manuals and documentation.

SOLID MODEL

Solid models have both edges and surfaces, plus computer-calculated mass under their surfaces. The model on the left in Figure 1.3 is a solid model of the same object we've shown as a wireframe and as a surface model. Though it appears very similar to a surface model, it can be sliced in half, as demonstrated in the center model, to show that it is truly solid (in a computer-generated sense). Also, AutoCAD can report mass property information—such as volume, center of gravity, and mass moments of inertia—for the solid. As with surface models, solid models look like wireframes, as shown on the right, unless a hidden line removal command is in effect.

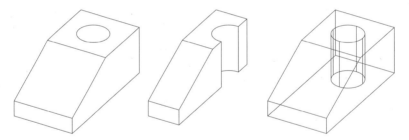

Figure 1.3

All solid models, no matter how simple or complex, are composed of simple geometric shapes or primitives. These primitives consist of boxes, cylinders, cones, wedges, spheres, and so on. Once created, these primitives are either merged or subtracted to form the final model. For the model at "A" in Figure 1.4, the solid modeling process begins by constructing the profile of the slab at "B" and then extruding this shape to form the base. A solid box is created and moved on top of the base at "C", where both shapes are joined to form a single solid. Another box is created and moved into position at "D". However instead of joining this primitive, a cutout is created by removing it from the solid. Holes are created by first creating a cylinder, moving it into position, and removing from the base at "E". Finally another cylinder is created and moved into position before it is removed creating the finished solid object at "F". Primitives such as these boxes and cylinders can actually be constructed right on the solid model. The moving of the primitives into position in this example were only used for illustrative purposes.

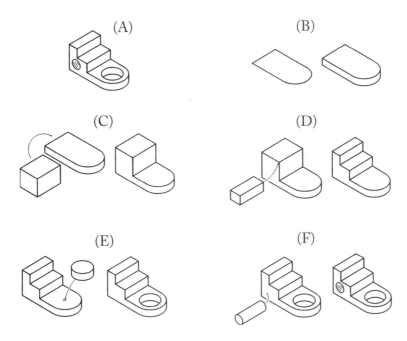

Figure 1.4

RENDERING

A shaded, realistic-looking picture of a surface or solid model is called a *rendering*. AutoCAD has a full set of commands for making renderings, including commands for installing and controlling lights and for manipulating the characteristics of a model's surfaces. An example rendering of a surface model is shown in Figure 1.5. Although this is a grayscale rendering, AutoCAD is fully capable of color renderings.

Figure 1.5

REASONS FOR USING 3D

The most obvious reason for making a 3D model is that it comes closer to representing the real object than a 2D drawing does. Even though a 3D computer model is a long way from reality, it is closer to reality than a 2D drawing.

Furthermore, you can transform 3D models into multiview, dimensioned 2D production drawings, thereby getting the best of both forms. Figure 1.6 shows a production drawing made from the surface model shown as a rendered image in Figure 1.5.

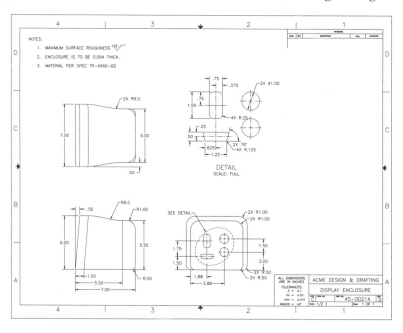

Figure 1.6

AutoCAD models can also be used directly in building objects without using a drawing. Solid models are generally better for this than surface or wireframe models; third-party programs are usually required to transform an AutoCAD model into a format for numerically controlled machining. AutoCAD's STLOUT command, however, exports a solid model to a file format compatible with a stereolithograph apparatus, often called rapid prototyping systems. These are machines that use computer data in making physical models from liquid resin.

A third reason for making 3D models is for the realistic renderings that can be made from them. Production drawings are invaluable, but renderings often show a design more clearly. They are good for spotting flaws and verifying a design, as well as for use in presentations and documentation.

3D CAPABILITIES OF AUTOCAD

AutoCAD started as a 2D drafting program. As the product continued to mature in 2D, a move to 3D began with the addition of a few 3D features. With an established 3D database, significant additions such as solid modeling and rendering were added. Now, in AutoCAD 2002, the program is a full-featured 3D modeling system.

- AutoCAD has a complete 3D coordinate system for specifying points and drawing objects anywhere in space.

- To assist in point input and for working in local areas, AutoCAD has a movable user coordinate system.

- You can set viewpoints from any location in space that can look in any direction.

- The computer screen can be divided into multiple viewports for simultaneously viewing 3D space from different viewpoints and different directions.

- AutoCAD has a good assortment of surface entities for making surface models that have a variety of shapes.

- Solid models of most objects typically manufactured in machine shops can be made within AutoCAD.

- 3D models can be transformed into standard multiview, dimensioned production drawings.

- AutoCAD has a built-in renderer with lights and surface materials, capable of making realistic-looking shaded pictures from 3D models.

3D LIMITATIONS OF AUTOCAD

The 3D features of AutoCAD have had the reputation of being difficult to learn and use, with few practical applications. The main reason for this reputation was probably due to the limited 3D capabilities of earlier versions of AutoCAD. At first, input was awkward and tedious, and there were few real 3D entities. Building even the simplest 3D model required major effort. Furthermore, output was so limited and primitive that even after a model was finished, there was little that could be done with it.

Although that reputation is no longer deserved, there are still some problems and limitations in working in 3D. First of all, virtually our entire interface with 3D models is through 2D devices. Our pointing devices—for example, whether we use a mouse or a digitizer tablet—are restricted to moving on a flat, 2D surface. Although there are some 3D digitizers, they are intended for obtaining point data on existing physical objects—there are no general input devices for pointing in 3D space.

A more severe restriction is the 2D computer screen we must use. Figure 1.7 is an example of what you might see on your computer screen as you work on the model shown in Figures 1.5 and 1.6. You have almost no sense of depth, and it is extremely difficult to discern which objects are in front and which are behind. The object may be 3D, but the image on your screen certainly is not.

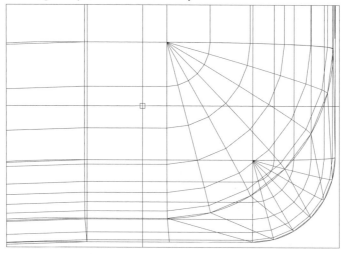

Figure 1.7

Problems related to 2D interface also extend to output. It is possible to create a physical 3D object directly from an AutoCAD 3D model, but the vast majority of output is on 2D paper. Multiview, dimensioned drawings are still the most used medium for transferring data from a 3D model into production. Although AutoCAD has a good assortment of tools for making standard engineering and architectural drawings from 3D models, the process is not completely seamless. More problems are associated with surface models than with solid models due to their surface mesh lines, and sometimes considerable effort is required to transform them into an acceptable format.

Figure 1.7 also illustrates how crowded the entities that make up 3D models can be. Entities are not only close together but some are even on top of others. This adds to visualization problems, and it sometimes makes it difficult to select objects. It is helpful to use layers so that only the objects you are currently working with are visible, but this requires extra steps, and you must be systematic and well organized.

Aside from visualization problems, AutoCAD surface objects themselves leave something to be desired. At their heart they are all three- and four-sided flat planes. Even rounded surfaces, such as the corners on the model shown in Figure 1.5, are faceted, as shown in Figure 1.8. You can control the size of these flat planes so that rounded surfaces can appear reasonably smooth, but the surface is still an approximation of a rounded surface. These facets not only affect the appearance of AutoCAD surfaces but also are a hindrance to using surface models for numerically controlled machining and other direct output.

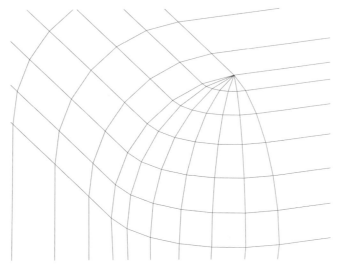

Figure 1.8

AutoCAD also has very few tools for editing or modifying surface objects. You cannot trim, extend, break, chamfer, or fillet them. There is no way, for example, to make a hole, either round or rectangular, through an existing surface. You will have to erase the surface, then make an edge for the hole, and build a new surface around that edge.

Although AutoCAD's solid modeling capabilities are good, they are not as good as those of several other programs. AutoCAD, for example, cannot create solid models based on true helix or spiral curves; it cannot create a solid model from a set of profiles, such as those of a boat hull or a propeller blade; and it cannot create variable radius fillets. Moreover, AutoCAD's editing capabilities of solids are limited and not intuitive. For example, a fillet can be deleted. However, you can change its radius only by offsetting its surface, and offsetting the surface outward to make the radius larger is not allowed. Also, you cannot use surfaces in AutoCAD to work on solids or visa versa. They can co-exit on the same drawing but only as individual objects.

AutoCAD does not have parametric solid modeling capabilities, as do the Autodesk Mechanical Desktop and Inventor products. The difference between parametric and nonparametric solid models is that dimensions *control* sizes and geometry in parametric models, whereas dimensions merely *report* sizes and geometry in nonparametric models. Parametric models are edited by changing dimensions. Therefore, if you wanted to increase the size of a round hole, you would increase the value of its diameter dimension. The size of the hole would then automatically change to match the new dimension value. While parametric models are flexible and easily modified; they must be constructed according to a complicated set of rules called constraints.

Finally, whereas AutoCAD's renderer has enough features to make realistic images of 3D models, it does not have any animation capabilities. If you need movement in renderings—perhaps to show how a certain mechanism works or for a walk-through presentation of a building design—you must use another rendering program that works within AutoCAD or export the AutoCAD model to another program.

Of all these 3D limitations, those related to 2D interfaces and visualization will probably be the most frustrating. Nevertheless, with practice, you will soon become proficient. AutoCAD has all of the tools you need to build useful, practical 3D models that have almost any geometry. After you learn to use these tools, using AutoCAD to make 2D drawings will seem as backward as using a drafting board and pencils seems to you now.

CHAPTER REVIEW

Directions: Circle the letter corresponding to the correct response in each of the following.

1. When you construct 3D models, you do not use any of the AutoCAD commands commonly used in making 2D drawings.

    a. true

    b. false

2. 3D models do not need to be as precise as 2D drawings.

    a. true

    b. false

3. It is not possible to create a 3D model that is part wireframe and part surface model.

    a. true

    b. false

4. A surface model has mass.

    a. true

    b. false

5. Although AutoCAD has full 3D capabilities internally, one's interface with 3D models is generally through 2D devices.

    a. true

    b. false

6. Rounded surfaces on AutoCAD surface models are approximated by small three- and four-sided flat faces called facets.

    a. true

    b. false

7. One of the important capabilities of AutoCAD surface models is that they can easily be modified and edited.

    a. true

    b. false

8. The removal of hidden lines is not possible on a wireframe model.

    a. true

    b. false

9. You must implement a special mode within AutoCAD before you can create 3D models.

   a. true

   b. false

10. You can create parametric solid models with AutoCAD.

    a. true

    b. false

Directions: Fill in the blanks as indicated for each of the following.

11. Identify the following characteristics with the appropriate 3D model type—wire frame, surface, or solid.

   _____ a. This type of model has opaque faces yet is considered hollow.

   _____ b. This model type can be cut into pieces.

   _____ c. This type of model cannot be rendered.

   _____ d. Models of this type are represented only by edges.

   _____ e. This model type has volume and mass associated with it.

Working in 3D Space

LEARNING OBJECTIVES

Chapter 2 will describe and show you how to use AutoCAD's tools for working in 3D space. When you have completed this chapter, you will:

- Know the properties of AutoCAD's world coordinate system (WCS), how to specify point locations in it, and how to orient 3D models in it.

- Know how to manage and interpret AutoCAD's user coordinate system (UCS) icon.

- Be able to move around in 3D space and establish viewpoints that look in any direction.

- Be able to construct 3D models with 2D drawing techniques through AutoCAD's user coordinate system (UCS).

- Know how to set up and work with multiple viewports.

3D COORDINATE SYSTEMS

Ordinary 1D points are at the heart of every AutoCAD object. A line starts at one point and ends at a second point. The center of a circle is a point. Even complex curves twisting and turning through space are defined by equations based on points.

AutoCAD uses a 3D rectangular coordinate system for designating the locations of these points in space. It is often called the Cartesian coordinate system, after the French mathematician René Descartes (1596–1650), who is credited with its development.

In this 3D coordinate system:

- There are three axes labeled X, Y, and Z, which are perpendicular to one another.

- These axes meet at a common point called the origin.

- Each axis is double ended, having a positive and negative direction beginning at the origin.
- The positive direction of the Z axis is determined by the right-hand rule, explained shortly.
- The location of a point in space is stored as three numbers, separated by commas in the form: x,y,z. Where:

  x is the point's distance from the origin in the X direction.

  y is the point's distance from the origin in the Y direction.

  z is the point's distance from the origin in the Z direction.

- The origin is often referred to as the 0,0,0 point.

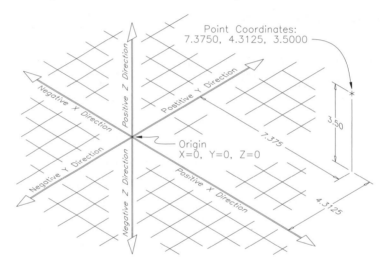

Figure 2.1

As shown in Figure 2.1, a point having the coordinates of:

7.375, 4.3125, 3.50

means:

- 7.375 units in the X direction from the origin;
- 4.3125 units in the Y direction from the origin; and
- 3.50 units in the Z direction from the origin.

The plane used for 2D drafting in AutoCAD is simply one plane—the XY plane—within this 3D coordinate system. Nevertheless, the XY plane is important; many AutoCAD prompts and messages for 3D commands still refer to it. Furthermore, AutoCAD's dot grid and snap mode work only on the XY plane. Some AutoCAD objects, such as circles and 2D polylines, can only be drawn on the XY plane or on a plane parallel to it.

THE RIGHT-HAND RULE

When you look straight toward the XY plane, you know that because it is perpendicular to the XY plane, one end of the Z axis is pointed directly toward you. But how do you know whether it is the axis's positive or negative end that is pointed toward you? Physically, it could point in either direction. Therefore, some widely accepted convention must be used to establish the direction of the positive end of the Z axis relative to the XY plane. The convention AutoCAD uses is called the right-hand rule.

Of course there are mathematical definitions of the right-hand rule, but it is easiest to visualize the rule by using your own right hand. One popular technique for illustrating the right-hand rule is, as shown in Figure 2.2, to point the thumb of your right hand in the positive direction of the X axis, the index finger of your right hand in the positive direction of the Y axis, and bend your other fingers in the positive direction of the Z axis.

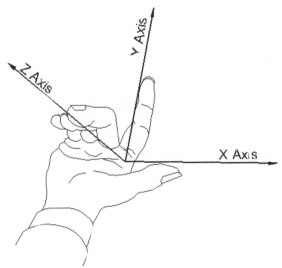

Figure 2.2

Although it is not especially important for you to be able to identify right- and left-handed coordinate systems, it is important for you to easily recognize the direction of the Z axis. In other words, you need to know which end is up. As we'll discuss later, AutoCAD provides some help in determining the orientation of the Z axis through its coordinate system icon.

MODEL SPACE COMPARED TO PAPER SPACE

AutoCAD refers to the 3D space we've just described as model space. AutoCAD has another form of space called paper space, which has only a 2D XY plane, rather than a complete 3D X,Y,Z coordinate system. The main purpose of paper space is to make 2D drawings from 3D models. Paper space will be discussed completely in Chapter 8, but we mention it here because you are likely to see references to it in some AutoCAD messages and documentation.

SPECIFYING POINTS IN 3D SPACE

Because all AutoCAD objects—lines, text, circles, splines, or surfaces—are drawn by specifying their key points, AutoCAD offers several different methods to enter point locations.

POINTING DEVICES

The most convenient way to enter point locations is to point to them with a mouse or a digitizer puck. Pointing devices, however, can only locate points that lie on the XY plane. They can break out of the XY plane only through object snaps, such as an entity endpoint, on an existing object. Consequently, in 3D modeling, you will end up physically entering some point locations by typing them in with the keyboard.

ENTERING X,Y,Z COORDINATES

When entering point locations from the computer keyboard, you will use X,Y,Z coordinates most of the time. The three numbers must be separated with commas, although you can type in just two numbers. When the third number is omitted from the coordinates of a point, AutoCAD assumes the point is on the XY plane, and sets the value of the Z coordinate to 0 or on a plane parallel to the XY plane if an elevation other than 0 has previously been established. Either absolute coordinates (based on the coordinate system origin) or relative coordinates (based on the last point entered) may be used. Relative coordinates are preceded with the @ symbol.

USING POINT FILTERS

Most of the time, you will specify all three coordinates of a point simultaneously either by entering the coordinates on the command line, by using an object snap, or by picking a location with your pointing device. You can, however, use point filters to specify one or two of the three coordinates separately. These filters can be used any time that AutoCAD expects a point by entering a period followed by the coordinate you want to filter out. Thus, you would type .x to filter the X coordinate. AutoCAD will respond by prompting for a point that can supply the filtered coordinate. Often, you will use an object snap to do that. Then, AutoCAD will prompt for the missing coordinates, and you can use filters for them also. In addition to the .x, .y, and .z filters to filter one coordinate, you can use .xy, .xz, and .yz to filter two points at a time.

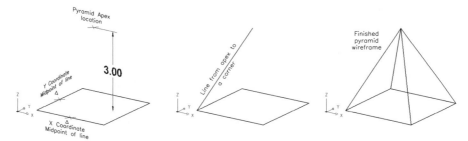

Figure 2.3

Suppose, for example you are constructing a wireframe pyramid. The apex of the pyramid is to be centered within the base and 3.0 units above it. You have drawn the base of the pyramid, as shown on the left in Figure 2.3. You will now use the following command line input to draw a line from the apex to one of the base corners:

Command: LINE (*Press ENTER.*)

Specify first point: .x (*Press ENTER.*)

of (*Use a midpoint object snap on the horizontal line.*)

of (*need YZ*): .y (*Press ENTER.*)

of (*Use a midpoint object snap on the vertical line.*)

of (*need Z*): 3.0 (*Press ENTER.*)

Specify next point or [Undo]: (*Use an endpoint snap on any of the existing lines.*)

Specify next point or [Undo]: (*Press ENTER.*)

Notice that AutoCAD uses the word of to prompt for the filtered coordinate, and after you specify a filtered coordinate, AutoCAD prompts for the remaining coordinates. Now that one line has been drawn from the apex of the pyramid to one corner, you can use endpoint object snaps to draw the three remaining edges of the pyramid.

ENTERING CYLINDRICAL COORDINATES

The points for some 3D objects—those based on helixes and spirals for instance—are more easily entered using cylindrical coordinates. Cylindrical coordinates do not represent a different coordinate system; they are merely another way to specify 3D point locations.

The format for absolute cylindrical coordinates is:

D<A,Z

where   D is the point's distance on the XY plane from the origin;

A is the point's angle on the XY plane from the X axis;

Z is the point's distance from the XY plane.

You will no doubt notice that cylindrical coordinates are 2D polar coordinates with the addition of a Z dimension. Figure 2.4 shows the location of a point having the absolute cylindrical coordinates of 3<45,2. Of course, you may precede cylindrical coordinates with the @ symbol to specify that the point is relative to the last point entered.

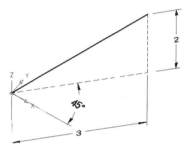

Figure 2.4

ENTERING SPHERICAL COORDINATES

Spherical coordinates use a distance and two angles in specifying a point location. Their format is:

<HA<VA

where   D is the point's straight line distance from the origin;

HA is the point's angle on the XY plane from the X axis; and

VA is the point's vertical angle from the XY plane.

A point with the spherical coordinates of 4<45<35 is shown in Figure 2.5. Notice that distance is measured from the origin to the point—it is not a horizontal distance on the XY plane as it is for cylindrical coordinates. Spherical coordinates can be useful when you want to establish a point that is a set distance and angle from an existing point.

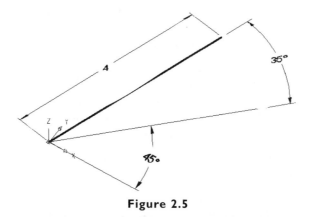

Figure 2.5

THE ROLE OF THE USER COORDINATE SYSTEM

Although it is necessary to have a global coordinate system—in which every point in space is tied to a single origin—when constructing 3D models, it is often convenient to have a local coordinate system that can be tied to a particular object in space. Consider, for instance, a box located in 3D space and twisted relative to the global coordinate system, as shown in Figure 2.6. Each corner of this box will have coordinates relative to the global coordinate system origin. However, if you were to measure objects on or in the box, you would probably ignore the global coordinate system and base your measurements on one of the box's corners. This would be a local coordinate system.

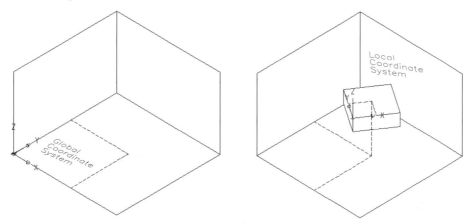

Figure 2.6

AutoCAD has a local coordinate system. It has the same features as the global coordinate system, but it can be moved and twisted in space to suit your needs. AutoCAD calls this movable, local coordinate system the user coordinate system (UCS) and refers to the fixed, global coordinate system as the world coordinate system (WCS). We'll fully explain the UCS and the commands that manage it later in this chapter, but in the meantime, you will see references to the WCS and UCS in some AutoCAD command prompts and options.

THE USER COORDINATE SYSTEM ICON

AutoCAD provides an icon to help you stay oriented in 3D space. This icon, often located in the lower left-hand corner of the viewport, is invisible to virtually all AutoCAD operations. It cannot be plotted or printed directly or included in an entity selection set.

A surprising amount of information, as shown in Figures 2.7, 2.8 and 2.9, is contained in the icon:

- The direction of the X and Y axes are shown by the straight lines with arrows labeled X and Y.

- When the UCS is exactly the same as the WCS, a box appears at the intersection of the X and Y axes. If the box is not shown, then the UCS has been moved or twisted in relation to the WCS.

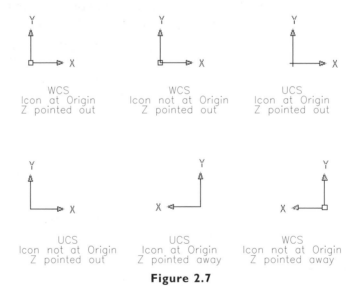

Figure 2.7

When viewing the UCS from different viewpoints in Figure 2.8, it is much easier to determine whether the Z axis is pointing toward or away from you. A positive Z direction displays the UCS icon with a solid line for the Z axis. The UCS icon identifies a negative Z direction (the Z axis pointing away from you) with a dashed line for the Z axis.

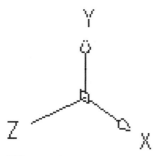
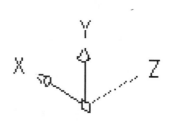

Z pointed out   Z pointed away
(Positive direction)   (Negative direction)

Figure 2.8

Illustrated in Figure 2.9 is an example of the UCS icon attached to a corner of an object. Since a solid line identifies the Z axis, the icon is pointing towards you.

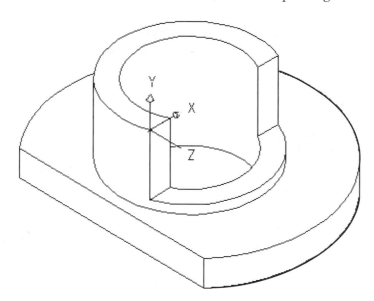

Figure 2.9

In addition to the XY arrows and the broken pencil forms, the UCS icon assumes a drafting triangle form when AutoCAD operates in paper space, and it looks like a 3D cube when AutoCAD is in the perspective mode. Also, the icon becomes three colored arrows in certain viewing modes. We will describe these icon forms later when we cover their subjects.

THE UCSICON DIALOG BOX

The command that controls the UCS icon is UCSICON. This command turns the icon on and off and controls whether or not the icon is positioned at the coordinate system origin. It does not set the location of the UCS. When the graphics screen is divided into several viewports (which we will cover later in this chapter), the settings for the UCS icon can vary by viewport. This command can also be selected from the View pulldown menu in Figure 2.10.

Figure 2.10

The command line format for the command is:

Command: UCSICON

Enter an option [ON/OFF/All/Noorigin/Origin/Properties] <ON>:
  (Select an option or press ENTER.)

ON

Causes the UCS icon to be displayed. Select this option by pressing ENTER or by typing in the entire word, ON.

OFF

Turns the icon off—it will not be displayed. At least the first two letters of OFF must be typed in for AutoCAD to distinguish it from ON.

ALL

This option causes changes to the icon to apply to all viewports. When it is selected, AutoCAD will display the following prompt:

Enter an option [ON/OFF/All/Noorigin/ORigin] <ON>: (*Select an option or press ENTER.*)

The effects of these options are the same as those in the main UCSICON menu.

NOORIGIN

Forces the icon to be displayed in the lower left-hand corner of the viewport, regardless of the actual location of the UCS origin.

ORIGIN

Locates the icon at the UCS origin whenever possible. If the origin is not in the viewport or if it is so near the edge that the icon will not fit, the icon will be shown in the viewport's lower left-hand corner.

Compare Figure 2.11 (Noorigin) with Figure 2.12 (ORigin) for the results of these options.

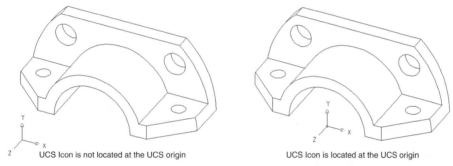

UCS Icon is not located at the UCS origin   UCS Icon is located at the UCS origin

Figure 2.11   **Figure 2.12**

Tip: Most of the time you will want the UCS icon to be displayed and positioned at the UCS origin. However, if you are working close to the origin and the icon obscures your work, temporarily turn it off or force it away from the origin with the Noorigin option of UCSICON.

PROPERTIES

Displays the UCS Icon dialog box in Figure 2.13. Through this dialog box, you can control the line width of the icon along with its size. You can even change its color if desired. Illustrated in the Preview panel is the current state of the icon. The UCS Icon can be displayed in either a 3D or 2D state. The 2D state of the UCS icon will appear flat compared to its 3D counterpart. Examples of both icon states are also displayed in Figure 2.13.

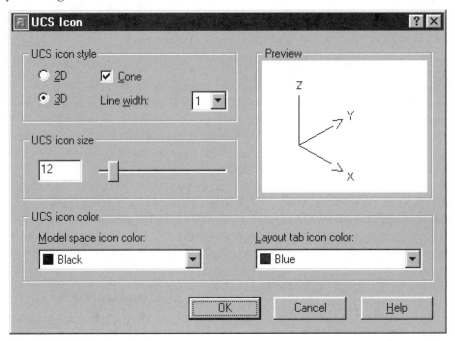

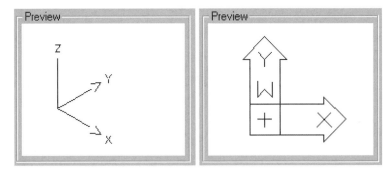

Figure 2.13

ORIENTING MODELS IN 3D SPACE

In the real world you can orient objects in almost any direction you want. You can, for instance, have the front of a house face directly east, directly south, or any angle in between. Gravity and the earth, however, control the orientation of the house's roof and foundation. In AutoCAD's 3D world you can also orient objects in any way you want, but you don't have gravity or the earth to contend with.

Nevertheless, you should be consistent in the orientation of your 3D models and follow some established conventions for positioning them. An important reason for this is that AutoCAD's 3D commands, menus, and documentation are often in terms of front, top, side, and elevation. A less obvious reason is that most AutoCAD commands for setting view direction are relative to the WCS. It is easy to become disoriented if your model is not set up with the WCS in mind. Figure 2.14 shows examples of proper and improper orientation of a 3D model.

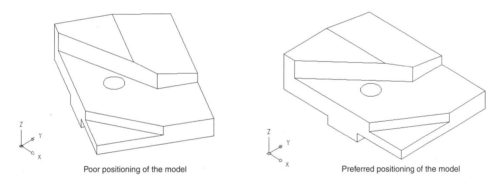

Poor positioning of the model   Preferred positioning of the model

Figure 2.14

These orientation conventions are:

- Position the model so that as many of its flat sides as possible are parallel with the X,Y,Z axes. If the model is cylindrical, position its axis parallel with one of the WCS axes.

- The top of a model is that portion seen when looking straight down toward the XY plane from the positive Z direction. In other words, up is in the positive Z direction; down is in the negative Z direction. AutoCAD often uses the word elevation when referring to the Z direction (see Figure 2.15).

- The front of a model is the portion seen when looking in the positive Y direction.

- The right side of a model is that side seen when looking in the negative X direction.

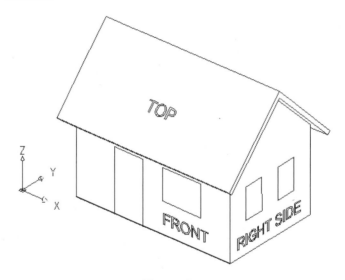

Figure 2.15

AutoCAD menus occasionally use compass direction terms—with north equal to the positive Y direction, south equal to the negative Y direction, east equal to the positive X direction, and west equal to the negative X direction (see Figure 2.15).

It is very important to properly position the model in the beginning stages of construction. For one, viewing commands will be difficult to use if the model is not constructed in the correct orientation. The top or bottom of the model should be constructed on the WCS (World Coordinate System). To work on a side of the model not parallel to the top or bottom (front or side), the UCS must be re-positioned. Constructing of the right side of the house in Figure 2.15 is an example of where this would be used. The UCS icon would be rotated 90 degrees about the X and Y axes.

This will be explained later on in this chapter.

SETTING VIEWPOINTS IN 3D SPACE

In 2D drafting with AutoCAD, virtually everything is done while looking straight down on the XY plane. In 3D modeling, however, you need a variety of viewpoints. Sometimes you need to see the left side of your model, at other times you need to see its front, and much of the time you will work from a viewpoint that allows you to see three sides at once. To accommodate this need, AutoCAD has an assortment of commands for setting viewpoints from any direction in 3D space.

- VPOINT sets general viewpoints in 3D space. Generally, command line input is used to set the viewpoints.
- PLAN sets viewpoints that are perpendicular to the XY plane.
- VIEW restores previously set viewpoints that have been named and saved.
- DVIEW sets general viewpoints in 3D space, plus perspective views and views with clipping planes that hide portions of a 3D model.
- 3DORBIT dynamically sets general viewpoints in space.

We will discuss the VPOINT, PLAN, and VIEW commands in this chapter. Because DVIEW is more useful in surface modeling than in wireframe modeling, we will discuss it in Chapter 4.

As you will see later in this chapter, AutoCAD also has the ability to divide your computer screen into several different viewports, each showing the model from a different viewpoint. All of the commands for setting viewing directions apply only to the viewport in which you are currently working when multiple viewports exist.

THE VPOINT COMMAND

The VPOINT command sets 3D viewing angles in the current viewport. The viewpoint is relative to the world coordinate system, even if a user coordinate system (UCS) is currently being used. If a UCS is being used, AutoCAD will switch to the WCS for the duration of the VPOINT command. The only exception to this is when the Worldview system variable is set to 0; then the viewpoint is set according to the UCS. The format for the VPOINT command is:

Command: VPOINT

Current view direction: VIEWDIR=current view direction coordinates

Specify a viewpoint or [Rotate] /<display compass and tripod>: (*Enter three numbers, R, or press ENTER.*)

COORDINATES

The view direction coordinates, which are three numbers separated by commas, represent the X,Y,Z coordinates of a direction from the model to the viewing point. The coordinates for the current view direction are shown on the command line, even if the Rotate or Compass and Tripod options were used to set the last view direction. Since the coordinates represent a direction rather than a point in space, the relationship of the three coordinate numbers to each other is important, but their magnitude is not. As shown, for example, in Figure 2.16 the coordinates of 5.2654,-6.2751,5.7358 result in the same view as the coordinates of 0.5265,-0.6265,0.5736. These numbers have nothing to do with zoom levels.

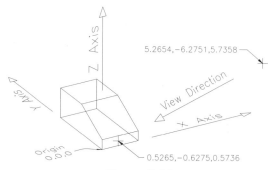

Figure 2.16

ROTATE

Pressing the R key initiates the Rotate option, bringing up the following prompts:

Enter angle in XY plane from X axis <current>: (*Specify an angle.*)

Enter angle from XY plane <current>: (*Specify an angle.*)

Although input to both prompts can be by pointing, you will probably prefer to type in the angles. The first angle is measured horizontally on the XY plane from the X axis, whereas the second angle is measured vertically from the XY plane. Since angle measurements wrap around, angle inputs of -45°, 315°, and 675° are the same. Angles from the existing viewpoint are offered as defaults, even if the previous viewpoint was set by the Coordinate option or by the Compass and Tripod option.

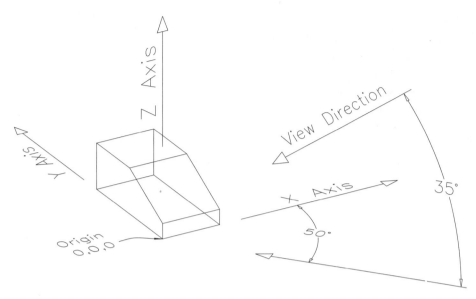

Figure 2.17

Figure 2.17 shows a viewing direction based on the angles of -50° (clockwise) in the XY plane and 35° from the XY plane. This particular pair of rotation angles is equivalent to the view direction coordinates used in the previous example. The resulting view of the model is shown in Figure 2.18.

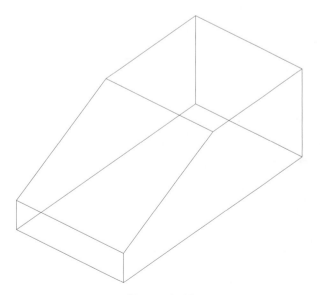

Figure 2.18

COMPASS AND TRIPOD

When you press ENTER in response to VPOINT's initial prompt, the screen switches to a display, shown in Figure 2.19, of an icon representing the coordinate system's X,Y,Z axes; plus another icon, in the upper right-hand corner of the screen, consisting of a circle with a smaller circle inside, both of which are divided into four quadrants. There is also a small crosshair cursor that you can move about with your pointing device. The UCS icon is also shown; however, this icon is static and plays no role in setting the viewpoint. Autodesk calls this screen the Compass and Tripod display. Moving the crosshair cursor within the compass (the concentric circles) sets a viewpoint as shown in Figure 2.20.

Figure 2.19

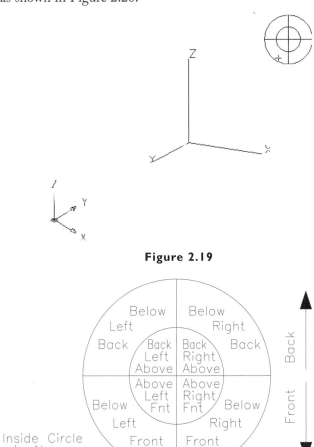

Figure 2.20

As the cursor is moved about, the axis tripod twists in space to indicate the corresponding viewpoint. When you have the viewpoint you desire, press the pick button on your pointing device to complete the command and return to the AutoCAD editing screen.

Tips: It is impossible to set an exact viewpoint using the Compass and Tripod. Consequently, you are not likely to find this option of VPOINT nearly as useful as the other two.

The Coordinate option of VPOINT is especially convenient for setting an exact viewpoint in a cardinal direction that shows a model's top, front, or side. Rotation angles can also be used, though they do require a little more typing. Table 2.1 lists the VPOINT X,Y,Z coordinates along with the VPOINT rotation angles for setting view directions for the six standard orthographic views.

Table 2.1 Coordinates and Angles for Orthographic Viewpoints

View	X,Y,Z Coordinates	Rotation Angles From X Axis	From XY Plane
Plan (Top)	0,0,1	270°	90°
Front	0,-1,0	270°	0°
Back	0,1,0	90°	0°
Right side	1,0,0	0°	0°
Left side	-1,0,0	180°	0°
Bottom	0,0,-1	270°	-90°

Setting exact isometric viewpoints, in which the horizontal angles appear to be slanted 30°, can be done using either the coordinates or the rotation angles shown in Table 2.2. Although these viewpoints give a view that appears to be identical to AutoCAD's pseudo-3D isometric snap mode, you do not need to use isometric ellipses to make circles and arcs or use the ISOPLANE command when you work in 3D.

Table 2.2 Coordinates and Angles for Standard Isometric Views

View	X,Y,Z Coordinates	Rotation Angles From X Axis	From XY Plane
Upper front-right	1,-1,1	315°	35.2644°
Upper front-left	-1,-1,1	225°	35.2644°
Upper back-right	1,1,1	45°	35.2644°
Upper back-left	-1,1,1	135°	35.2644°
Under front-right	1,-1,-1	315°	-35.2644°
Under front-left	-1,-1,-1	225°	-35.2644°
Under back-right	1,1,-1	45°	-35.2644°
Under back-left	-1,1,-1	135°	-35.2644°

In AutoCAD 2002, the View/3D Views pull-down menu and the View toolbar use the VIEW command, rather than VPOINT, to set orthographic and isometric viewpoint. As a result, the UCS will sometimes be automatically reoriented to match the viewpoint. The relationship of the VIEW command to viewpoints will be discussed later in this chapter.

Although AutoCAD has numerous tools for setting orthographic views, these views are seldom a good choice as you build 3D models. They provide no sense of depth, and objects will often be stacked on top of one another, impairing both visualization and object selection. Exact isometric views are also awkward to work with sometimes, especially when the model has square cross-sections, as the one in Figure 2.21 has.

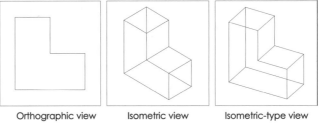

Orthographic view   Isometric view   Isometric-type view

Figure 2.21

THE DDVPOINT COMMAND

The DDVPOINT command is a dialog box version of the Rotate option of VPOINT, combined with the PLAN command. Its dialog box is shown in Figure 2.22. The command can be invoked from the View menu by choosing 3D Views and then Viewpoint Presets. The dialog box can also be displayed by typing in DDVPOINT on the command line.

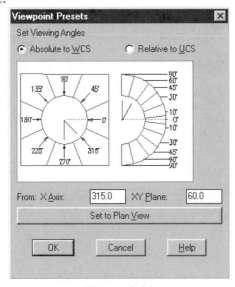

Figure 2.22

Set Viewing Angles:

Absolute to WCS   Sets the view direction relative to the world coordinate system.

Relative to UCS   Sets the view direction relative to the current user coordinate system.

From:

X Axis: The rotation angle on the XY plane from the X axis. You can type in an angle or use the image tile on the left.

XY Plane: The vertical rotation angle from the XY plane. You can type in an angle or use the image tile on the right.

The red (dotted) arm in each image tile indicates the current angle while the black (solid) arm indicates the new angle. If you pick in one of the outer bounded areas, the angle snaps to the labeled value. Picking within the inner areas will set a viewing angle between the labeled angles, although it is difficult to set a precise angle by pointing. The angle boxes below the image tiles show the selected angles.

Set to Plan View:

This button will set a plan view relative to the coordinate system selected by Set Viewing Angles.

THE 3DORBIT COMMAND

3DORBIT allows you to use your pointing device to dynamically set viewpoints in real time. When you start 3DORBIT, the UCS icon will change from its flat X and Y axes form to that of three cylindrical arrows, the screen cursor will change to one of four types that will be described shortly, and a large circle will appear in the center of the current viewport, as shown in Figure 2.23. AutoCAD refers to the large circle as the arcball. Also, if the AutoCAD grid is turned on, its appearance will change from a matrix of dots to a grid of lines on the XY plane.

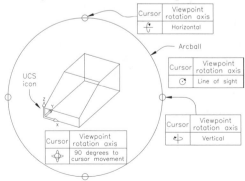

Figure 2.23

You set viewpoints with 3DORBIT by rotating the line of sight about an axis; and you rotate the line of sight by moving the screen cursor as you hold down the pick button of your pointing device. When you have the viewpoint you desire, release the pick button to set the viewpoint. The angle of the viewpoint rotation axis, as well as the appearance of the screen cursor, depends on the location of the cursor when you press the pick button.

- If you press the pick button when the cursor is within the arcball, the screen cursor's appearance will change to a sphere surrounded by a horizontal circle, and the viewpoint rotation axis will be perpendicular to the movement of the cursor. For instance, if you drag the cursor up and to the right 45°, the viewpoint will rotate about an axis that is tilted 135° from the horizontal edges of the viewport.

- When you press the pick button as the cursor is outside the arcball, the screen cursor will change to a dot surrounded by an arc-shaped leader, and the viewpoint rotation axis will be the line of sight.

- When you press the pick button as the cursor in within either of the small circles at the top and bottom of the arcball, the screen cursor will change to a horizontal line that serves as the axis of an arc-shaped leader. The viewpoint rotation axis is parallel with the horizontal edges of the viewport.

- Pressing the pick button when the cursor is within either of the small circles on the right and left quadrants of the arcball will cause the screen cursor to change to a vertical line encircled by an arc-shaped leader. The viewpoint rotation axis is parallel with the vertical edges of the viewport.

3DORBIT has a right-click shortcut menu for initiating commands and selecting options related to views and visualization. The items in this menu that are applicable to wireframe models are:

Exit	**Exits the 3DORBIT command. You can also exit 3DORBIT by pressing the ENTER or ESC keys.**
Pan	**Initiates a real-time pan operation. Press the pick button of your pointing device, drag the screen cursor to relocate the position of the 3D model within the viewport, and release the pick button to set the position.**
Zoom	**Initiates a real-time zoom operation. Press the pick button of your pointing device and drag the screen cursor up to enlarge the image of the model, or down to decrease the size of the image.**
Orbit	**Returns to the 3DORBIT command after performing a menu operation, such as Pan or Zoom.**
Reset View	**Restores the viewpoint, zoom level, and image location to that which exist-ed when the 3DORBIT command was initiated.**
Preset Views	**Sets one of the six orthographic views, or one of the four isometric**

views that look down on the XY plane.

The other items in this menu are more applicable to solid and surface models than to wireframe models, so we will not describe them until Chapter 4.

 Tip: When you first begin using 3DORBIT, you may find it best to use only the rotation axes defined by the small circles on the quadrants of the arcball, to avoid skewing the viewpoint into a confusing angle. You could, for example, start from one of the small circles on the right and left side of the arcball to rotate the viewpoint relative to the X axis, then start from one of the small circles on the top and bottom of the arcball to rotate the viewpoint relative to the XY plane.

USING THE 3DCORBIT COMMAND

3DCORBIT initiates a continuous rotation of the viewpoint. The screen cursor will change to a small circle, encircled by two arc-shaped leaders. You begin the rotation by pressing the pick button of your pointing device, dragging the cursor in any direction, and then releasing the pick button. The speed at which you drag the cursor sets the relative speed at which the viewpoint rotates. Click the pick button to end the rotation, and press the ENTER or ESC keys to exit the command.

USING THE 3DSWIVEL COMMAND

In setting viewpoints dynamically, you can imagine that you are looking at your 3D model through the viewfinder of a camera. When you use 3DORBIT, the camera is always pointed directly at the model as you move it about in 3D space, looking for a good viewpoint. On the other hand, when you use 3DSWIVEL the camera remains in one point and is swiveled, or rotated, about that one point. Therefore, as you rotate the camera to the left, the model will shift to the right in the viewfinder. And, as you rotate the camera down, the model shifts upward in the viewfinder.

THE COMPASS SYSTEM VARIABLE

This system variable can have a value of either 0 or 1. When it has a value of 1, three sets of short lines, with each set arranged in a circle whose plane is perpendicular to that of the other two sets, will be displayed within the 3DORBIT arcball. The planes of the circles represent the XY, ZX, and YZ planes. The compass is illustrated in Figure 2.24.

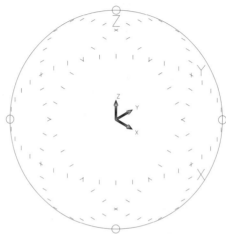

Figure 2.24

TRY IT! - SPECIFYING 3D POINTS AND USING THE VPOINT COMMAND

In this exercise, you will use your knowledge of the WCS and the VPOINT command to build the model shown in Figure 2.25. Building this model will demonstrate how to specify 3D points, as well as how to set up 3D viewpoints.

Begin a new drawing using the standard AutoCAD prototype or template drawing, DRAW-ING.DWG. This starts with a drawing area that is blank except for the coordinate system icon positioned in the lower left corner of the screen. As you can tell from the right-hand rule, the Z axis is pointed out of the screen directly toward you. Therefore, the viewpoint has the direction coordinates of 0,0,1.

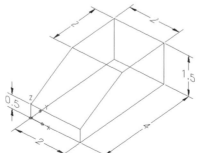

Figure 2.25

Using the dimensions in Figure 2.25, draw the base of the wireframe, which is a rectangle composed of lines:

Command: LINE *(Press ENTER.)*

Specify first point: 0,0

Specify next point or [Undo]: @2,0

Specify next point or [Undo]: @0,4

Specify next point or [Close/Undo]: @-2,0

Specify next point or [Close/Undo]: C *(Press ENTER.)*

Even though drawing this rectangle doesn't involve three dimensions, it does demonstrate two important points:

- You can use your pointing device to locate points, as long as those points are on the drawing plane.
- Whenever just two coordinates are typed in, AutoCAD sets the Z coordinate to 0.

Next, draw the top rectangular part of the wireframe. Because it is not on the XY plane, you'll have to type all three coordinates of each of the four points.

Command: LINE *(Press ENTER.)*

Specify first point: 0,2,1.5 *(Press ENTER.)*

Specify next point or [Undo]: @0,-2 *(Press ENTER.)*

Specify next point or [Undo]: @-2,0 *(Press ENTER.)*

Specify next point or [Close/Undo]: @0,2 *(Press ENTER.)*

Specify next point or [Close/Undo]: C *(Press ENTER.)*

It will not seem like you accomplished very much by drawing these last four lines because three of them are exactly over lower lines. Therefore, use the Rotate option of VPOINT to switch to a viewpoint, which will better show what you have drawn. The 3DORBIT could also be used to perform this task of viewing the model in 3D.

Command: VPOINT *(Press ENTER.)*

Current view direction: VIEWDIR= 0.0000,0.0000,1.0000

Specify a viewpoint or [Rotate] <display compass and tripod>: R *(Press ENTER.)*

Enter angle in XY plane from X axis <270>: -50 (or 310) *(Press ENTER.)*

Enter angle from XY plane <90>: 35 *(Press ENTER.)*

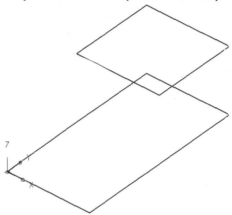

Figure 2.26

The resulting view will be similar to one shown in Figure 2.26, though you will probably need to adjust the view with the ZOOM command. AutoCAD changes the zoom level every time VPOINT changes the view direction. Usually AutoCAD zooms in so that the model completely fills the screen—an equivalent of zoom-extents. As a result, you will usually do a realtime zoom immediately after VPOINT.

From this viewpoint, the last four lines you drew appear to overlap the original four lines. Actually, they are in different planes, but this is not obvious. This lack of depth perception due to the 2D computer screen is one of the problems you'll always encounter when you work in 3D. With this model it is easy to imagine that the two rectangles are in different planes; however, on complicated models it is not so easy.

The single line in the wireframe that represents the top of the front vertical face can be made using several methods. One way is to use a copy of the existing line, with a displacement of one-half unit in the Z direction.

Command: COPY *(Press ENTER.)*

Select objects: *(Select the 2-unit-long line that is on the X axis.)*

Specify base point or displacement, or [Multiple]: 0,0,.5 *(Press*

ENTER.)

Specify second point of displacement or <use first point as displacement>: (*Press ENTER.*)

Now your wireframe model will look like the one shown in Figure 2.27.

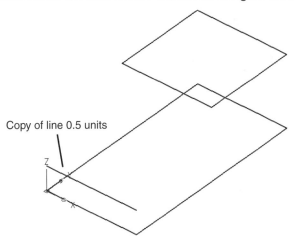

Copy of line 0.5 units

Figure 2.27

Since all of the wireframe's points are now established, the easiest way to draw the remaining six lines is to use object endpoint snaps. Therefore, set up a running endpoint OSNAP and draw lines between the points connected with the dashed lines shown in Figure 2.28. That finishes the construction of the wireframe model. Compare your model with the one in the electronic file 3d_ch2_01.dwg.

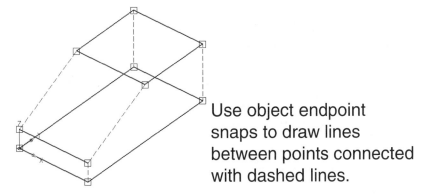

Use object endpoint snaps to draw lines between points connected with dashed lines.

Figure 2.28

Now use the VPOINT command to see the wireframe model you just completed from four different view directions:

1.   0,-1,0 to see the front

2.   1,0,0 to see the right side

3.   1,1,1 to look down on the top, back, and right side.

4.   Rotate 120° from X axis and 60° from XY plane to see the top, back, and left side of the model.

The resulting views are shown in Figure 2.29.

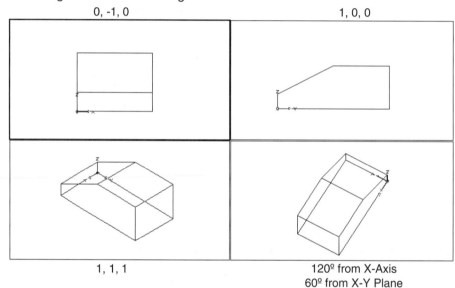

0, -1, 0

1, 0, 0

1, 1, 1

120º from X-Axis
60º from X-Y Plane

Figure 2.29

THE PLAN COMMAND

The PLAN command sets the viewpoint in the current viewport to one looking straight toward the XY plane from the positive Z direction. The command format is:

Command: PLAN

Enter an option [Current ucs/Ucs/World] <Current>: (*Select an option, or press ENTER.*)

All three options zoom to the equivalent of a Zoom-Extents.

CURRENT UCS

The Default option, selected by entering C or by pressing ENTER, sets the plan view relative to the current user coordinate system. If no UCS is being used, this option is equivalent to the VPOINT command with view direction coordinates of 0,0,1.

UCS

Sets the plan view relative to a named user coordinate system. (Named user coordinate systems will be fully explained later in this chapter.) AutoCAD will display the following follow-up prompt:

Enter name of UCS or[?] : (*Enter a UCS name or ?*)

_ Name of UCS

   Enter the name of the UCS system.

_ ?

Typing in a ? will bring up a list of named coordinate systems.

WORLD

Sets the plan view relative to the WCS. This option results in the same viewpoint as VPOINT with the view direction coordinates of 0,0,1.

Figure 2.30 illustrates the difference between the Current UCS and the World options of PLAN.

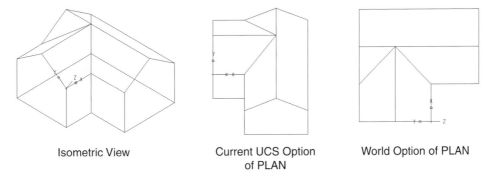

Isometric View   Current UCS Option   World Option of PLAN
                              of PLAN

Figure 2.30

 Tip: 3D models are often more easily visualized from isometric viewpoints, but sometimes you do need to look straight down on the drawing plane. The PLAN command is more convenient than VPOINT to set up such a view.

These examples of the PLAN command use a wireframe model similar to the one in the previous exercise. As shown on the left in Figure 2.31, the UCS has been rotated so that it is XY plane in on the WCS YZ plane, and the viewpoint is rotated 45° from the WCS X axis and 30° from the WCS XY plane.

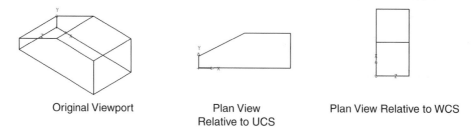

| Original Viewport | Plan View Relative to UCS | Plan View Relative to WCS |

Figure 2.31

The following command line input sets the plan view relative to the UCS, as shown in the center of Figure 2.31.

Command: PLAN *(Press ENTER.)*

Enter an option [Current ucs/Ucs/World:] <Current>: *(Press ENTER.)*

To set the plan view relative to the WCS, use the following command line input. The results are shown on the right in Figure 2.28.

Command: PLAN *(Press ENTER.)*

Enter an option [Current ucs/Ucs/World:] <Current>: W *(Press ENTER.)*

THE USER COORDINATE SYSTEM

Specifying points in 3D space can be awkward because pointing devices, except with object snaps, are restricted to a flat plane. To draw efficiently, you must have some way to point to locations easily and accurately in 3D space. Early versions of AutoCAD could only move the XY plane up or down from the WCS XY plane (with the ELEV command). In Release 10, though, AutoCAD introduced a local coordinate system called the User Coordinate System (UCS) that can be both moved and twisted relative to the WCS. Although pointing devices have the same XY plane restrictions, there is no restriction in the orientation of the XY plane itself.

When a UCS is in use, the WCS drops out of sight, so that all references to points and directions are relative to the current UCS. The only time the WCS reappears is when you are setting viewpoints. Generally, AutoCAD sets view direction according to WCS—temporarily restoring the WCS whenever you initiate the VPOINT command. You can, however, use the Worldview system variable to change this characteristic.

THE ELEV COMMAND

The ELEV command was an early attempt by AutoCAD to break out of the XY plane. It is now an obsolete but still present command that performs two unrelated functions. One is to move the drawing plane up or down in the Z direction so that you can use a pointing device to draw in a plane parallel to the XY plane. The second function is to give objects thickness, or height. Thickness is a property that is often called extrusion thickness because wireframe objects appear to have been extruded, or stretched, in the Z direction. Because thickness acts much like a surface, we will explain it more thoroughly when we cover surface modeling in Chapter 4. The format for the ELEV command is:

Command: ELEV

Specify a new default elevation <0.0000>: (*Specify a distance.*)

Specify a new default thickness <0.0000>: (*Specify a distance.*)

NEW CURRENT ELEVATION

Enter a new elevation for the drawing plane. A positive number moves the drawing plane in the positive Z direction (up), while a negative number moves the drawing plane in the negative Z direction (down). AutoCAD's grid (if it is turned on) moves with the elevation change, and the Z coordinate in the status bar indicates the new elevation. The UCS icon, however, will not move.

THICKNESS

Specify a new thickness. All entities that accept a thickness will be drawn using this value. This option has nothing to do with elevation. We will discuss thickness in Chapter 4.

Tip: The UCS command does a much better job of managing the drawing plane and should be used rather than the ELEV command. Furthermore, AutoCAD has few built-in indicators to show that the drawing plane has been moved away from the XY plane, making it easy to mistakenly draw in a plane offset from the one you thought you were on.

TRY IT! - USING THE ELEV COMMAND

Begin a new drawing file. Set the current viewing position using the VPOINT command and a coordinate position of 1,-1,1. The top 2-by-2 rectangle of the wireframe model shown back in Figure 2.25 could be drawn using ELEV. After drawing the large base rectangle, which is on the XY plane, the drawing plane could be raised 1.5 units in the Z direction with the following command line input:

Command: ELEV (Press ENTER)

Specify new default elevation <0.0000>: 1.5 (*Press ENTER.*)

Specify new default thickness <0.0000>: (*Press ENTER to accept the default.*)

Next, set AutoCAD's grid snap to one; then, as shown in Figure 2.32, the top of the wireframe could be drawn by pointing:

Command: LINE (*Press ENTER.*)

Specify first point: (*Select 0,2 with the pointing device.*)

Specify next point or [Undo]: (*Select 2,2 with the pointing device.*)

Specify next point: or [Close/Undo]: (*Select 2,4 with the pointing device.*)

Specify next point or [Close/Undo]: (*Select 0,4 with the pointing device.*)

Specify next point or [Close/Undo]: C (*Press ENTER.*)

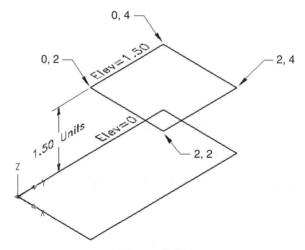

Figure 2.32

After drawing this rectangle, the drawing plane could be returned to the XY plane by changing the Elevation system variable, rather than by using the ELEV command, with the following command line input.

Command: ELEVATION (*Press ENTER.*)

Specify new value for ELEVATION <1.5000>: 0 (*Press ENTER.*)

THE UCS COMMAND

UCS is the command that manages the user coordinate system (UCS). As this movable coordinate system allows you to use 2D drawing techniques to create 3D models, UCS will be one of your most often used commands. It pays to learn the command to the extent that its use is almost automatic. Options of the UCS command can be selected from the Tools pulldown menu illustrated in Figure 2.33. Toolbars are also available in Figure 2.34 to assist in picking options of the UCS command.

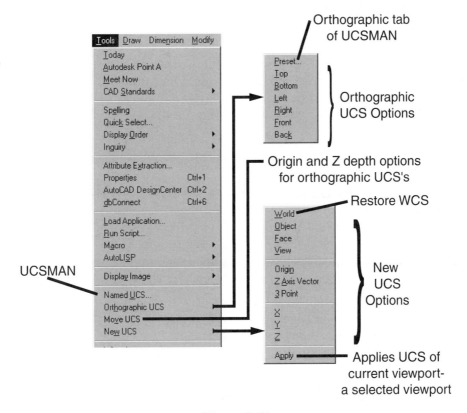

Figure 2.33

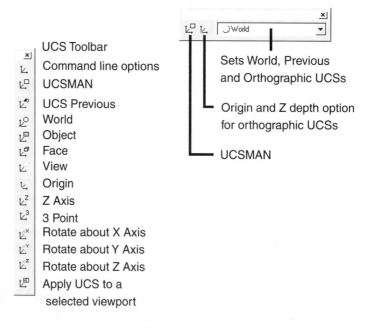

UCS Toolbar

Command line options

UCSMAN

UCS Previous

World

Object

Face

View

Origin

Z Axis

3 Point

Rotate about X Axis

Rotate about Y Axis

Rotate about Z Axis

Apply UCS to a
selected viewport

Sets World, Previous
and Orthographic UCSs

Origin and Z depth option
for orthographic UCSs

UCSMAN

Figure 2.34

The command line format of UCS is:

Command: UCS

Current UCS name: *NO NAME*

**Enter an option
[New/Move/orthoGraphic/Prev/Save/Del/Apply/?/World]<World>:
(*Enter an option or press ENTER.*)**

Each viewport (viewports are discussed later in this chapter) can have a different UCS, and the UCS command applies to the current viewport. Notice that AutoCAD displays the name of the current UCS on the command line, before issuing the prompt for establishing a new UCS. As with all AutoCAD command line options, you can select an option by typing in the entire option name or by typing in just the uppercase letters in the option name.

NEW

When you select the New option, the following command line options will be offered:

**Specify origin of new UCS or [ZAxis/3point/OBject/FAce/View/X/Y/Z]
<0,0,0>: (*Enter an option or press ENTER.*)**

Origin

This option, which is the default option, moves the origin of the UCS without changing the current directions of the X, Y, and Z axes, as shown in Figure 2.35. Pressing ENTER will result in no change to the UCS. Any of AutoCAD's standard methods for selecting a new origin may be used—including pointing, object snaps, and typed-in coordinates. All input is relative to the current UCS.

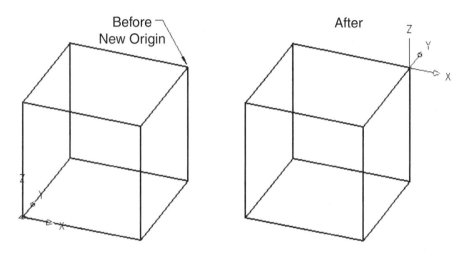

Figure 2.35

ZAxis

Moves the origin and orients the UCS relative to the direction of the Z axis. The follow-up prompts are:

Specify new origin point <0,0,0>: (*Specify a point or press ENTER.*)

Specify point on positive portion of Z-axis <current>: (*Specify a point.*)

Both points may be selected by any of AutoCAD's standard methods for specifying points. The XY plane of the UCS will be located at the first point selected and will be perpendicular to a line from the first point to the second point, as shown in Figure 2.36. The X axis will be parallel to the WCS XY plane.

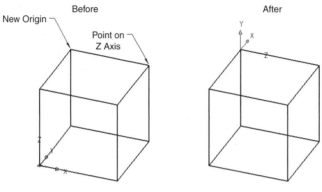

Figure 2.36

3point

Moves the origin and orients the UCS relative to X and Y axes. The follow-up prompts are:

Specify new origin point <0,0,0>: *(Specify a point or press ENTER.)*

Specify point on positive portion of the X-axis <current>: *(Specify a point.)*

Specify point on positive-Y portion of the UCS XY plane <current>: *(Specify a point.)*

All three points, which define a plane, may be selected using any of AutoCAD's methods for specifying points. The first point sets the origin of the plane, whereas the second point establishes the direction of the X axis relative to this origin. The third point determines how the XY plane is rotated in space. This last point does not have to be on the Y axis—it can be anywhere off the line between the first two points. See Figure 2.37 for an example.

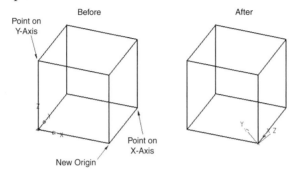

Figure 2.37

OBject

Moves and reorients the UCS based on an existing object. The follow-up prompt for this option is:

Select object to align UCS: *(Select an object.)*

You must select the object by picking a point on it. Any object type can be selected, except for the following: 3D solid, 3D polyline, 3D mesh, spline, viewport, mline, region, ellipse, ray, xline, leader, and mtext.

The direction of the Z axis will be the same as the extrusion direction of the selected entity, whereas the location of the origin depends on the type of entity selected. Also, the directions of the X and Y axes depend on the location of the entity selection point. Table 2.3 shows the resulting UCS origin and orientation for some common object types.

Table 2.3 UCS Origin and Orientation

Object Type	Origin Location	UCS Orientation
Arc	Center of arc	The X axis points in the direction of the arc endpoint nearest the point picked.
Circle	Center of circle	The X axis is aimed toward the point picked.
Line	Endpoint nearest the point picked	The line lies in the XZ plane of the new UCS while the line's other endpoint will have a Y coordinate of 0.
2D Polyline	Polyline start point	The X axis points toward the next vertex.
3D Face	Face start point	The XY plane will be in the plane of the 3D face, with the X axis pointing toward the second point, and the Y axis toward the third or fourth points.
Text and Blocks	Insert point	The X axis points in the object's 0 degree direction.

Prior to Release 13, this option was named Entity, and even AutoCAD 2002 will accept that name, or just a typed-in letter E.

Face

This option places the UCS on the flat face of a 3D solid. It issues the prompt:

Select face of solid object: *(Select the planar face of a 3D solid.)*

You can select a face by picking a point on its edge, or by picking a point on its surface. The edges of the selected face will be highlighted, and the UCS will be placed on the face. Its origin will be in the face's corner nearest the face selection point, and its X axis will be on the edge nearest the face selection point, as shown in Figure 2.38. (If there are no straight edges, the UCS origin will be on the perimeter of the face.) Then AutoCAD will issue the prompt:

Enter an option [Next/Xflip/Yflip] <accept>: *(Enter an option or press ENTER.)*

The UCS will move to the next suitable face when the Next option is selected. The

Xflip and Yflip options rotate the XY plane of the UCS 180° about its X or Y axis.

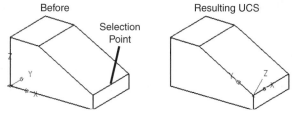

Figure 2.38

View

Reorients the UCS so that the XY plane is perpendicular to the current view direction, with the X axis parallel to the bottom of the viewport and the Y axis pointed vertically. The Z axis will be pointing out of the screen, toward the viewer. The origin of the UCS is unchanged. There are no follow-up prompts.

X/Y/Z

Each of these three options rotates the UCS around the specified axis. The follow-up prompt is:

Specify rotation angle about N axis <90>: (*Specify an angle.*)

N in this prompt is the initially selected axis—either X, Y, or Z. Angles may be specified by typing in an angle or by pointing. Rotation direction is according to the right-hand rule. An easy way to visualize rotation direction is to mentally grasp the axis you want the UCS rotated around with your right hand so that your thumb points away from the origin, as shown in Figure 2.39. Positive rotation angles will then be in the direction of your curled fingers (counterclockwise). Negative rotation angles will be in the opposite direction of your curled fingers (clockwise).

MOVE

This option in the main UCS menu is for moving the origin of UCSs that have been created by the orthoGraphic option. The option's follow-up prompt is:

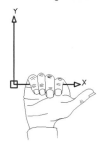

Figure 2.39

Specify new origin point or [Zdepth]<0,0,0>: *(Specify a point, enter a Z, or press ENTER.)*

If you specify a point, the UCS origin will move to that point. The Zdepth option moves the UCS origin up or down from the current XY plane along the Z axis. You will be prompted to specify a distance. Positive distance values move the origin in the positive Z axis direction, whereas negative values move the origin in the negative Z axis direction

orthoGRAPHIC

This option rotates the XY plane of the UCS so that it is parallel with the XY, ZX, or YZ plane of a base coordinate system. It displays the prompt:

Enter an option [Top/Bottom/Front/BAck/Left/Right]<Top>: *(Enter an option or press ENTER.)*

The Top and Bottom options place the UCS XY plane on the base coordinate system's XY plane, the Front and BAck options place it on the base coordinate system's ZX plane, and the Left and Right options place it on the base coordinate system's YZ plane. The Top, Front, and Right options point the Z axis of the UCS in the positive Z axis direction of the base coordinate system, while the Bottom, BAck, and Left options point the Z axis of the UCS in the negative Z axis direction of the base coordinate system. See Figure 2.40 for examples of the six orthographic UCSs.

The base coordinate system is the one that is stored in the Ucsbase system variable. By default, the WCS is the base coordinate, but you can store the name of any previously saved UCS in Ucsbase.

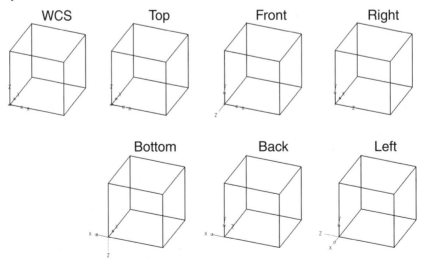

Figure 2.40

PREV

Returns the UCS to its previous location and orientation. This option can be repeated to step back through the last ten UCS settings.

RESTORE

Changes the UCS to a position and orientation set by a previously saved UCS. The follow-up prompt is:

Enter name of UCS to restore or [?}: *(Enter a ? or a name.)*

The question mark option allows you to see a list of named user coordinate systems. It displays the following prompt:

Enter UCS name(s) to list<*>: *(Enter a name list or press ENTER.)*

You can use wildcard characters (such as ? and *) to bring up a filtered list of named UCSs, or press ENTER to see the names of all saved UCS.

SAVE

Saves the current UCS configuration to a specified name. The follow-up prompt is:

Enter name to save current UCS or [?] : *(Enter a ? or a name.)*

The rules for naming UCSs are the same as for layers, views, and other AutoCAD objects. The question mark option brings up the prompt:

Enter UCS name(s) to list<*>: *(Enter a name-list or press ENTER.)*

Press ENTER to see a list of all previously saved UCS names or use wildcards to see a filtered list of UCS names.

The construction process of a 3D model can become very time consuming if you have to keep setting up previously created User Coordinate Systems. For this reason, as you create a UCS, it is considered good practice to save it. Typical names for UCS positions include Front, Top, Side, Auxiliary, etc. It then becomes much easier to display a previously used UCS with the Restore option.

DEL

Deletes a saved UCS. The follow-up prompt is:

Enter UCS name(s) to delete <none>: *(Enter a name list.)*

You can enter as many names as desired and use wildcard characters to delete several named user coordinate systems, even the current one.

APPLY

This option copies the UCS from one viewport to another viewport. It displays the prompt:

Pick viewport to apply current UCS or [All] <current>: *(Select a viewport, enter A, or press ENTER.)*

The UCS that will be copied is the one for the viewport that was the current viewport when the UCS command is initiated. (See the discussion of viewports for an explanation of the term current viewport.) If you choose the All option, the UCS of the current viewport will be applied to all viewports. If you select a viewport, the UCS of the initial viewport will be applied to that viewport when you press ENTER.

?

The question mark option displays a table of saved UCSs, showing their origin and axes directions relative to the current UCS. The follow-up prompt is:

Enter UCS name(s) to list<*>: *(Enter a name list or press ENTER)*

Press ENTER to see a list of the names of all saved UCSs or use wildcard characters to see a filtered list of UCS names.

WORLD

This option, selected by entering W or pressing ENTER from the main UCS prompt, restores the world coordinate system.

Tips: When you are beginning a 3D model or you are starting to work on a section of the model in which few or no entities exist, the Origin, ZAxis, X,Y, and Z options are especially useful.

As your model progresses and you want to set the UCS relative to existing objects, the options of Origin, ZAxis, and 3point will be used often. Use object snaps to position the UCS precisely.

Although you do not have to use negative angles when rotating the UCS about the X,Y, or Z axis, it is often easier to visualize results with them. For instance, if you are revolving the UCS about the Z axis it will probably be easier for you to think in terms of a -45° rotation, rather than a 315° rotation, even though they are equivalent.

The View option can be useful for precisely setting a compound angle for the XY plane relative to the WCS. Suppose, for instance, you want to set up a drawing plane whose X axis is pointed 120° from the WCS X axis and whose Y axis is tilted 45° from the WCS XY plane. You could do this by using the Rotate option of VPOINT to set a view direction 30° from the X axis and 45° from the XY plane as shown in Figure 2.41, and then using the View option of the UCS command to set the XY plane perpendicular to this view direction.

As 3D models become more confusing due to complexity, use the UCS-World and PLAN-World commands to reduce return to a base position for the UCS and for viewing the model.

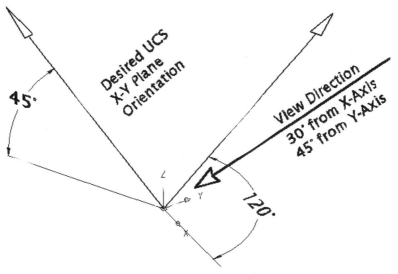

Figure 2.41

TRY IT! - USING THE UCS COMMAND

This exercise will give you experience in moving and orienting the UCS. In the figures, the original position of the UCS will be shown on the left, and the resulting position from a specific UCS option will be shown on the right.

Before you begin manipulating the UCS, draw the three one-unit-long lines and the 0.5-radius arc shown in Figure 2.42 on the XY plane of the UCS. The actual location of these objects on the XY plane is not important. Then use the VPOINT command to set a viewpoint that has the direction coordinates of 1,-1,1 and zoom back as necessary to give yourself working room. Lastly, copy the arc to a relative destination of 1.0 units in the positive Z direction.

Once you have created these wireframe objects, use them to try out the following options of the UCS command. Notice that you can access all of these options from the main UCS prompt, even though they are not listed in the prompt.

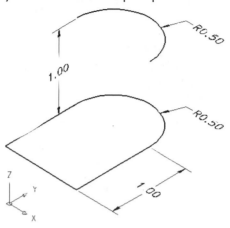

Figure 2.42

The Origin option:

Command: UCS (*Press ENTER.*)

Current UCS name: *WORLD*

Enter an option
  [New/Move/orthoGraphic/Prev/Restore/Save/Del/Apply/?/World]<W orld>: O (*Press ENTER.*)

Specify new origin point <0,0,0>: (*Pick the arc endpoint as shown.*)

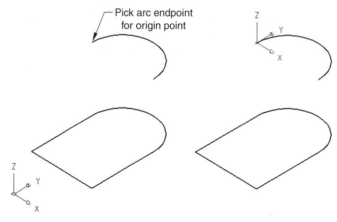

Pick arc endpoint for origin point

Figure 2.43

The View option:

Command: UCS *(Press ENTER.)*

Current UCS name: *NO NAME*

Enter an option
[New/Move/orthoGraphic/Prev/Restore/Save/Del/Apply/?/World]<W
orld>: V *(Press ENTER.)*

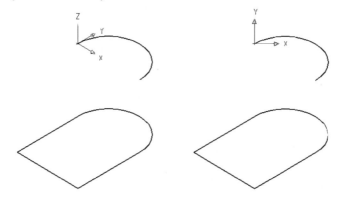

Figure 2.44

The 3point option:

Command: UCS *(Press ENTER.)*

Current UCS name: *NO NAME*

Enter an option
[New/Move/orthoGraphic/Prev/Restore/Save/Del/Apply/?/World]<W
orld>: 3 *(Press ENTER.)*

Specify new origin point <0,0,0>: *(Pick the line endpoint as shown.)*

Specify point on positive portion of X-axis <2.0234,0.4082,-0.5774>:
   (Pick the line endpoint as shown.)

Specify point on positive-Y portion of the UCS XY plane
   <1.7304,1.2743,

-0.5744>: *(Pick either endpoint of the arc.)*

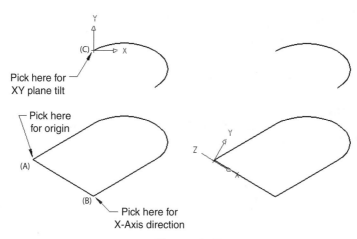

Pick here for
XY plane tilt

Pick here
for origin

(A)

(B)

Pick here for
X-Axis direction

Figure 2.45

The OBject option:

Command: UCS (*Press ENTER.*)

Current UCS name: *NO NAME*

**Enter an option
[New/Move/orthoGraphic/Prev/Restore/Save/Del/Apply/?/World]<World>: OB (*Press ENTER.*)**

Select object to align UCS: (*Pick arc near one end as shown.*)

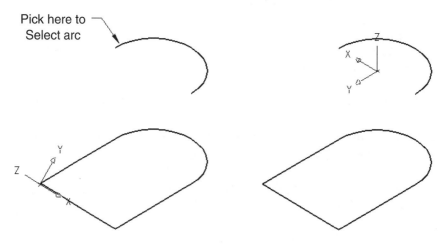

Pick here to
Select arc

Figure 2.46

The ZAxis option:

Command: UCS (*Press ENTER.*)

Current UCS name: *NO NAME*

Enter an option
[New/Move/orthoGraphic/Prev/Restore/Save/Del/Apply/?/World]<W orld>: ZA (*Press ENTER.*)

Specify new origin point <0,0,0>: (Pick the line endpoint as shown.)

Specify point on positive portion of Z-axis <0.5000,1.0000,0.0000>:
(*Pick the line endpoint as shown.*)

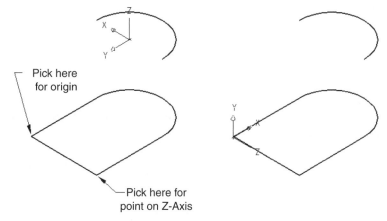

Pick here for origin

Pick here for point on Z-Axis

Figure 2.47

Rotate about X-axis option:

Command: UCS (*Press ENTER.*)

Current UCS name: *NO NAME*

Enter an option
[New/Move/orthoGraphic/Prev/Restore/Save/Del/Apply/?/World]<W orld>: X (*Press ENTER.*)

Specify rotation angle about X axis: <90>: -90 (*Press ENTER.*)

3D

EXERCISES

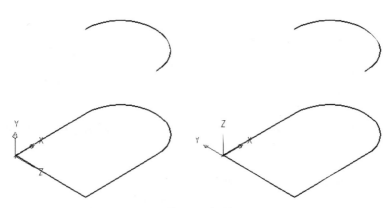

Figure 2.48

Rotate about Z-axis option:

Command: UCS (*Press ENTER.*)

Current UCS name: *NO NAME*

Enter an option
[New/Move/orthoGraphic/Prev/Restore/Save/Del/Apply/?/World]<W
orld>: Z (*Press ENTER.*)

Specify rotation angle about Z axis <90>: (*Press ENTER to accept the*
default rotation angle.)

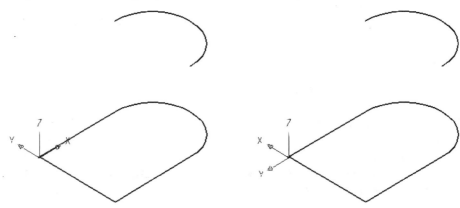

Figure 2.49

Rotate about Y-axis option:

Command: UCS (*Press ENTER.*)

Current UCS name: *NO NAME*

**Enter an option
[New/Move/orthoGraphic/Prev/Restore/Save/Del/Apply/?/World]<World>:Y** (*Press ENTER.*)

Specify rotation angle about Y axis: <90>: -45 (*Press ENTER.*)

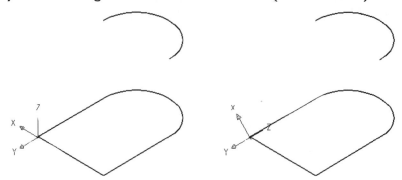

Figure 2.50

THE UCSMAN COMMAND

The UCSMAN command displays a dialog box, which is titled UCS, that has three tabs for you to use in managing named UCSs, for setting orthographic UCSs, and for controlling the parameters of some system variables related to UCSs.

NAMED UCSS

The names of all existing UCSs will be displayed in a list box. This list box will always have World as an entry. If the current UCS is unnamed, it will be listed as Unnamed; if more than one UCS has been used, a UCS named Previous will be listed. You can highlight any name in the list by clicking on it. Then, you can click on the Set Current button to make the highlighted UCS the current UCS. You can also right-click the UCS name to bring up a shortcut menu having options to make the UCS the current UCS, delete it, or rename it (see Figure 2.51).

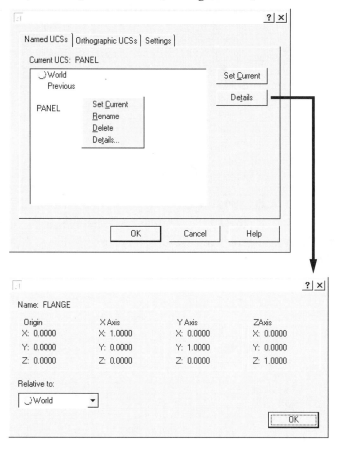

Figure 2.51

If you click on the Details button or select Details from the shortcut menu, a secondary dialog box showing the origin location and the axes directions of the highlighted UCS relative to a base UCS will be displayed. By default, the base UCS is the WCS, but you can select another UCS through the dialog box's pull-down list box labeled Relative To.

OrthoGRAPHIC UCSS

The actions of this dialog box are similar to those of the Move and Orthographic options of the UCS command (see Figure 2.52). All six orthographic UCSs will be displayed in a list box. When you click on a UCS, it will be highlighted, and you can click on the Set Current button to have the UCS be the current one. The Details button brings up the same secondary dialog box displaying origin and axes data that the Named UCSs tab displays.

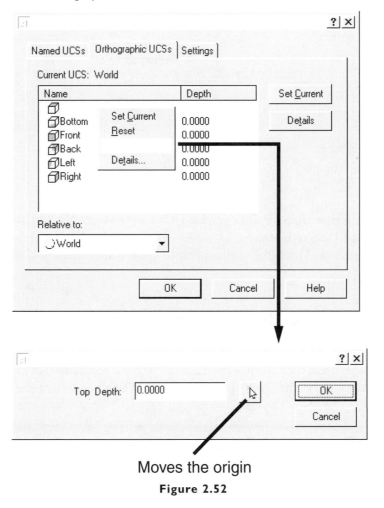

Moves the origin

Figure 2.52

The shortcut menu for each orthographic UCS also allows you to set the UCS to be the current UCS and to display details about the UCS. The Depth option of the shortcut menu displays a secondary dialog box titled Orthographic UCS Depth. Just as in the Depth option of the UCS command, setting a depth in this dialog box moves the UCS origin along the Z axis from the base UCS origin. The button to the right of the Depth edit box allows you to specify an origin that is off of the Z axis of the base UCS (see Figure 2.53). The dialog boxes will be temporarily dismissed, and you will be prompted from the command line to specify a point for the orthographic UCS origin.

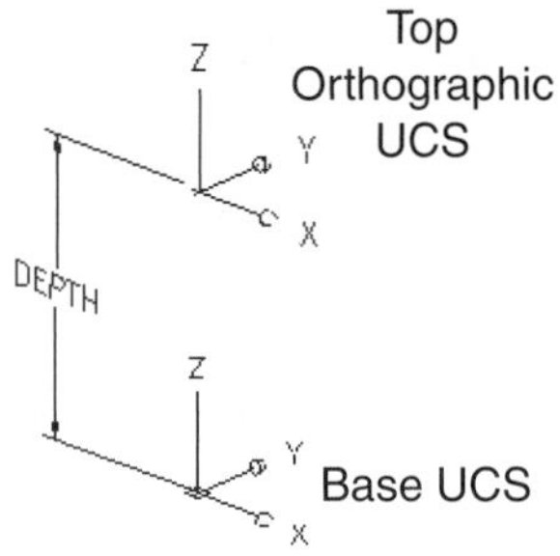

Figure 2.53

By default, the base UCS for orthographic UCSs is the WCS. You can, though, specify any named UCS to be the base UCS by selecting it from the pull-down list box labeled Relative to. The name of the specified UCS is stored in the Ucsbase system variable.

SETTINGS

The cluster of checkboxes labeled UCS Icon Settings (see Figure 2.54) produce the same results as the options of the UCSICON command, which was discussed earlier in this chapter. When the checkbox labeled Update View to Plan When UCS is Changed is selected, the view direction automatically switches to that of a plan view whenever the UCS is changed. The setting of the Ucsfollow system variable is controlled by this checkbox. The checkbox labeled Save UCS with Viewports controls the setting of the Ucsvp system variable. This system variable will be described when multiple viewports are discussed later in this chapter.

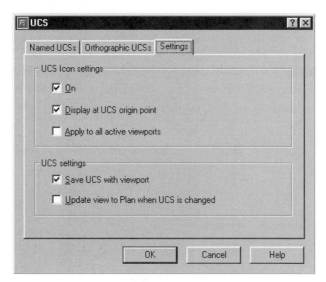

Figure 2.54

THE VIEW COMMAND

The VIEW command has eight preset views. Six of them set an orthographic view, while each of the other four set an isometric view that looks down on the XY plane of the UCS. To set one of these preset views, click on its name to highlight it, and then click on the Set Current button of the dialog box. You can also right-click the high-lighted name and select Set Current from the shortcut menu.

The view will be relative to the named UCS selected in the drop-down list box labeled Relative To (see Figure 2.55). Just as with the corresponding drop-down list box of the Orthographic UCSs tab of the UCSMAN command, the name of the selected UCS is stored in the Ucsbase system variable.

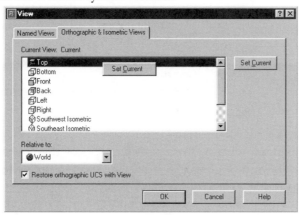

Figure 2.55

The AutoCAD menus and toolbars used for setting viewpoints are shown in Figures 2.56 and 2.57. In this figure, the viewpoint directions are not given for the menu options for setting orthographic and isometric views, because the VIEW command, rather than VPOINT, is used to set those viewpoints in AutoCAD 2002.

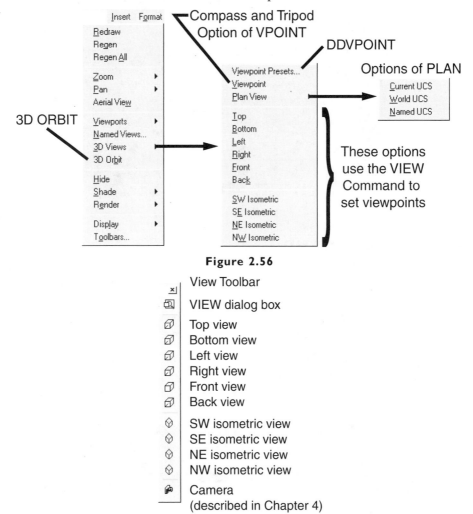

Figure 2.56

View Toolbar
VIEW dialog box
Top view
Bottom view
Left view
Right view
Front view
Back view

SW isometric view
SE isometric view
NE isometric view
NW isometric view

Camera
(described in Chapter 4)

Figure 2.57

When the check box labeled Restore Orthographic UCS with View is selected, the UCS will automatically change to match the UCS of the corresponding orthographic view, as shown in Figure 2.58. This checkbox controls the contents of the Ucsortho system variable. When it is checked, Ucsortho is set to a value of 1, and when it is cleared, Ucsortho is set to a value of 0. This system variable affects only orthographic views.

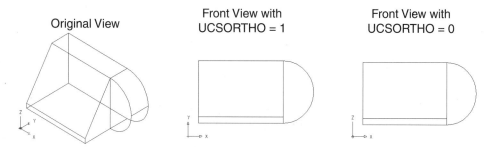

Figure 2.58

Tip: The orthographic and isometric view options of the View toolbar buttons and the View/3D Views pull-down menu in AutoCAD 2002 use the VIEW command rather than the VPOINT command to set viewpoints. As a result, setting an orthographic view will also set the UCS when Ucsortho has a value of 1. If you prefer that these menu and toolbar options not set the UCS as they set orthographic views, assign a value of 0 to Ucsortho.

MULTIPLE TILED VIEWPORTS

Viewing 3D space on a 2D computer screen causes major visualization problems when making 3D models. You have virtually no spatial perception, which makes it difficult to determine which lines are in front planes and which are in back planes. Furthermore, there are a lot of lines to contend with. Even though a 3D wireframe model of an object probably contains fewer entities than a multiview drawing of the object, the entities are crowded together. Surface models, with their mesh lines and grid lines, are even more cluttered than wireframes, as you will see in Chapter 4.

As a visualization aid, AutoCAD gives you the ability to divide your computer screen into rectangular sections, called viewports, so you can see your model from several different view directions at the same time. One viewport, for instance, can show the back of your model, while a second shows a plan view of it.

Figure 2.59 illustrates how multiple viewports can help. The screen is divided into one large viewport on the right and two smaller viewports on the left. The upper left viewport shows a plan view of the current UCS. Even though the UCS icon shows the XY origin of the UCS, it is not possible in this view to determine the coordinate system's Z location. The XY plane could be on the upper flat surface of the model, on the lower flat surface, or even floating in space somewhere, not even close to the model. However, by looking at the large viewport on the right, you can see that the UCS origin is located on the model's upper flat surface.

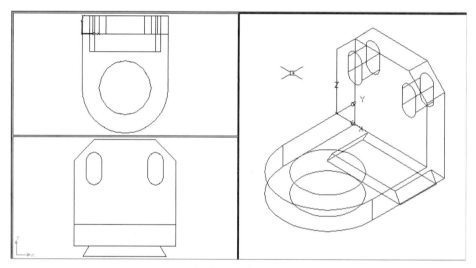

Figure 2.59

CHARACTERISTICS OF TILED VIEWPORTS

Multiple viewports are almost like having multiple computer screens. Each viewport can have a different zoom level as well as a different view direction. Grid, snap, view resolution, and the coordinate system icon can also be set by viewport. And, beginning with AutoCAD 2002, each viewport can even have its own UCS. Furthermore, some features that we'll cover later in the book—hide, shade, render, and the perspective view mode—can also be set individually for each viewport. But, there is still just one 3D model, and any change to it instantly shows up in all viewports. (Provided that the change is in a location shown in all viewports.)

The maximum number of viewports you can have depends on your computer's video system. However, no matter what system you have, AutoCAD will allow you to have more viewports than you are ever likely to need. You will sometimes have to compromise between the number of viewports you want and their relative size because it can be hard to see details and make selections in small viewports. Obviously, the size of your computer screen is a big factor—you'll be able to have more usable viewports on a large screen than on a small one.

Autodesk calls these tiled viewports because they are comparable to ceramic tiles on a floor, in that:

- Viewports must completely fill the computer screen graphics area.
- Viewports cannot overlap.
- There can be no space between viewports.
- Viewports cannot be moved.
- A viewport's size and shape cannot be changed.

Tiled viewports only work in model space. In Chapter 8 we'll discuss AutoCAD's paper space, which has another type of viewport, called floating viewports, that can overlap, have spaces between viewports, be moved, and have their size and shape changed. The Tilemode system variable determines whether AutoCAD operates in the model space mode or the paper space mode. When Tilemode is set to 1, its default setting, model space is in effect and viewports are tiled. When Tilemode is set to 0, AutoCAD operates in paper space mode and uses floating viewports. Paper space and floating viewports are intended for output.

USING TILED VIEWPORTS

Even though several viewports may be on the screen, there is only one viewport in which commands take place—the current viewport. The current viewport will be the only viewport in which the cursor crosshairs show up, and it will also have a heavier border around it than the other viewports. In Figure 2.66, the large viewport on the right is the current viewport. When you move the cursor to another viewport, the crosshairs change to a small arrow to signify that it is not the current viewport.

Any viewport can be made to be the current viewport by moving the cursor to it and pressing your pointing device's pick button. The first press of the pick button in a viewport is a signal to AutoCAD to change the current viewport. You can even start most commands in one viewport and end them in another. In Figure 2.59, for instance, you could start a line in the large viewport, move the cursor to either of the small viewports, click the pick button to make it the current viewport, and pick a point for the line ending. You can also change current viewports by simultaneously pressing the CTRL and R keys.

In Release 14 and earlier versions of AutoCAD, commands that affected the view—such as ZOOM, PAN, SNAP, GRID, and VPOINT—did not allow you to change viewports in the middle of the command. When any of those commands were invoked, the cursor was locked in the current viewport and could not be moved to another one. In AutoCAD 2002, though, only the PAN command locks the cursor in the current viewport.

Viewport setups can be named and saved within a drawing file. Autodesk calls them viewport configurations because, in addition to the number and layout of viewports, AutoCAD saves for each viewport:

- The UCS
- The GRID and SNAP settings
- VIEWRES setting
- ZOOM level
- UCSICON settings
- View direction and target location
- Settings for perspective and clipping planes (These will be explained in Chapter 4 when the DVIEW command is discussed.)

Consequently, when a viewport configuration is restored, each viewport has the same appearance settings as when it was saved. Of course, any changes that were made on the model show up in the newly restored viewports.

Names for viewport configurations must follow AutoCAD's rules for named objects. In AutoCAD 2002, names can have up to 255 characters and have spaces within the name. Characters that are regularly used for control and filtering purposes—such as question marks, colons, commas, and asterisks—are not allowed.

The REDRAW and REGEN commands affect only the current viewport. If you want to clean up the display in all viewports, issue the REDRAWALL command. If you want to force a regeneration in all viewports, issue the REGENALL command.

VIEWPORTS AND THE UCS

Each viewport can have its own UCS in AutoCAD 2002. Moreover, the UCS command affects only the current viewport. You can copy the UCS of one viewport to another by way of the Apply option of the UCS command. To do this, you would make the viewport having the UCS you wanted to duplicate be the current viewport. Then, you would invoke UCS, choose the Apply option, and pick the destination viewport for the UCS by picking a point it in. The Apply option also has a provision for applying the UCS of the current viewport to all viewports.

Another, more roundabout, way to change the UCS of one viewport to match that of another is with the Ucsvp system variable. This system variable can have a value of either 0 or 1. Also, the value of Ucsvp can vary from one viewport to another. When Ucsvp is set to 0, the UCS of the viewport will automatically change to match the UCS of the current viewport. When Ucsvp is set to 1, which is the default value, the UCS of the viewport it is locked to the viewport and is independent of the UCS of the current viewport.

An example of how the Ucsvp system variable works is shown in Figure 2.60 which shows the same viewport configuration three times. Ucsvp is set to 0 in the large viewport on the right. The initial location and orientation of the large viewport's UCS is shown in Figure 2.60A. Notice in Figure 2.60B that as the upper left viewport becomes the current viewport, the UCS in the large viewport changes to match the UCS of the upper left viewport. And notice in Figure 2.60C the large viewport's UCS changes again to match the lower left viewport when it becomes the current viewport.

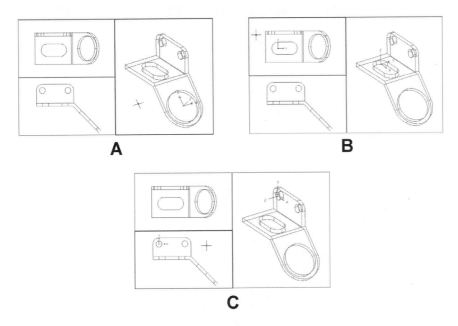

Figure 2.60

THE VPORTS COMMAND

VIEWPORTS, which is almost always abbreviated to VPORTS, is the command that creates and manages tiled viewports. The command displays a dialog box that has two tabs—New Viewports and Named Viewports.

NEW VIEWPORTS

You will use the New Viewports tab, which is shown in Figure 2.61, to create tiled viewports.

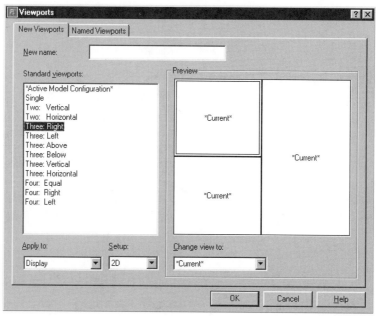

Figure 2.61

STANDARD VIEWPORTS

The names of AutoCAD's twelve standard viewport arrangements, plus the current arrangement, are shown in this list box. Click on a name to highlight and select it.

PREVIEW

The selected viewport arrangement is displayed in this pane. The words within each viewport, such as Current and SE Isometric, refer to the viewpoint that each viewport will have.

NEW NAME

To save the selected viewport arrangement, enter a name in this edit box.

APPLY TO

This drop-down list box contains two options—Display and Current Viewport. When you select Display, then entire AutoCAD graphics area will be divided into the selected viewport arrangement. When you select Current Viewport, the selected viewport arrangement will apply only to the viewport that was current when VPORTS was invoked.

When Single has been selected as the viewport arrangement, the entire graphics area will revert to a single viewport, regardless of the Apply To setting (see Figure 2.62).

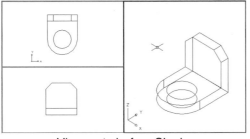
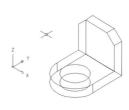

Viewports before Single   Viewports after Single

Figure 2.62

SETUP

Two options—2D and 3D—are in this drop-down list box. When 2D is selected, the viewpoints in all of the new viewports will be the same as that of the current viewport. When 3D is selected, each viewport will have one of the six orthographic views or one of the four isometric views that look down on the XY plane (see Figure 2.63). If the Ucsortho system variable is set to 1, the UCS in viewports for orthographic views will automatically match the line of sight of the viewport.

Existing Viewport and Viewpoint

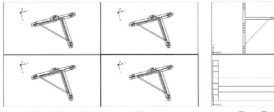
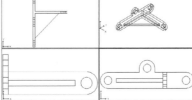

Four Equal Viewports, 2D setup   Four Equal Viewports, 3D setup

Figure 2.63

CHANGE VIEW TO

When 3D has been selected as the Setup, this drop-down list box will contain the names of the six orthographic views and the four isometric views that look down on the XY plane, as well as viewpoint of the current viewport. You can assign any of these views to any of the viewports that are shown in the preview pane. For instance: Suppose you have selected the Three: Right viewport arrangement and the 3D Setup. The default viewpoints are a top view in the upper left viewport, a front view in the lower left viewport, and an SE isometric viewpoint in the right viewport, as shown on the left in Figure 2.64. If you wanted the lower left viewport to have a left view, rather than a front view; and the right viewport to have an SW isometric view, rather than an SE isometric view, you could change the viewpoints as shown on the right in Figure 2.64.

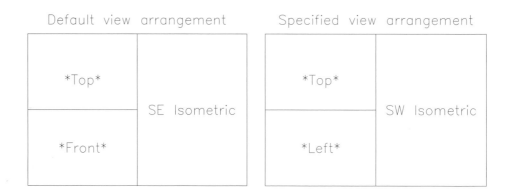

Figure 2.64

NAMED VIEWPORTS

The Named Viewports tab, which is shown in Figure 2.65, of the Viewports dialog box is for managing viewport configurations that have been saved. The names of saved viewport configurations are shown in the list box on the left side of the dialog box. You can select one of these viewport configurations by clicking on it. The Preview image pane will display the viewport arrangement. You can also activate a shortcut menu by right-clicking on a name. This shortcut menu has two options: one is the rename the viewport configuration, and the other is to delete the named viewport configuration.

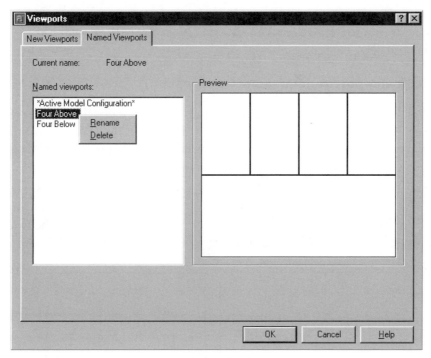

Figure 2.65

COMMAND LINE OPTIONS

If you start VPORTS from the command line and precede the name with a hyphen, a command line prompt for creating and managing tiled viewports will be displayed.

Command: -VPORTS (or -VIEWPORTS)

Enter an option [Save/Restore/Delete/Join/SIngle/?/2/3/4]<3>: (*Enter an option or press ENTER.*)

The first three options (Save, Restore, and Delete) are for named viewport configurations. The next three (Join, SIngle, and ?) are for managing existing viewports. The last three (2, 3, and 4) make new viewports. The command line options for creating viewports (the 2, 3, and 4 options) always divide the current viewport, rather than the entire graphics area. Except for Join and ?, all of these options are available from the dialog box version of VPORTS, and therefore, only the Join and ? options will be described here.

JOIN

Joins two adjacent viewports to create a larger viewport. The follow-up prompts are:

Select dominant viewport <current>: (*Press ENTER or select a viewport.*)

Select a viewport to join: (Select a viewport.)

The resulting viewport will take on the view direction, zoom level, and other appearance characteristics of the dominant viewport. Pressing ENTER will select the current viewport (the one that currently contains the crosshair cursor). To select another viewport, move the cursor to it and press the pick button.

At the second prompt, move the cursor to an adjacent viewport and press the pick button. The resulting viewport must be rectangular—it is not possible to make L-shaped or T-shaped viewports. Thus, in Figure 2.66 you can join only the two small viewports. You cannot join the large viewport with either of the small ones.

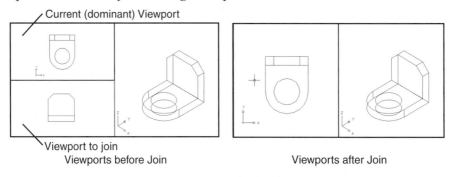

Figure 2.66

?

The question mark option brings up a list of viewports, giving each viewport's identification number and its corner coordinates, along with the names and coordinates of any saved viewport configurations. For example, the screen configuration shown in Figure 2.67 would be listed as:

Current configuration:

id# 3

    Corners: 0.0000,0.5000 0.5000,1.0000

id# 2

    Corners: 0.5000,0.0000 1.0000,1.0000

id# 4

    Corners: 0.0000,0.0000 0.5000,0.5000

Configuration QUAD:

    0.5000,0.5000 1.0000,1.0000

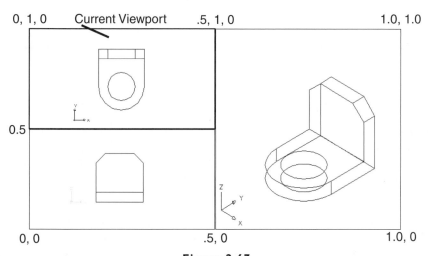

Figure 2.67

Each viewport has an identification number, along with corner coordinates, which are a fraction of the total length and height of the screen. The lower left corner of the screen has the coordinates of 0,0, whereas the upper right corner is 1,1, as shown in the previous figure. The current viewport is listed first. In this example it has an identification number of 3 and its corner coordinates are 0.0,0.5 and 0.5,1.0. AutoCAD reserves viewport id# 1 for the main paper space, so model space viewports will never have an id# lower than 2.

The last item shown in the viewport list are the names and coordinates of saved viewport configurations. In this example there is one saved configuration, which is named QUAD.

 Tips: Named viewport configurations can be saved in a prototype or template drawing to use in quickly setting up viewports with viewpoints you often use. You could, for example, have a named viewport configuration of four viewports showing the model's top, front, and right sides, plus an isometric viewpoint.

Even though AutoCAD 2002 permits object names up to 255 characters long, you may want to keep your viewport configuration names to 31 or fewer characters and use only numbers, letters, underscores, hyphens, and dollar signs in the names to ensure their compatibility with earlier versions of AutoCAD.

Since the VPORTS options for creating viewports can divide the current viewport, rather than the entire screen, you can create viewport arrangements that go beyond the standard options—especially when combined with the Join option. Suppose, for example, you wanted to see your model in one large viewport with three small viewports below it. First, make three horizontal viewports as in A of Figure 2.68. Next, combine the two upper viewports as in B of the figure. Finally, create three vertical viewports in the bottom viewport, as in C of the figure.

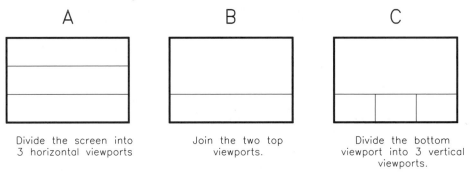

A — Divide the screen into 3 horizontal viewports

B — Join the two top viewports.

C — Divide the bottom viewport into 3 vertical viewports.

Figure 2.68

Although viewport identification numbers are not especially useful from AutoCAD's command line, they are useful in AutoLISP and ADS programming for manipulating viewports. AutoCAD stores the identification number of the current viewport in the Cvport system variable. Programs can use this variable both in finding and setting the current viewport.

Figure 2.69 shows pull-down menus and toolbars for Tiled Viewports.

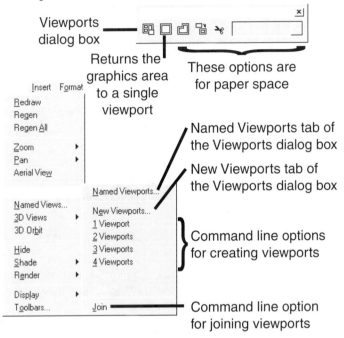

Figure 2.69

TRY IT! - CONSTRUCTING A 3D WIREFRAME MODEL

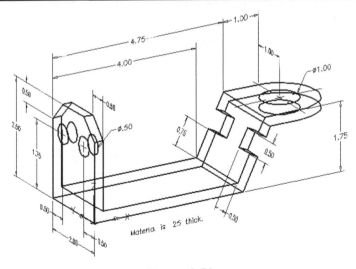

Figure 2.70

In this exercise, you can put your knowledge of the UCS to work in building the 3D wireframe shown in the Figure 2.70. Although it would be possible to build this model without using the UCS, it would be awkward and slow. However, by moving and twisting the UCS, this model can be made easily and quickly using the 2D commands you are used to.

The technique you will use in building this model will be to pick out a plane to draw on and position the UCS to allow drawing in that plane with a pointing device. When everything in that plane is drawn, you will pick out another plane and move and reorient the UCS to draw objects in that plane.

Although the techniques used in constructing this model will be efficient, most of them could be done using other, equally efficient, techniques. This is especially true when setting a new UCS—there will usually be several ways to accomplish the same results.

Drawing Setup:

1. Start a new drawing named BRACKET.DWG. Use the basic ACAD.DWT template or DRAWING.DWG drawing file with English measurement units.

2. Use the ORigin option of UCSICON to have the UCS icon located at the UCS origin whenever possible.

3. Turn on the grid and set its spacing to one unit. Although this setup is not absolutely necessary, you are likely to find the grid helpful in visual izing the location of the XY plane.

4. Activate the snap mode and set the snap setting to 0.25 units. The dimensions of this model are in multiples of 0.25, so drawing in the snap mode will be efficient.

5. Set the number of digits to the right of the decimal point to two to cut down on the number of zeros displayed in coordinates.

6. Make and start using a new layer named WF01. This layer should have a continuous linetype in any color you like.

Building the Wireframe:

The only 3D commands you will use to construct this 3D model will be UCS and VPOINT. The lines, circles, and arcs will be drawn just as if the 3D model were a 2D drawing.

In the following steps, we will describe only the 3D commands in detail.

1. Set up a 3D view in preparation for drawing the model's front and back sides:

Command: VPOINT (*Press ENTER.*)

Current view direction:
VIEWDIR=0.00,0.00,1.00

Specify a view point or [Rotate/] <Display compass and tripod>: R (*Press ENTER.*)

Enter angle in X-Y plane from X-axis   **<270>: 240 or –120**
(Press ENTER.)

Enter angle from X-Y plane <90>: 20 (Press ENTER.)

> PAN and ZOOM as necessary to show the coordinate system's origin plus approximately six or seven units in the X direction. Even though you could draw the model's front and back sides from a plan view, this isometric viewpoint will give you a better picture of what you are doing.

2. Set up the XY plane for drawing the model's front side by rotating the UCS 90° about the X axis:

Command: UCS (Press ENTER.)

Current UCS name: *NO NAME*

Enter an option
[New/Move/orthoGraphic/Prev/Restore/Save/
Del/Apply/?/World]<World>: X (Press ENTER.)

Specify rotation angle about X axis: <90>: (Press ENTER to accept
the default 90° rotation angle.)

> The UCS icon will flip up to its new orientation, the W in the icon will disappear, and the dot grid will fill the entire screen.

3. Use the LINE command and your pointing device to draw the six lines on the front face of the model using the X,Y coordinates shown in the Figure 2.71.

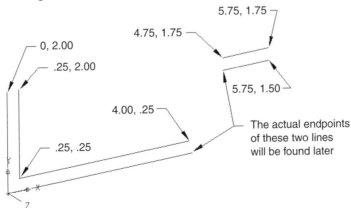

Figure 2.71

Notice that you can draw these lines using AutoCAD's ORTHO and SNAP modes. You will add the slanted lines later because there is a notch in that edge. Also, the actual coordinates of one end of both bot tom lines will be determined later—for now, use the X coordinates of the upper lines.

4. Although you could make the back of the model by moving the UCS two units in the minus Z direction and drawing six lines identical to those on the front, it will be quicker to copy the existing six lines two units in the negative Z direction. You know that the direction is the neg ative Z direction because the UCS icon has a square drawn around its origin. This means that the positive end of the Z axis is pointed out of the computer screen.

Command: COPY (*Press ENTER.*)

Select objects: (*Select the six lines.*)

Specify base point or displacement, or [Multiple]: 0,0,-2 (*Press ENTER.*)

Specify second point of displacement: or <use first point as displacement>: (*Press ENTER.*)

Your model will look like the one shown in Figure 2.72.

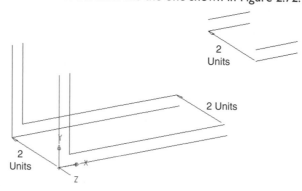

Figure 2.72

5. Now you are ready to move on to the left side of the model. First, use the ZAxis option of the UCS command to set a new UCS.

Command: UCS (*Press ENTER.*)

Current UCS name: *NO NAME*

Enter an option [New/Move/orthoGraphic/Prev/Restore/Save/ Del/Apply/?/World]<World>: ZA (*Press ENTER.*)

Specify new origin point (0.00,0.00,0.00): 0,0,-2 (*Press ENTER.*)

Specify point on positive portion of Z-axis <0.00,0.00,-1.00>: -1,0,-2 (*Press ENTER.*)

> Notice that the coordinates used for the new origin and the new Z axis direction were relative to the current origin. Although the 3point option of UCS would have been more straightforward to use here, we wanted to demonstrate the ZAxis option.

6. Change the viewpoint using the Rotate option of VPOINT to get a better perspective of the model's left side.

Command: VPOINT (*Press ENTER.*)

*** Switching to the WCS ***

Current view direction: VIEWDIR=-0.47,-0.81,0.34

Specify a view point or [Rotate/]l<display compass and tripod>: R (*Press ENTER.*)

Specify angle in XY plane from X axis (240): 210 (*Press ENTER.*)

Specify angle from XY plane <20>: 15 (*Press ENTER.*)

*** Returning to the UCS ***

> Notice in the command line messages that AutoCAD switched to the WCS during the VPOINT command. The model will completely fill the screen, so you may want to zoom back a little for more working space.

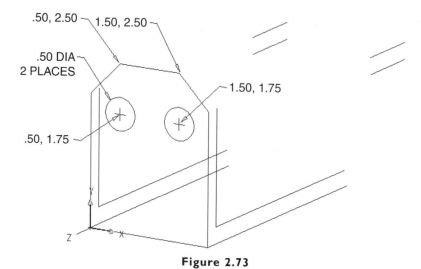

.50, 2.50

1.50, 2.50

.50 DIA
2 PLACES

1.50, 1.75

.50, 1.75

Figure 2.73

7. The vertical edges of the left side of the model were drawn in step 3. Complete this side by drawing the bottom line plus the three top lines, as well as the two circles (which represent round holes), using the coordinates shown in Figure 2.73.

8. Copy the three top lines and the two circles 0.25 units in the minus Z direction.

9. Draw the inside line using the object endpoint snaps shown in Figure 2.74. Although these objects could be made by moving the UCS 0.25 units in the minus Z direction and drawing them in the XY plane, it is probably easier to copy the existing objects and then draw the horizontal line that represents the inside edge of the model with object end point snaps.

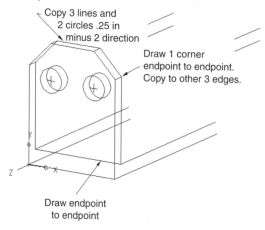

Copy 3 lines and
2 circles .25 in
minus 2 direction

Draw 1 corner
endpoint to endpoint.
Copy to other 3 edges.

Draw endpoint
to endpoint

Figure 2.74

10. Also use object endpoint snaps to draw any one of the corners of the chamfers. Then use three copies of that line to make the remaining three corners. Here, too, the UCS could be moved and oriented to per mit drawing in the XY plane, but it is easier to draw just one line using object endpoint snaps and then make copies of it. Now the model will look like the figure above.

11. We will now move to the slanted area of the model. First, align the UCS using the 3point option of the UCS command in conjunction with the points shown in Figure 2–75. The viewpoint has been rotated back to 240° from the X axis in the XY plane, and 20° from the XY plane. You can set this viewport by stepping back through previous ZOOMs, or by using the VPOINT command. The command line sequence of prompts and input to set the UCS is:

Command: UCS (*Press ENTER.*)

Current UCS name: *NO NAME*

Enter an option [New/Move/orthoGraphic/Prev/Restore/Save/ Del/Apply/?/World]<World>: 3 (*Press ENTER.*)

Specify new origin point (0.00,0.00,0.00): (*Pick point A, using an object endpoint snap.*)

Specify point on positive portion of X-axis <1.00,0.25,-4.25>: (*Pick point B, using an object endpoint snap.*)

Specify point on positive-Y portion of the UCS XY plane <0.00,1.25,-4.25>: (*Pick point C or D, using an object endpoint snap.*)

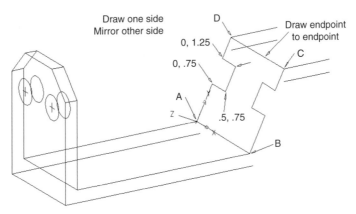

Figure 2.75

12. Now draw the six lines comprising the bottom and one side of the slanted, notched area. Use the coordinates shown in the above figure for drawing the side. As the top of the slanted area does not match your snap setting, you'll have to use an object endpoint snap there. Then mirror the five edge lines to make the other half of the slanted area face. Even though the UCS is tilted relative to the WCS, the MIRROR command works here exactly as it does in 2D.

13. Copy the lines on the slanted area 0.25 units in the minus Z direction. Next, use endpoint snaps to draw a line from one side of the notch to the other. Then make seven copies of it to connect the edges of the notched area, as shown in Figure 2.76.

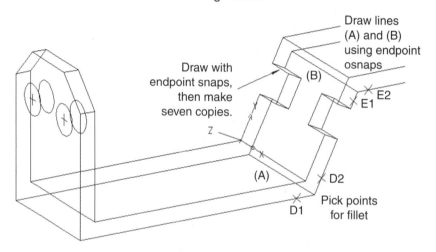

Figure 2.76

14. Connect the lines that make up the edges of the bottom side of the slanted area with the FILLET command, using a 0 radius. Pick near the points labeled D1 and D2 and E1 and E2 in the figure at the top of this column. In AutoCAD versions later than Release 12 these fillets can be made with the UCS oriented as it is. With earlier versions, however, the UCS must be positioned so that the XY plane is parallel to the plane of the objects being filleted. Now your model should look like the one shown in Figure 2.76.

The only section left to make on the wireframe model is the arc and circle area. Before you start drawing in this section, change the view point to one looking down at a steeper angle, from the other direction.

Command: VPOINT (Press ENTER.)

*** Switching to the WCS ***

Current view direction: **VIEWDIR=-0.47,-0.81,0.34**

Specify a view point or **[Rotate/]I<display compass and tripod>: R** (*Press ENTER.*)

Specify angle in **XY** plane from **X** axis (20): **295** (*Press ENTER.*)

Specify angle from **XY** plane <20>: **45** (*Press ENTER.*)

*** **Returning to the UCS** ***

15. Move the UCS to the top of the wireframe with the 3point option of UCS, and the points shown in Figure 2.77.

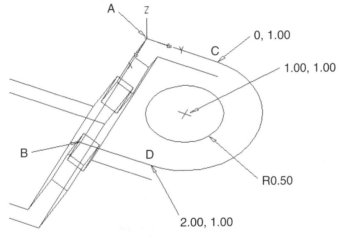

A   Z

C

0, 1.00

1.00, 1.00

B

D

R0.50

2.00, 1.00

Figure 2.77

Command: UCS (*Press ENTER.*)

Current UCS name: *NO NAME*

Enter an option **[New/Move/orthoGraphic/Prev/Restore/Save/Del/Apply/?/World]<W orld>: 3** (*Press ENTER.*)

Specify new origin point (0.00,0.00,0.00): (*Pick point A, using an object endpoint snap.*)

Specify point on positive portion of X-axis <1.00,0.25,-4.25>: (*Pick point B, using an object endpoint snap.*)

Specify point on positive-Y portion of the UCS XY plane <0.00,1.25, -4.25>: (*Pick point C or D, using an object endpoint snap.*)

16. Draw the arc using the Start-Center-End method of the ARC command with the coordinates shown in Figure 2.77. Then draw the one-unit-diameter circle, which represents a round hole, centered on the coordinate shown in Figure 2.77.

17. The last step in this wireframe is to copy the arc and circle 0.25 units in the minus Z direction, as shown in Figure 2.78. Then, your wireframe should be similar to the one shown in Figure 2.70 at the beginning of this exercise, minus the dimensions.

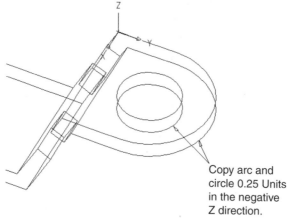

Copy arc and circle 0.25 Units in the negative Z direction.

Figure 2.78

You can compare your version of this wireframe with the one in file 3d_ch2_02.dwg on the CD-ROM included with this book. In Chapter 6 you will make this same model as a 3D solid.

COMMAND REVIEW

3DORBIT

Dynamically sets general viewpoints in 3D space. Moreover, some commands associated with 3DORBIT allow you to use shaded images of 3D surface and solid models in setting viewpoints.

DDVPOINT

Uses a dialog box for setting 3D viewpoints, including plan views.

DVIEW

An enhanced but more complicated version of VPOINT. DVIEW can set views in which distant objects appear to be smaller than foreground objects (perspective views), views that have an invisible wall (a clipping plane) for hiding portions of a 3D model, as well as set viewpoints similar to **VPOINT**.

ELEV

This command moves the drawing plane away from, but always parallel to, the **XY** plane.

MVIEW

This command creates and manages paper space's floating viewports.

PLAN

Sets a viewpoint looking straight down toward the **XY** plane. It is similar to the **VPOINT** coordinates of 0,0,1.

PROPERTY

The **PROPERTY** window has provisions for changing the thickness of existing objects, but not their elevation.

REDRAWALL

This command forces a **REDRAW** in all viewports.

REGENALL

This command forces a regeneration in all viewports.

UCS

Controls the location and orientation of the **UCS**—not the **UCS** icon.

UCSICON

This command controls the icon that shows the location and orientation of the current **UCS**.

UCSMAN

This command uses a dialog box for managing saved and orthographic **UCS**s.

VIEW

Saves and restores views. **3D** viewpoints can be saved. In AutoCAD 2002, **VIEW** has preset views for the six orthographic views and the four isometric views that look down on the **XY** plane.

VPOINT

Sets general viewpoints within **3D** space.

SYSTEM VARIABLE REVIEW

CVPORT

This variable stores the identification number of the current viewport.

ELEVATION

This variable stores the distance that the current drawing plane is offset from the XY plane.

THICKNESS

This variable stores the current extrusion thickness of an object.

TILEMODE

This variable controls whether AutoCAD operates in model space mode or in paper space mode. When AutoCAD is in paper space mode, the UCS icon is shaped like a 30–60° drafting triangle.

UCSICON

This variable contains the UCS icon settings for the current viewport.

UCSAXISANG

The default rotation angle offered in the prompts for the UCS options that rotate the UCS about the X, Y, or Z axes is stored in this system variable.

UCSBASE

This system variable stores the name of the base UCS for defining the origin and axes orientation of the UCS orthographic options.

UCSFOLLOW

When Ucsfollow is set to 1, AutoCAD will switch to a plan view whenever the UCS is changed.

UCSNAME

This variable stores the name of the current UCS. It is a read-only variable, so it cannot be used to select a named UCS.

UCSORG

This read-only variable stores the WCS coordinates of the current UCS.

UCSORTHO

When Ucsortho is set to 1, the 3D setup in the New Viewports tab of the Viewports dialog box will cause the UCS in viewports having orthographic views to match the viewpoint.

UCSVP

This system variable controls whether the UCS in a viewport remains fixed or changes to match the UCS of the current viewport.

UCSXDIR

This read-only variable stores the direction of the current UCS X axis relative to the WCS.

UCSYDIR

This read-only variable stores the direction of the current UCS Y axis relative to the

WCS.

VIEWDIR

This variable stores the **X,Y,Z** viewpoint coordinates.

WORLDVIEW

When **Worldview** is set to its default value of 1, the **VPOINT** command uses the **WCS** for angles and view direction coordinates. If a **UCS** is in effect, **AutoCAD** temporarily switches back to the **WCS** for the duration of the **VPOINT** command. When **Worldview** is set to 0, AutoCAD does not switch to the **WCS** when setting a viewpoint.

CHAPTER PROBLEMS

Use the tools and knowledge you have acquired in this chapter to draw the 3D wireframe models shown in the following figures. Notice that in each figure one point is labeled A and another is labeled B, and that the distance between those two points is listed below the figure. You should verify the accuracy of each of your completed models by measuring the distance between those two points (with AutoCAD's DISTANCE command and endpoint and quadrant object snaps) on your model.

   Completed versions of these five wireframe models are in file 3d_ch2_03.dwg on the CD-ROM.

Problem 2-1

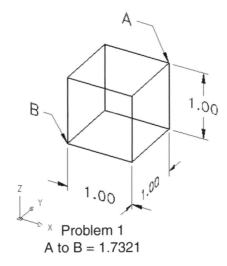

Problem 1
A to B = 1.7321

Figure 2.79

Problem 2-2

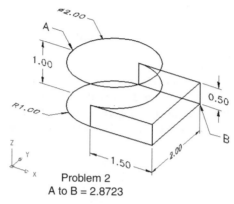

Problem 2
A to B = 2.8723

Figure 2.80

Problem 2-3

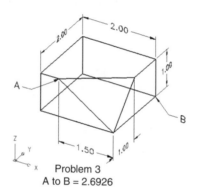

Problem 3
A to B = 2.6926

Figure 2.81

Problem 2-4

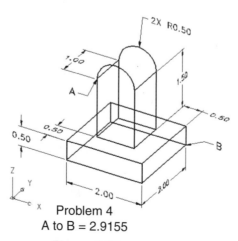

Problem 4
A to B = 2.9155

Figure 2.82

Problem 2-5

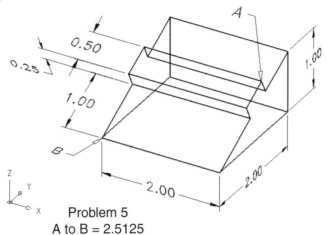

Problem 5
A to B = 2.5125

Figure 2.83

CHAPTER REVIEW

The answers to the following questions are in the Instructor's Guide.

1. Match the UCS option from the list on the left with an action from the list on the right.

_____ a. Object 1. Moves and orients the UCS according to an existing object.

_____ b. Origin 2. Moves and orients the UCS relative to the direction of the Z axis.

_____ c. Previous 3. Moves, but does not reorient, the UCS to a new location.

_____ d. Restore 4. Restores a named UCS.

_____ e. View 5. Restores the last UCS position and orientation.

_____ f. Z 6. Rotates the UCS about the Z axis.

_____ g. Zaxis 7. Rotates the XY plane to face the view direction.

_____ h. Face 8. Places the XY plane on a planar surface on a 3D solid.

_____ g. Apply 9. Copies the UCS within the current viewport to a selected viewport.

Directions: Answer the following questions.

2. A line's start point is at the origin, and its endpoint is established by the spherical coordinates of 3.5<56.4<-25. How long is the line? Is the end of the line on the positive or negative side of the XY plane, as defined by the Z axis?

3. If you were to start a line at the WCS origin and use the cylindrical coordinates of 5<90,2 to establish its endpoint, what would be the absolute X,Y,Z coordinates of its endpoint?

4.   Refer to Figure 2.84 as you answer these questions:

_____   a.   Which coordinate system is in effect, the WCS or the UCS?

_____   b.   Is the UCS icon located at the origin?

_____   c.   Is the viewpoint looking from the positive Z direction toward the XY plane?

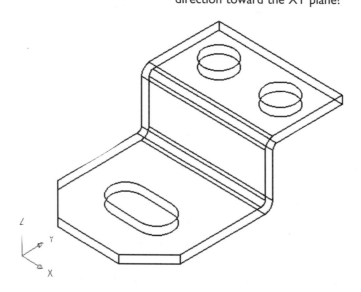

Figure 2.84

5.   Name four UCS options that only rotate the UCS about the origin but do not move the origin.

Directions: Circle the letter corresponding to the correct response in each of the following.

6.   If you save the current UCS, the current viewpoint will be saved as well and will be restored when the UCS is restored.

a. true

b. false

7.   The VIEW command has no relationship with UCSs.

a. true

b. false

CHAPTER 3

Building Wireframe Models

LEARNING OBJECTIVES

This chapter will explain how to draw and modify wireframe objects in 3D space. When you have completed Chapter 3, you will:

- Know how the AutoCAD object types you have used to make 2D drawings can be used in 3D space to make wireframe models.

- Be familiar with AutoCAD's specialized 3D wireframe modification commands.

- Be able to draw complex curves that twist and turn through 3D space.

BUILDING WIREFRAME MODELS

Now you have a good background in AutoCAD's 3D coordinate systems, along with some experience in setting viewpoints from space and using multiple viewports. Therefore, we can move on to creating and modifying objects in 3D space. First, we will begin a detailed look into how the objects you have always used in 2D drawings are handled in 3D models. We will refer to them as 2D objects because each individual object is confined to one plane; even through that plane is not necessarily parallel to the XY plane. Then we will move on to AutoCAD's 3D curves. These are objects than are not confined to a single plane—they can twist and turn through 3D space. Both 2D- and 3D-type entities are used to make wireframe models.

Wireframes, in which objects are represented only by their edges, are the simplest kind of 3D models. These models appear to be made of thin sticks or wire, and they can never have a realistic appearance because there is no surface between their edges. They cannot be rendered, and they cannot hide objects that are behind them. Despite these limitations, wireframes are an important type of 3D model.

- First of all, wireframes are the basis—the skeleton—for virtually all surface models (see Figure 3.1). As we will see in Chapter 4, most AutoCAD surfaces require one or more wireframe boundary objects to define the shape and extent of the surface.

- Several programs for stamped sheet metal products are able to make unfolded patterns from AutoCAD 3D wireframe models (see Figure 3.2). Some of these unfolding programs run within AutoCAD, whereas others are stand-alone programs that can import AutoCAD models as DXF files. These programs are needed because product designers work with a model of a completed (folded) sheet metal part, but manufacturers of the part need a flat, unfolded version of it because they must cut out and shape the part from flat sheets of metal.

- Wireframe models are also useful for preliminary design layouts in which you are seeking to establish sizes, locations, and distances. For example, to begin the design of a pump room that will be crowded with equipment and piping, you would first use just pipe center lines, along with envelope outlines of key equipment and structure locations, to establish positions. At this stage in the design, you are more interested in how things will fit together than how they will look.

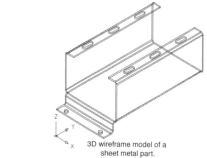

3D wireframe model of a
sheet metal part.

Figure 3.1

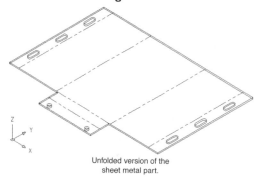

Unfolded version of the
sheet metal part.

Figure 3.2

- The first stages in designing buildings and structures will also be done using wireframes. Here, too, you are primarily interested in locating and positioning objects, and 3D wireframe models are not only easy to work with but they can provide accurate dimensions in three dimensions, often eliminating the need for distance calculations.

2D OBJECTS IN 3D SPACE

Most of the objects used in 3D wireframe models are the same 2D objects used in 2D drawings. Therefore, we will explore the behavior of these objects in 3D space before we move on to 3D curves. We will even cover some AutoCAD specialty objects, such as solids and traces, that could be used in the construction of wireframes (but seldom are).

POINT

Although, strictly speaking, a point is a 1D object having only location—no length, width, or height—points can be placed anywhere in space. AutoCAD's special formats for points, such as X's and boxes (set by the Pdmode system variable), however, are always drawn parallel to the XY plane.

LINE

Although a line, being the trace of a point moving in one direction, is confined to a single plane, they are fully 3D in that there is no restriction in locating them in space. Their endpoints can have virtually any combination of X,Y,Z coordinates, although you need to use either object snaps or typed-in coordinates to establish line endpoints off the current XY plane. You will use lines frequently as you build wireframes.

RAY

Rays are a special form of lines introduced with Release 13 that extend for an infinite distance from their starting point. They are intended for construction purposes, so you don't normally use them as part of a wireframe model. Like lines, rays can start from any point in space and point in any direction.

XLINE

Xlines were also introduced with Release 13 and are intended for construction purposes. They are lines that have no endpoints, extending an infinite distance in both directions. There are no restrictions in their directions, except the Hor and Ver options of the XLINE command places the xline only on the current XY plane.

MLINE

Mlines can only be drawn in a plane parallel to the current XY plane. If the starting point is off of the XY plane, AutoCAD will keep the rest of the mline at that same elevation. If you pick a vertex point (using an object snap) that has a different elevation, it will be projected onto the original plane. Also, AutoCAD will not accept a typed-in Z coordinate.

CIRCLE

AutoCAD always draws circles in a plane parallel to the current XY plane. Usually, you will select a center point on the current XY plane, and the resulting circle is in that plane. You can, however, specify a center point that is off the current XY plane by either typing in coordinates or using an object snap. The resulting circle will be drawn parallel to the XY plane at the Z elevation of its center point. Any pointing you do to specify a circle radius merely inputs a distance, not direction. Therefore, there is no way to draw a circle that is tilted relative to the current XY plane.

ARC

Like circles, arcs are a 2D entity, always lying in a plane parallel to XY plane. This is to be expected since an arc is simply a segment of a circle. If you select points that have different Z coordinates as you draw an arc, AutoCAD projects those selected points onto a plane that is parallel to the current XY plane. The plane on which the arc is drawn has the first point's Z elevation. Consequently, the surest way to draw an arc in 3D space is to first orient the user coordinate system (UCS) so that the XY plane is positioned to match the plane of your desired arc.

POLYLINE

Polylines, made with the PLINE command, must also be drawn in a plane parallel to the current XY plane. AutoCAD will accept a point that is off the current XY plane as the first point of a polyline, but all subsequent points will then have the same Z coordinate as the initial point. AutoCAD will ignore typed-in Z coordinates for the remaining vertex points and will project points that are selected through object snaps onto the original Z elevation plane. The AutoCAD manuals often refer to these polylines as 2D polylines to distinguish them from 3D polylines, which we will cover shortly. As with arcs and circles, the best way to draw 2D polylines is to position the XY plane in the plane in which you want the polyline to lay. This is also true for the other entities in the 2D polyline family—polygons, donuts, and ellipses.

HATCH

AutoCAD always draws hatch patterns on the current XY plane. Furthermore, if you use the Pick Points option of the BHATCH command, the areas picked must be on the XY plane. If you use HATCH or the Pick Objects option of BHATCH, the objects selected can be off the XY plane, but the hatch will be projected onto the XY plane rather than be within the selected objects (see Figure 3.3).

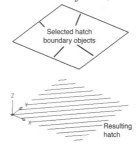

Figure 3.3

MODIFYING 2D OBJECTS IN 3D SPACE

It is not unusual to spend as much time editing and modifying objects as you do creating them. This is not necessarily due to errors or design changes. Some objects cannot be created in the form you want, so you must draw them and then modify them. At other times you may use copies of existing objects, which must be modified to create new ones. Consequently, AutoCAD has a rich assortment of tools for modifying objects.

In this section, we will describe how the familiar AutoCAD editing commands—MOVE, MIRROR, BREAK, FILLET, CHAMFER, TRIM, and EXTEND—work in 3D space. Because you undoubtedly have used these commands extensively in 2D drafting, we will not go into detail as to how to initiate the commands or even how they work, except when they work differently in 3D space than they do in 2D.

MOVE AND COPY

These two commands work the same in 3D space as they do in 2D. You will use them frequently as you build 3D models. Both of these commands first ask for a base point and then for a displacement point. You may use any of the methods that were discussed in Chapter 2 to specify these points.

MIRROR

MIRROR makes a mirror-image copy of an object by reflecting it behind a plane, although it is not obvious that you are using a plane because you pick just two points. The reflection plane is always perpendicular to the XY plane. Consequently, in 3D you must have the object you intend to mirror on the XY plane, or on a plane that is parallel to the XY plane (see Figure 3.4). Later in this chapter we will discuss another AutoCAD mirror command, MIRROR3D, which allows more flexibility in specifying a reflection plane.

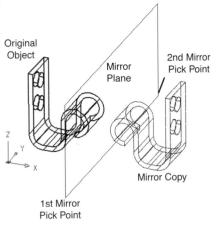

Figure 3.4

ROTATE

This command rotates objects around an axis that is perpendicular to the XY plane (in other words, the rotation axis always points in the Z direction). The rotation point selected establishes the XY plane location of the axis. The objects selected to be rotated are revolved around this axis regardless of how they are oriented relative to the axis or to the XY plane (see Figure 3.5). Later in this chapter we will cover a 3D specialty command, ROTATE3D, that gives you more choices in orienting the rotation axis.

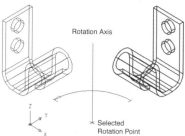

Figure 3.5

ARRAY

The ARRAY command activates a dialog box for making multiple copies of objects. The copies can be arranged in rows and columns with the Rectangular option or in a circular pattern with the Polar option. Both options work well in 3D space, though they are always relative to the XY plane. For rectangular arrays, rows are always in the Y direction and columns are always in the X direction; polar arrays are always around an axis that is perpendicular to the XY plane. The objects to be arrayed, however, do not have to be on the XY plane, nor do they have to be parallel to it.

AutoCAD has another array command, 3DARRAY, which allows you to make rectangular arrays in the Z direction as well as in the X and Y directions and to specify a direction of the axis for polar arrays. We will discuss that command later in this chapter.

BREAK

AutoCAD's BREAK command is used to break an object, such as a line or arc, by picking two points. The object then becomes two separate objects; if the two points are spaced apart there will be a gap between the two objects. BREAK can also be used to shorten one end of an object by picking the first point on the object and picking the second point beyond the object.

FILLET AND CHAMFER

Both of these related commands require two open objects, such as lines and arcs. (An option for both commands allows them to also work on a polyline.) FILLET connects the end of the objects with an arc, while CHAMFER connects their ends with a beveled, or angled, line. The two objects do not have to touch, and if you specify a fillet radius or a chamfer distance of 0, AutoCAD will simply connect the two objects. You can even fillet two lines that are parallel. AutoCAD will connect them with an arc having a radius equal to half the distance between the two lines.

AutoCAD will also perform filleting and chamfering operations regardless of how the objects are oriented relative to the UCS. The only requirement is that the two objects be in the same plane. Thus, in Figure 3.6, AutoCAD will chamfer lines A and G, as well as A and F. AutoCAD will fillet those same pairs, plus the pairs of A and B, A and C, and A and D. However, you cannot fillet lines A and E because they are not coplanar.

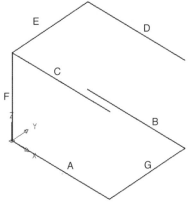

Figure 3.6

EXTEND AND TRIM

These two commands perform similar, but opposite operations. EXTEND lengthens an object to meet a boundary object. TRIM, on the other hand, trims objects back to the point where they intersect a boundary object. Before Release 13 all of the objects had to be in the same plane, and that plane had to be parallel to the current XY axis. Furthermore, the objects actually had to intersect the boundary for TRIM and potentially intersect for EXTEND.

Also, the system variable Projmode controls both the EXTEND and TRIM commands. This variable, which acts as a three-way switch, can have a value of 0, 1, or 2.

When Projmode is set to 0, objects to be extended must be pointed so that they will actually run into the boundary, whereas objects to be trimmed must actually intersect the boundary. This means that EXTEND and TRIM work as they did in earlier AutoCAD releases, except that the current orientation of the UCS is of no consequence. The objects, however, cannot be in different planes, even if those planes intersect. The line and the two circles in Figure 3.7 are perpendicular to the current UCS XY plane. Nevertheless, the line can be used to trim either circle, and either circle can be used to trim the line. However, neither circle can be used to trim the other.

When Projmode is set to 1, its default setting, the boundary object does not have to have any points in the same plane as the objects being trimmed or extended. The boundary is projected in a perpendicular direction, relative to the current XY plane, onto the plane of the objects to be extended/trimmed. In the left side of Figure 3.8, the line labeled B is at a higher elevation than the circle. Nevertheless, if it is selected as a trim boundary, it will be projected down to the circle's plane to trim the circle as shown on the right side of Figure 3.8.

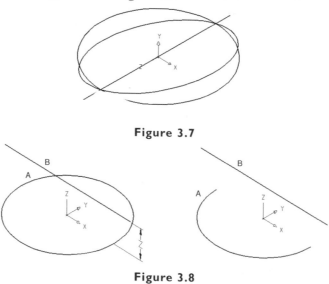

Figure 3.7

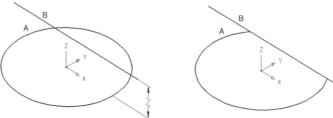

Figure 3.8

Lastly, when Projmode is set to 2, objects are trimmed or extended according to their apparent intersection with the boundary in the current view, as shown in Figure 3.9, which uses the same circle and line shown in the previous figure. The effects of this setting of Projmode should not be confused with those of the Edgemode system variable. In effect, Edgemode can extend boundaries so that objects are extended or trimmed according to their apparent intersection with the lengthened boundary. Projmode, on the other hand, projects the boundary in the view direction onto the plane of the object to be extended or trimmed.

Figure 3.9

These results from the setting of Projmode are summarized in the following table. You can also control the projection mode directly through the EXTEND and TRIM commands. After you select a boundary, AutoCAD offers a Project option. When you select this option, AutoCAD will give three projection type options—None/Ucs/View—with the current setting of Projmode offered as a default. The None option gives the same results as a Projmode setting of 0; the Ucs option is equivalent to a Projmode setting of 1; and the View option is the same as when Projmode has a value of 2. These options of the EXTEND and TRIM command do not change the setting of Projmode; rather, they allow you to override Projmode.

Table 3.1 Projmode settings and results

Projmode Setting	Results
0	No projection. The boundary must be in the same plane as the objects to be trimmed or extended.
1	UCS plane projection. The boundary is projected perpendicularly, relative to the XY plane, onto the plane of the objects to be trimmed or extended.
2	View projection. Objects are extended or trimmed according to the apparent, view-dependent position of the boundary.

Suppose you are drawing a 3D wireframe model for a house that has a front extension with a roof that intersects the main roof of the house. You can easily start a line representing the peak of the secondary roof, as shown in Figure 3.10, but the endpoint of this line is unknown.

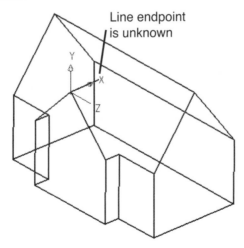

Figure 3.10

You can find the line's endpoint by switching to a viewpoint that shows the side elevation of the house. Then use the EXTEND command, with the View option, to extend the line to the profile of the main roof; see Figure 3.11. (If the line happened to be too long, you would use the TRIM command, rather than EXTEND).

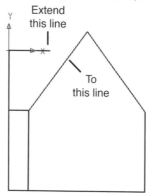

Figure 3.11

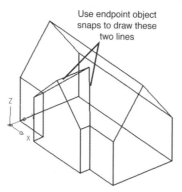

Figure 3.12

To complete the roof, you would change back to an isometric-type view and use object endpoint snaps to draw the two roof valleys, as shown in Figure 3.12.

SPECIALIZED 3D MODIFICATION COMMANDS

AutoCAD has four special modification commands for working in 3D space. Although these commands originate in AutoLISP and ADS (Autodesk Development System) programs, the programs are loaded automatically and are virtually indistinguishable from AutoCAD's built-in commands. Three of these commands—ROTATE3D, MIRROR3D, and 3DARRAY—are enhanced 3D versions of their corresponding 2D commands; a fourth, ALIGN, has no 2D counterpart. ALIGN has options that permit it to serve as both a 2D and a 3D command. All four of these commands can be used with surface and solid models as well as with wireframes. The commands can be started from the command line or the 3D Operation submenu in the Modify pull-down menu, shown in Figure 3.13.

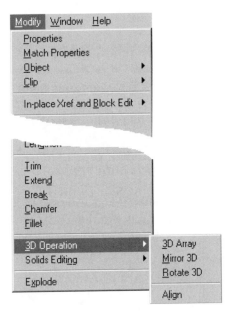

Figure 3.13

THE ALIGN COMMAND

ALIGN is a 3D move-and-rotate type of command. Beginning with Release 14, you can also scale the objects as they are moved and rotated. After selecting the objects to be moved, AutoCAD prompts you to specify up to three pairs of source and destination points. The first pair of points establishes the displacement for the move, the second pair rotates the object, and the third pair of points tilts the object. You can stop the command any time after the first pair of points has been entered by pressing ENTER when prompted for a source point. We will show four examples of the command using the same objects. The first example will stop the command after the first pair of points. It will simply move the box shown on the left in Figure 3.14 to the top of the wedge.

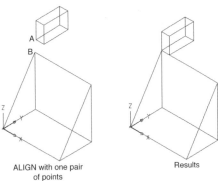

ALIGN with one pair   Results
of points

Figure 3.14

Command: ALIGN

Select objects: *(Select the box.)*

Specify first source point: *(Pick the indicated point on the box at "A".)*

Specify first destination point: *(Pick the indicated point on the wedge at "B".)*

Specify second source point: *(Press ENTER.)*

As you select the source and destination points, AutoCAD connects them with a line, which disappears when the object is moved. The results, which are the same as if the MOVE command was used, are shown on the right in Figure 3.14.

The next example will use two pairs of points to move the box shown on the left side of Figure 3.15 to the top of the wedge, and then rotate it.

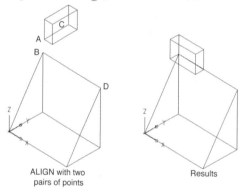

ALIGN with two
pairs of points

Results

Figure 3.15

Command: ALIGN

Select objects: *(Select the box.)*

Specify first source point: *(Pick the indicated point on the box at "A".)*

Specify first destination point: *(Pick the indicated point on the wedge at "B".)*

Specify second source point: *(Pick the indicated point on the box at "C".)*

Specify second destination point: (*Pick the indicated point on the wedge at "D".*)

Specify third source point or <continue>: (*Press ENTER.*)

Scale objects to alignment points? [Yes/No] <No>: (*Press ENTER.*)

The results are shown on the right side of Figure 3.15. If you respond to the prompt for scaling the objects with a Yes, the sizes of the objects are changed as they are moved and rotated. The scale factor used is the ratio of the distance between the two destination points and the distance between the two source points. The results when the Scale objects to alignment points option is selected with our example are shown in Figure 3.16.

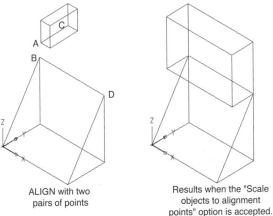

ALIGN with two
pairs of points

Results when the "Scale
objects to alignment
points" option is accepted.

Figure 3.16

Finally, we will repeat the command using all three pairs of source and destination points.

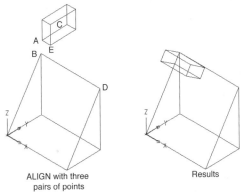

ALIGN with three
pairs of points

Results

Figure 3.17

Command: ALIGN

Select objects: *(Select the box.)*

Specify first source point: *(Pick the indicated point on the box at "A".)*

Specify first destination point: *(Pick the indicated point on the wedge at "B".)*

Specify second source point: *(Pick the indicated point on the box at "C".)*

Specify second destination point: *(Pick the indicated point on the wedge at "D".)*

Specify third source point or <continue>: *(Pick the indicated point on the box at "E".)*

Specify third destination point: *(Pick the indicated point on the wedge at "F".)*

The third pair of points tilted the moved object so that the plane defined by the three source points matches the plane defined by the three destination points.

Tip: Use object snaps when picking the source and destination points. This permits the command to operate independently of the UCS and lets you concentrate on the object being relocated.

THE ROTATE3D COMMAND

AutoCAD's standard ROTATE command always rotates objects around an axis that is parallel to the Z axis. ROTATE3D, on the other hand, allows you to use a rotation axis pointed in any direction. The command prompts you for an axis and then for the rotation direction.

The axis has positive and negative ends for controlling the direction of rotation, which follows the right-hand rule. An easy way to visualize the rotation direction is to mentally grasp the axis with your right hand so that your fingers are curled around the axis and your thumb points toward the positive end of the axis. Positive rotation will then be in the direction your fingers are curled.

The format for the command is:

Command: ROTATE3D

Current positive angle: ANGDIR=counterclockwise   ANGBASE=0

Select objects: (*Use any object selection method.*)

**Specify first point on axis or define axis by
[Object/Last/View/Xaxis/Yaxis/Zaxis/2points]: (*Specify an option, a
point, or press ENTER.*)**

2POINTS

This option, which is the default option, uses two points to define the rotation axis.
Pressing ENTER or typing in a 2 brings up the following prompts:

Specify first point on axis: (*Specify a point.*)

Specify second point on axis: (*Specify a point.*)

If you enter a point from the main prompt, AutoCAD will skip the first prompt. The
positive direction of the axis is from the first point to the second. See Figure 3.18 for
the result.

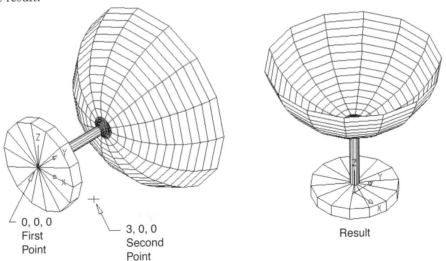

Figure 3.18

OBJECT

This option uses an existing object to define the axis. A line, circle, arc, or 2D polyline may be used as an axis object, with the axis location and direction as shown in
Table 3.2.

Table 3.2: Axis Objects

Object Type	Axis Location	Axis Orientation
Line	The line itself	Positive Z direction, or positive X or Y direction if the line is parallel to the XY plane
Circle	Center of circle	Extrusion direction of circle
Arc	Center of arc	Extrusion direction of arc
Polyline arc segment	Center of arc	Extrusion direction of polyline
Polyline line segment	The line segment	Same as for Line objects

LAST

This option uses the previous rotation axis. If there is no last axis, AutoCAD displays a message and repeats the main prompt.

VIEW

The View option sets up an axis parallel to the current viewing direction. A follow-up prompt asks for a point to position the axis.

Specify a point on the view direction axis <0,0,0>: *(Specify a point, or press ENTER.)*

XAXIS/YAXIS/ZAXIS

Each of these options orients the rotational axis parallel with one of the cardinal axes of the current UCS. A follow-up prompt asks for a point where N is either X, Y, or Z.

Specify a point on the N axis <0,0,0>: *(Specify a point, or press ENTER.)*

The rotation axis will run through the selected point, with its direction the same as the selected coordinated system axis. Selecting the Zaxis option is equivalent to using the standard ROTATE command.

After you define an axis, AutoCAD will issue the following prompt for specifying a rotation angle about the axis:

Specify rotation angle or [Reference]: *(Specify an angle, or enter R.)*

- Rotation angle
- The object will be rotated by the specified angle, which can be input either by pointing or by typing in a number.
- Reference
- This option uses two pairs of points to show AutoCAD a rotation angle—a base point and a point relative to it for the original angle, then another base point and a point relative to it for the new angle.

This first example will use the default two point option of ROTATE3D to rotate a wireframe box. The object before it is rotated is shown on the left in Figure 3.19, while after it has been rotated is shown on the right.

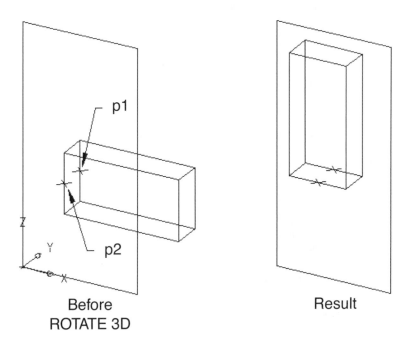

Before
ROTATE 3D

Result

Figure 3.19

Command: ROTATE3D

Current positive angle: ANGDIR=counterclockwise   ANGBASE=0

Select Objects: *(Select the box.)*

Specify first point on axis or define axis by
   [Object/Last/View/Xaxis/Yaxis/Zaxis/2points]: *(Pick point p1.)*

Specify second point on axis: *(Pick point p2.)*

Specify rotation angle or [Reference]: 90

The next example uses the Zaxis option to set up an axis, and the reference option to specify a rotation angle. The arc-shaped object on the left side of Figure 3.20 will be rotated into the position shown on the right side. Note the position of the User Coordinate System icon when performing this operation.

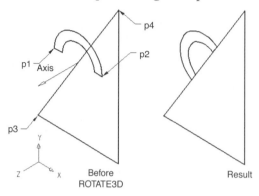

Figure 3.20

Command: ROTATE3D

Current positive angle: ANGDIR=counterclockwise   ANGBASE=0

Select Objects: *(Select the arc-shaped object.)*

Specify first point on axis or define axis by [Object/Last/View/Xaxis/Yaxis/Zaxis/2points]: Z

Specify point on the Z axis <0,0,0>: *(Pick midpoint of line between p3 and p4.)*

Specify the rotation angle or [Reference]: R

Specify the reference angle <0>: *(Pick point p1.)*

Specify second point: *(Pick point p2.)*

Specify the new Angle: *(Pick point p3.)*

Specify the second point: *(Pick point p4.)*

In this last example the rotation axis is based on an arc. The axis is located in the center of the arc shown on the left side of Figure 3.21, and is pointed in the arc's extrusion direction. The rotated keyhole-shaped object is shown on the right. In this figure, the extrusion direction of the arc is in the current Y direction.

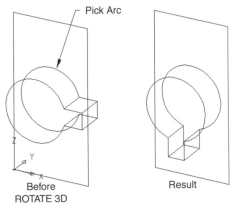

Figure 3.21

Command: ROTATE3D

Current positive angle: ANGDIR=counterclockwise   ANGBASE=0

Select Objects: *(Select the two arcs and 10 lines.)*

**Specify first point on axis or define axis by
    [Object/Last/View/Xaxis/Yaxis/Zaxis/2points]: O**

Select a line, circle, arc or 2d-polyline segment: *(Pick one of the arcs.)*

Specify the rotation angle or [Reference]: 90

THE MIRROR3D COMMAND

This command allows you to mirror objects to the opposite side of a plane positioned anywhere in space. Although the standard MIRROR command works well with 3D objects, the mirroring plane is always perpendicular to the XY plane. Like MIRROR, MIRROR3D gives you the choice of retaining or deleting the original objects.

When you invoke MIRROR3D, AutoCAD prompts you to select the objects to be mirrored, then displays a menu for specifying a mirroring plane. One or more prompts, depending on the mirror plane definition choice, will follow the main prompt.

Command: MIRROR3D

Select objects: *(Use any object selection method.)*

Specify first point of mirror plane (3 points) or

[Object/Last/Zaxis/View/XY/YZ/ZX/3points]<3 points>: *(Specify an option, a point or press ENTER.)*

3POINTS

This option, which is the default option, uses three points to define the mirror plane. Pressing ENTER or typing in a 3 brings up the following three prompts:

Specify first point on mirror plane: *(Specify a point.)*

Specify second point on mirror plane: *(Specify a point.)*

Specify third point on mirror plane: *(Specify a point.)*

If you entered a point from the main prompt, AutoCAD will skip the first point prompt.

PLANE BY OBJECT

This option uses an existing circle, arc, or 2D polyline to define a mirror plane. The mirror plane is in the same plane as the selected object, but it extends completely through space, so that the actual location of the object is of no consequence. Notice that although a line is not accepted for defining a mirror plane, a 2D polyline is, even if it consists of just a single line segment. 2D polylines, you will recall, are always made on (or parallel to) the XY plane. Therefore the plane that the 2D polyline was constructed on can be used as a mirror plane, even if it is not parallel to the current XY

plane. A follow-up prompt for this option will ask for an object to be used as a mirror plane.

Select a circle, arc or 2D-polyline segment: *(Select an object.)*

LAST

This option uses the previous mirror plane. If there is no previous plane, AutoCAD will repeat the menu for choosing a mirror plane.

ZAXIS

A single line that has one end on the plane and is pointed perpendicularly from the plane can define any plane. In geometry, such a line is called the normal of the plane. This option has two follow-up prompts for point locations. The first point establishes the location of the plane, and the second establishes the direction of the plane's normal.

Specify point on mirror plane: (*Specify a point.*)

Specify point on Z-Axis (*normal*) of mirror plane: (*Specify a point.*)

VIEW

This option, which uses the current view direction as a normal of the mirror plane, prompts for a point to set the plane's location.

Specify point on view plane <0,0,0>: (*Specify a point.*)

XY/YZ/ZX

Each of these three options use mirror planes that are parallel to the three principal coordinate system planes. The XY option uses a mirror plane parallel to the XY plane, the YZ option uses one that is parallel to the YZ plane, and ZX uses one parallel to the ZX plane. The follow-up prompt to these options is:

Specify point on MN axis <0,0,0>: (*Specify a point or press ENTER.*)

AutoCAD will replace MN with XY, YZ, or ZX, depending on your option choice. The mirror plane will pass through the point specified from this prompt, with the default location at the coordinate system origin. The YZ and the ZX options are equivalent to using the standard MIRROR command when mirror point pairs parallel to the X or Y axis are used.

Figure 3.22 illustrates XY, YZ, and ZX options using the same object, in the same position, with the coordinate system origin as the mirror plane location point.

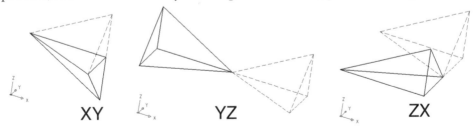

Figure 3.22

After you have defined a mirror plane, AutoCAD will ask if you want the original objects retained or deleted.

Delete source objects? [Yes/No] <N>: (*Enter Y or N, or press ENTER.*)

For the following example of the 3point option, the object before it is mirrored along with the three mirror plane points is shown on the left in Figure 3.23. The results are on the right, with the position of the original object shown in dashed lines.

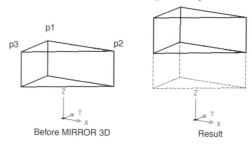

Before MIRROR 3D   Result

Figure 3.23

Command: MIRROR3D

Select objects: (*Select the wedge.*)

Specify first point of mirror plane (3 points) or

[Object/Last/Zaxis/View/XY/YZ/ZX/3points]<3 points>: (*Pick point p1.*)

Specify second point on mirror plane: (*Pick point p2.*)

Specify third point on mirror plane: (*Pick point p3.*)

Delete source objects? [Yes/No] <N>:Y

The order in which you pick the three points is not important.

In the next example, a circle is used to define a mirror plane. It is located behind the wedge that is to be mirrored and is in a plane perpendicular to the current XY plane. The results are shown on the right in Figure 3.24, with the original object shown in dashed lines.

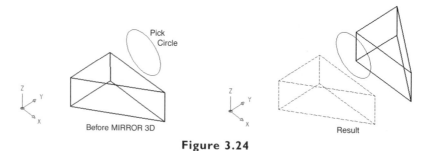

Before MIRROR 3D   Result

Figure 3.24

Command: MIRROR3D

Select objects: (*Select the wedge.*)

Specify first point of mirror plane (*3 points*) or

[Object/Last/Zaxis/View/XY/YZ/ZX/3points]<3 points>: O

Select a circle, arc or 2D-polyline segment: (*Select the circle.*)

Delete source objects? [Yes/No] <N>:Y

The final example will illustrate the Zaxis option. In Figure 3.25 the original object is shown on the left, and the mirrored object, with the original object shown in dashed lines, on the right.

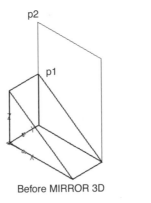
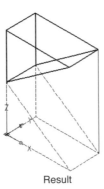

Before MIRROR 3D   Result

Figure 3.25

Command: MIRROR3D

Select objects: (*Select the wedge.*)

Specify first point of mirror plane (*3 points*) or

[Object/Last/Zaxis/View/XY/YZ/ZX/3points]<3 points>: Z

Specify point on mirror plane: (*Pick point p1.*)

Specify point on Z-axis (*normal*) of the mirror plane: (*Pick point p2.*)

Delete source objects? [Yes/No] <N>:Y

THE 3DARRAY COMMAND

The 3DARRAY command is an enhanced version of the standard ARRAY command for making both rectangular and polar arrays. Although the standard ARRAY command works well with 3D objects, it can only make rectangular arrays in the X and Y direction and can make polar arrays only about an axis perpendicular to the XY plane. 3DARRAY, on the other hand, can make rectangular arrays in the X, Y, and Z directions, as well as polar arrays around an axis pointed in any direction in 3D space (see Figure 3.26). Although 3DARRAY is an AutoLISP program, rather than a core component of AutoCAD, it is automatically loaded and can be invoked through the Modify pull-down menu, as well as from the command line.

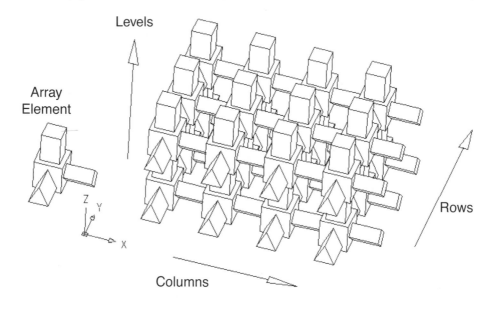

Figure 3.26

RECTANGULAR ARRAYS
Command: 3DARRAY

Select objects: (*Use any standard selection method.*)

Enter the type of array [Rectangular/Polar] <R>: R

Enter the number of rows (—) <1>: (*Enter a positive integer.*)

Enter the number of columns (||||) <1>: (*Enter a positive integer.*)

Enter the number of levels (...) <1>: (*Enter a positive integer.*)

Specify the distance between rows (—): (*Specify a distance.*)

Specify the distance between columns (||||): (*Specify a distance.*)

Specify the distance between levels (...): (*Specify a distance.*)

Rows are in the Y direction, columns are in the X direction, and levels are in the Z direction. The array shown above has three rows, four columns, and two levels. An input of 1 for a row, column, or level will make no copies in that direction. Even though 1 is the default in each of the direction prompts, an array with one row, one column, and one level is equivalent to a single item and is not allowed. Also, if the number of levels is one, the resulting array is equal to one made with the standard ARRAY command.

The prompts for distances between elements are issued only for directions in which more than one element was specified. Positive numbers generate the array in the positive axes' directions, whereas negative numbers generate the array in the negative axes' directions.

POLAR ARRAYS
Command: 3DARRAY

Select objects: (*Use any standard selection method.*)

Enter the type of array [Rectangular/Polar] <R>: P

Enter the number of items in the array: (*Enter a positive integer.*)

Specify the angle to fill (+=ccw, -=cw) <360>: (*Enter an angle or press ENTER.*)

Rotate arrayed objects? [Yes/No] <Y>: (*Enter Y, N or press ENTER.*)

Specify center point of array: (*Specify a point.*)

Specify second point on axis of rotation: (*Specify a point.*)

Although these prompts are similar to those of the standard ARRAY command's prompts for polar arrays, the center of the array requires two points rather than just one. The first point establishes one end of the array's rotational axis, and the second point sets the other end. The positive direction of this axis is from the first point to the second, with rotation following the right-hand rule. Negative numbers may be used to reverse the normal rotation direction.

On the left side of Figure 3.27, five fan blades will be arrayed a full 360° using points p1 and p2 as the rotational axis endpoints. The completed polar array is shown on the right.

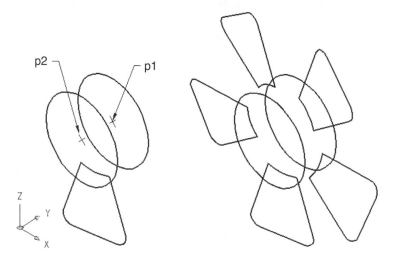

Figure 3.27

Command: 3DARRAY

Select objects: *(Select the fan blade.)*

Enter the type of array [Rectangular/Polar] <R>: P

Enter the number of items in the array: 5

Specify the angle to fill (+=ccw, -=cw) <360>: *(Press ENTER.)*

Rotate arrayed objects? [Yes/No] <Y>: *(Press ENTER.)*

Specify center point of array: *(Specify point p1.)*

Specify second point on axis of rotation: *(Specify point p2.)*

This array could also have been make with the standard -ARRAY command or the Array dialog box, provided that the UCS was first rotated 90° about the X axis.

3D CURVES IN 3D SPACE

Although you can go a long way in building 3D wireframes using only 2D entities and planar curves, some common objects—including screw threads, spiral staircases, and surface edges of consumer products—require curves that are fully 3D. AutoCAD has two object types that can be used to make 3D curves: 3D polylines and splines. The screen pull-down menus and toolbars for commands related to these objects are shown in Figure 3.28.

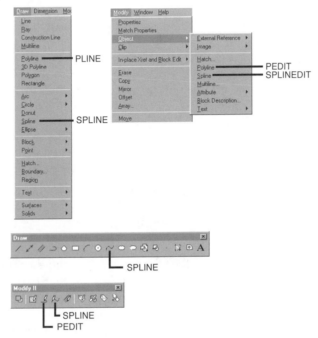

Figure 3.28

THE 3DPOLY COMMAND

3DPOLY, the command that makes 3D polylines, does not have any of the width and arc options available with the PLINE command that is used to make 2D polylines. 3DPOLY only makes polylines having line segments of zero width. In fact, the options and prompts for 3DPOLY are almost identical to those for the PLINE command.

Command: 3DPOLY

Specify start point of polyline: (*Specify a point.*)

Specify endpoint of line or [Undo]: (*Specify a point or enter a U.*)

Specify endpoint of line or [Undo]: *(Specify a point or enter a U.)*

Specify endpoint of line or [Close/Undo]: *(Specify an option, a point, or press ENTER.)*

The last prompt is repeated until you end the command by selecting the Close option or by pressing ENTER.

ENDPOINT OF LINE

Enter a point using any standard method, and AutoCAD will draw a line segment to it from the previous endpoint.

UNDO

Deletes the last segment and returns the cursor to the previous vertex point. You can use Undo to back through the polyline as far as the initial point.

CLOSE

Draws a line to the first point selected and ends the command. At least two segments must have been drawn before the Close option can be used. As with 2D polylines, there is a subtle difference between 3D polylines that have been completed with the Close option and those that have been completed by drawing a segment to the first point and pressing ENTER. We will discuss this further when we cover the PEDIT command.

Although 3DPOLY works just like the LINE command and the results look simply like connected lines, the line segments are part of a single entity that can be used as a boundary for surfaces, and can be copied, moved, and rotated. AutoCAD versions prior to Release 13 had limited editing abilities with 3D polylines—they could not BREAK, EXTEND, or TRIM 3D polylines. Beginning with Release 13, however, AutoCAD is able to perform all of those operations on 3D polylines.

TRY IT! - DRAWING A WIREFRAME 3D HELIX

In this exercise, you will try out the 3DPOLY command by making a helix—the curve formed when a straight line is wrapped around a cylinder. Screw threads, coil springs, spiral staircases, and barber poles are examples of objects based on a helix. Every helix has at least two dimensions—diameter and pitch. As shown in Figure 3.29, diameter is the width of the cylinder, and pitch is the distance along the cylinder from the start of the helix to the point at which one revolution is completed. Although neither diameter nor pitch must remain constant, we will keep them constant in this exercise.

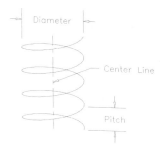

Figure 3.29

We will use a helix diameter of 2.0 units and a pitch of 1.0 unit. It will point, or grow, in the Z direction. The helix will consist of short, straight, 3D polyline segments, with the end of each segment offset from its start by a set angle from the helix center, plus an increase in elevation. We will make each angle offset 45°, which means that we'll step around the helix's center line eight times in one revolution (360° divided by 45°). Since the pitch is 1.0 units, the end of each segment will be 1/8 unit higher than its start (1 unit per revolution divided by eight segments per revolution). These dimensions are shown in Figure 3.30.

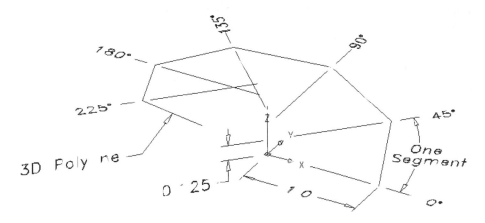

Figure 3.30

We will use typed-in cylindrical coordinates to draw the helix. You will recall from Chapter 2 that cylindrical coordinates are simply polar coordinates plus an elevation. Their form is:

radius<angle,elevation

As you type in the cylindrical coordinates for this helix, the radius will remain 1, each angle will increase 45°, and each elevation will be 0.125 units higher than the previous elevation. The center of the helix will be at the origin to avoid the need for relative coordinates. The command line sequence of prompts and input to draw the helix will be:

Command: 3DPOLY

Specify start point of polyline: 1<0,0

Specify end point of line or [Undo]: 1<45,.125

Specify endpoint of line or [Undo]: 1<90,.25

Specify endpoint of line or [Close/Undo]: 1<135,.375

Specify endpoint of line or [Close/Undo]: 1<180,.5

Specify endpoint of line or [Close/Undo]: 1<225,.625

Specify endpoint of line or [Close/Undo]: 1<270,.75

Specify endpoint of line or [Close/Undo]: 1<315,.875

Specify endpoint of line or [Close/Undo]: 1<0,1

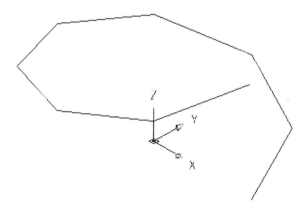

Figure 3.31

We stopped after one revolution, but we could have kept going. The resulting helix is shown in Figure 3.31. Even though it is extremely coarse and not very impressive, it can be easily smoothed by turning it into a spline curve with PEDIT. Figure 3.32 shows a helix made using these same parameters, except that it has two revolutions and has been turned into a spline curve with PEDIT.

Both the rough and smooth helix are in file 3d_ch3_01.dwg on the CD-ROM that accompanies this book. We will explain spline curves and PEDIT later in this chapter.

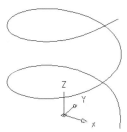

Figure 3.32

You can have a very coarse helix if you intend to smooth it into a spline curve. Even as few as four segments per revolution will work, although the two ends may not have a desirable shape. If you plan to use the helix as the boundary of an AutoCAD surface, there is little point in smoothing the helix because AutoCAD surfaces (which we will describe in Chapter 4) approximate curved boundaries with short, straight end segments. Also, as you will see when we cover solid modeling in Chapter 5, 3D polylines consisting of line segments may be used as a path for making extruded solids, whereas those transformed into spline curves cannot.

There are two ways to go wrong when using the technique of the previous exercise to make a helix. First, you can make an arithmetic mistake while setting up the segment offsets. Second, you can make a typing mistake while entering the cylindrical coordinates. Script files are a good way to reduce occurrences of the latter instance. They give you a chance to correct typing mistakes before they matter, and if the helix doesn't turn out as you expected, you may be able to modify the script file and redraw with minimum effort. Script files are also the easiest way to make special helix forms—such as those with flat or flaring ends or helixes with variable pitch and diameter.

Script files, however, do not eliminate the tedium of calculating the helix offsets and typing in coordinates. AutoLISP, on the other hand, is made for such repetitive tasks. While there are many AutoLISP programs in the public domain for drawing helixes and spirals, a simple one is listed below in case you do not have another available. This program is also on the CD-ROM that accompanies this book. This program makes a wireframe helix from a 3D polyline. It is able to make a helix having a diameter that increases with each revolution. The program prompts are for the beginning helix radius, radius change per revolution, number of revolutions, and number of line segments per revolution. Using a relatively large number of line segments per revolution produces a smooth helix, even without transforming it into a spline curve (see Figure 3.33).

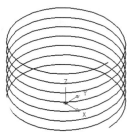

Figure 3.33

The program also asks for a point on which to set the helix center line (which can be anywhere on the XY plane), and for the direction from the centerline to the helix starting point. (For convenience, we have been starting our manually drawn helixes 0° from the center line.) The resulting helix extends in the Z direction.

Basically, the program draws the helix the way you did in the exercise with typed-in cylindrical coordinates. However, because AutoLISP does not have a function for cylindrical coordinates, the program must step up in the Z direction for each polyline segment and then use polar coordinates.

AUTOLISP PROGRAM FOR A WIREFRAME HELIX

```
; HELIX.LSP
; Draws a 3D wireframe helix using a 3D polyline
(defun C:HELIX ( / cm bm cen radius r_inc r_del pitch angle revs div
  angInc z_inc steps x y z n)
(setq cm (getvar "CMDECHO")            ; save current settings
    bm (getvar "BLIPMODE"))
(setvar "CMDECHO" 0)
                        ; get user input
(setq cen   (getpoint "\nCenter point for the helix: ")
    radius (getdist cen "\nInitial radius of helix: ")
    r_inc (getdist cen "\nRadius increment per revolution: ")
    pitch (getdist cen "\nPitch of helix: ")
    angle (getangle cen "\nDirection from center to helix start point: ")
    revs  (getreal "\nNumber of revolutions? ")
    div   (getint "\nNumber of divisions per revolution? ")
    angInc (/ 6.283185307 div)      ; 6.283185307 is 2 PI
    z_inc  (/ pitch div)
    steps  (* div revs)     ; the total number of segments in the helix
    x      (car cen)       ; get the X coordinate from the center point
    y      (cadr cen)      ; get the Y coordinate from the center point
    z      (caddr cen)     ; get the Z coordinate from the center point
    n      0
    r_del  (/ r_inc div)
)
(setvar "BLIPMODE" 0)
(command "3DPOLY" (polar cen angle radius)) ; begin drawing the helix
(while (< n steps)
  (setq angle (+ angle angInc)
      z (+ z z_inc)
      cen (list x y z)
      n (1+ n)                      ; increment n by one
      radius (+ radius r_del)
  )
  (command (polar cen angle radius))
) ; end while
(command "")                        ; finish drawing the helix
(setvar "BLIPMODE" bm)              ; restore settings
(setvar "CMDECHO" cm)
) ; end function helix
```

APPLYING THE PEDIT COMMAND TO 3D POLYLINES

PEDIT is a versatile command. Not only is it able to differentiate between polygon surface meshes, 2D polylines, and 3D polylines but it also offers appropriate options for each of these entity types. The command line format is:

Command: PEDIT

Select polyline: (*Select a 3D polyline.*)

Enter an option [Close/Edit vertex/Spline curve/Decurve/Undo/]:
  (*Enter an option or press ENTER.*)

This menu reappears after you have selected and finished an option. Press ENTER to finish the command. Notice that unlike the PEDIT prompt for 2D polylines, there is no option for joining 3D polylines.

CLOSE/OPEN

The first option in the PEDIT menu will be either Close or Open, depending upon the condition of the polyline. If the selected polyline is open, AutoCAD will display the Close option. Conversely, if it is a closed polyline, AutoCAD will display the Open option. A polyline is in a closed condition only when it is closed with the Close option of the 3DPOLY command or the Close option of PEDIT. Even if you finish drawing a 3D polyline by drawing a segment back to the first vertex, AutoCAD considers it to be open (see Figure 3.34).

When the Close option is selected, AutoCAD will draw a line segment from the last vertex to the first vertex, and the PEDIT menu will switch to show the Open option. If the last vertex is in the same point as the first vertex so that there is no need for a closing segment, AutoCAD will merely change Close to Open in the PEDIT menu. On the other hand, when you select the Open option, AutoCAD will remove the last segment from the polyline, and will change the Open option to Close.

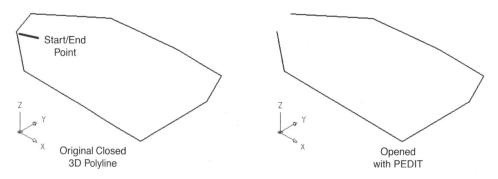

Figure 3.34

EDIT VERTEX

This option is for detailed editing of the polyline, allowing you to relocate, delete, and add vertices. It brings up the follow-up prompts:

**Enter a vertex editing option
[Next/Previous/Break/Insert/Move/Regen/Straighten/eXit]<N>:**
(Enter an option or press ENTER.)

AutoCAD will place an X marker on the first vertex of the polyline, or on the first visible vertex if part of the polyline is off the screen. You will use the Next and Previous options to move this marker to the vertex you are interested in, and then select one of the other options.

- Next

- Moves the X marker to the next vertex toward the end of the polyline. Notice that the default option for the Edit vertex menu is Next, so you can also press ENTER to step through the vertices. The X marker will stop on the last visible vertex; it will not wrap around to the start vertex even if the polyline is closed.

- Previous

- Moves the X marker to the previous vertex and changes the default menu option to Previous. The marker will stop when it reaches the first visible vertex toward the beginning of the polyline.

- Break

- Removes one or more segments from the interior of the polyline or from one of its ends. AutoCAD stores the present location of the X marker as the first break point and displays the following prompt for you to select the other break point and initiate the break:

Enter an option [Next/Previous/Go/eXit]<N>: *(Enter an option or press ENTER.)*

- The Next and Previous options are for moving the new X marker to the vertex of the second break point. Then, select the Go option for AutoCAD to perform the break, and return to the Edit vertex menu. The eXit option is an escape, allowing you to return to the Edit vertex menu without breaking the polyline.

- Insert

- Adds a new vertex to the polyline after the X marker. AutoCAD will display the prompt:

Specify location for new vertex: *(Specify a point.)*

- As you move your pointing device, AutoCAD uses a rubberband line from the X marker to help you position the location of the new vertex (see Figure 3.35).

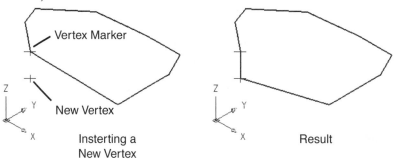

Figure 3.35

- Move
- Moves the marked vertex to a new position. AutoCAD will display the prompt:

Specify new location: for marked vertex (*Specify a point.*)

- As you move your pointing device, AutoCAD uses a rubberband line from the X marker to help you locate the new position.
- Regen
- Regenerates the polyline.
- Straighten
- This option removes segments between two vertices, replacing them with a single line segment. AutoCAD stores the present location of the X marker and displays the following prompt for you to select the other vertex and perform the operation:

Enter an option [Next/Previous/Go/eXit] <N>: (*Enter an option or press ENTER.*)

- The Next and Previous options move the new X marker to the other straighten vertex. Then, select the Go option for AutoCAD to perform the operation and return to the Edit vertex menu. The eXit option is an escape, allowing you to return to the Edit vertex menu without removing segments to straighten the polyline.

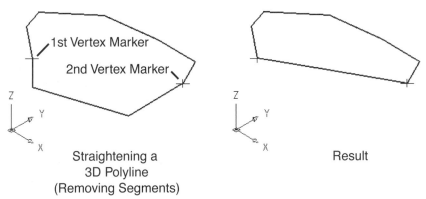

Straightening a
3D Polyline
(Removing Segments)

Result

Figure 3.36

- eXit
- Returns to the main PEDIT menu.

SPLINE CURVE

Smoothes the polyline into a B-spline curve. We will discuss the properties and characteristics of spline curves later in this chapter. The order of the spline is either quadratic or cubic, depending on the setting of the Splinetype system variable (see Figure 3.37 for examples).

Splinetype Setting	*B-spline Curve Order*
5	quadratic
6	cubic

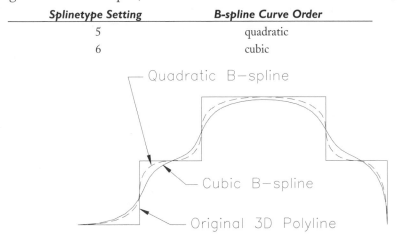

Figure 3.37

The default setting of Splinetype is 6. Whereas cubic B-splines tend to be slightly smoother and straighter than quadratic B-splines, one is not better than the other; rather, it depends on the general shape you are after. The original line segment ends (the vertices) are used as control points, so the smoothed 3D polyline will not necessarily pass through them. In fact, the endpoints are likely to be the only vertices that match those of the original polyline. Usually, quadratic splines will come closer to the original vertexes than cubic splines.

Spline-fit polylines are still made of line segments, with the relative length of these line segments determined by the value in the Splinesegs system variable. Although spline-fit 2D polylines are completely curved when Splinesegs is given a negative value, AutoCAD ignores negative values for 3D polylines. Nevertheless, even the default setting of Splinesegs, which is 8, results in a very smooth curve.

DECURVE

Returns a spline-fit polyline to its original condition. If a spline fit 3D polyline has been edited with the BREAK or TRIM commands, it cannot be decurved.

UNDO

Cancels the most recent operation. You can undo back to the beginning of the PEDIT session.

EXIT

Ends the PEDIT session.

Tips: You can also move vertices with grips and with the STRETCH command. 3D polylines can also be edited with the BREAK and TRIM commands.

Because the number of vertices and line segments in spline curved polylines increase significantly from those of the original polyline, you should think twice before you use the Spline curve option. Also, AutoCAD cannot use spline curves for some purposes, such as a path for extruded solids (Chapter 5). Furthermore, when polylines are used as the boundaries for surface meshes (Chapter 4), smooth curves are not needed since AutoCAD surfaces always approximate curved edges using short, straight lines.

SPLINE CURVE BASICS

We made several references to spline curves in the previous section without exactly explaining what they are. Also, the next command we will cover makes a 3D AutoCAD entity that has the word spline as a name. Therefore, we will now spend some time discussing splines. This information will help you understand how spline curves attain their shape, as well as the meaning behind many of AutoCAD's prompts related to splines.

Complex curves, such as those forming the cross-sections for automobile body parts, stylish telephones, and ship hulls, have always been a problem in drafting and drawing. Simple curves—such as lines and arcs—can be drawn with reasonable precision on drawing boards using straightedges and compasses. The results are basically the same regardless of who makes the drawing. Complex curves made with French curves and drafting splines, however, are more subjective and are likely to be drawn differently by different individuals.

Problems of preciseness and repeatability in complex, flowing curves continue in computer modeling. Simple curves, made of arcs and lines, are easily managed with relatively simple, standard equations, but complex, flowing curves that twist and turn through space require special treatment.

The most successful representation of complex curves within computer graphics goes by the name of B-splines. Although there are several different types of B-splines, they all have three components:

1. The curve is divided into segments, with points called knots, separating the segments. This allows relatively simple equations to be used on segments of the curve, rather than complicated equations applying to the entire curve. Often, these knots will be evenly spaced along the curve. Some types of B-splines, however, use knots that are unevenly spaced. They are referred to as nonuniform B-splines. AutoCAD does not display the knots.

2. Points, usually off the curve, are used for pulling the curve into shape. These are called control points. They perform a function analogous to the weights used to pull and hold flexible steel or plastic drafting splines in place (which is the basis for the name of computer B-spline curves). Usually, these control points have an equal weight, but in some instances some control points have a stronger effect in shaping the curve than others. Splines that have unequally weighted control points are known as rational B-splines.

    Splines that have irregularly spaced knots plus the ability to give some control points more pull than others are known as nonuniform, rational B-splines. This name is invariably shortened to the catchy acronym of NURBS. The splines that result from smoothing 2D and 3D polylines are not NURBS. Although this entity type can have unequally weighted control points, most of the time they are not. Usually, rational B-splines are needed only to exactly represent certain shapes from conic sections.

3. Mathematical functions are used to establish the curve's shape within the segments. These functions have a property called order, which controls the maximum number of times the curve segment can change curvature. A curve with an equation of order two will have no curvature—it is a straight line—whereas an equation of order three will have a con-

stant curvature—it is an arc. Fourth-order equations can have one curvature change, fifth-order equations can have as many a two curvature changes in a segment, and so on (see Figure 3.38).

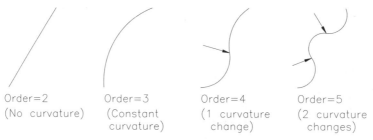

Order=2	Order=3	Order=4	Order=5
(No curvature)	(Constant curvature)	(1 curvature change)	(2 curvature changes)

Figure 3.38

Sometimes the word degree is used rather than order. The degree of an equation is its order minus one. Furthermore, equations of the second degree are often also referred to as quadratic, and third-degree equations as cubic. We have already seen that AutoCAD uses these two terms in its options for spline-fit 3D polylines.

AutoCAD does allow you to control the order of splines. Polylines, both 2D and 3D, can have an order of either three (quadratic) or four (cubic). Spline objects (made with the SPLINE command, which we will discuss next) have a default order of 4, but this can be increased as high as 26. Generally, fourth-order equations produce smooth curves and are the best choice because higher-order equations require more control points, sometimes resulting in curves having oscillations and undesirable inflections.

Now that we have covered the components of spline curves, we will briefly go over the properties of individual curves within a spline (see Figure 3.39):

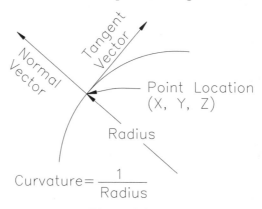

Figure 3.39

1. Each curve will have a position—a location in space determined by its control points and its curve equation.

2. There will be a direction for every point on the curve. This direction is the curve's tangent vector at that point. Perpendicular to the curve's tangent vector is its normal vector.

Each curve will have a radius. The reciprocal of the curve's radius is called curvature. Thus, curves with a high curvature have a small radius, whereas curves with a low curvature will have a large radius. Curves with zero curvature are lines; they have an infinite radius (see Figure 3.40).

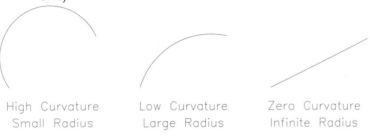

High Curvature
Small Radius

Low Curvature
Large Radius

Zero Curvature
Infinite Radius

Figure 3.40

Because spline curves are actually sets of individual curves, the relationship of each curve's properties to those of its neighbors is important. These relationships, referred to as curve continuity (see Figure 3.41 for examples), are also important when you are trying to join individual curves. If two curves do not touch, there is no continuity at all. If the ends of two curves do touch, then there is positional continuity. If their tangent, where they touch, is the same, then there is both positional and tangent continuity. If the two curves have the same curvature where they join, then they have positional, tangent, and curvature continuity. These curves blend together so smoothly, you cannot tell where one curve ends and the other begins.

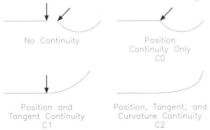

No Continuity

Position
Continuity Only
C0

Position and
Tangent Continuity
C1

Position, Tangent, and
Curvature Continuity
C2

Figure 3.41

These three levels of continuity have been given special labels. Two adjacent curves having position continuity only have a C0 level of continuity. Adjoining curves with the same tangent have C1 continuity, whereas those with equal curvature have C2 continuity. Although AutoCAD does not use these abbreviations, some 3D modeling programs (such as Mechanical Desktop) do, and they also apply to surfaces. AutoCAD B-splines will always have internal C2 continuity.

THE SPLINE COMMAND

Spline is the name of a command and of an object type. The command, SPLINE, makes nonuniform rational B-spline (NURBS) curves, which are shown as a spline object in output from the LIST command. A spline is similar to a 3D polyline in that it has zero width, can twist and turn through all three dimensions in space, and can be open or closed. Unlike a 3D polyline, however, a spline is always smoothly curved. It is also a more sophisticated entity than the 3D polyline. Its shape, especially in the end segments, can be controlled better than a 3D polyline's. You also have control over some subtle curve properties, such as tolerance, direction, control point weight, and order (although some of these properties are controlled through the SPLINEDIT command rather than with SPLINE).

Furthermore, spline objects are easier for AutoCAD to manage than spline-fit polylines, requiring less memory and disk space. Consequently, AutoCAD uses splines for ellipses and for curved leaders.

The command line format for the SPLINE command is:

Command: SPLINE

Specify first point or [Object] (*Enter an O, or specify a point.*)

ENTFIRST POINT

Although this is the default option, you must initiate it by selecting the starting point of the spline. If you press ENTER, AutoCAD will exit the command. After you specify the first point, AutoCAD draws a rubberband straight line from it and displays a prompt for the second point:

Specify next point: (*Specify a point.*)

AutoCAD refers to these user-selected points as "fit points." After you have specified the second point AutoCAD switches to a prompt offering additional options for the remaining points:

Specify next point or [Close/Fit Tolerance] <start tangent>: (*Specify a point or enter an option.*)

Even though it is not listed in the prompt, you can type in undo or just the letter u to step back through the spline.

ENTER POINT

This option adds another segment to the spline. AutoCAD draws a straight rubberband line from the previous point and simultaneously displays the overall shape of the spline as you select the point. Press ENTER to signal the end of the spline. AutoCAD will then prompt for the spline's start and end tangents.

Specify start tangent: (*Specify a point or press ENTER.*)

This point establishes the direction of the spline at its beginning.

Specify end tangent: (*Specify a point or press ENTER.*)

This establishes the direction of the spline's end. Pressing ENTER will leave the spline as it is drawn.

CLOSE

Draws a curved segment from the last entered point to the first point of the spline. The shape of this closing segment will be such that it will be tangent to both the last point and first point of the spline, but you can set the tangent direction.

Specify tangent: (*Specify a point or press ENTER.*)

AutoCAD draws a rubberband line from the first point and shows you the shape of the spline as you select a point. Pressing ENTER will leave the shape of the spline as AutoCAD has drawn it. Specifying a tangent direction allows you to make the spline tangent to an existing object.

Splines that have been completed with the Close option are called periodic by AutoCAD. Splines which have been completed by pressing ENTER, are classed as nonperiodic, even if the last point is in the same location as the first point.

FIT TOLERANCE

By default, tolerance is zero, which forces the spline to pass through every point selected. Giving tolerance a value will cause the spline to bypass the points, resulting in a straighter, less curvy spline (see Figure 3.42). The effects of tolerance values are relative; a value of 0.5, for instance, does not mean that the spline will stay one-half unit from the input points. Tolerance affects the entire spline, not just selected points. In

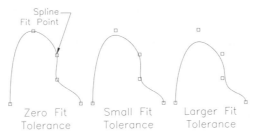

Figure 3.42

addition, tolerance values are retained until you change them, even when you begin another spline and even when you exit and reopen the drawing file.

OBJECT

This option converts existing spline-fit polylines, either 2D or 3D, into spline objects. The shape of the existing polyline is retained. Whether the original polyline is kept or deleted depends on the setting of the system variable Delobj. When Delobj is set to 1, its default setting, the original polyline spline is deleted. When Delobj is set to 0, the original polyline remains.

Tips: You should use just a few key points when making a spline curve and let the curve shape itself. Too many points may result in curves with undesirable inflections.

The start and end tangent directions have a significant impact on the shape of the curve. Object snaps are often useful when you want the spline to blend with an existing object.

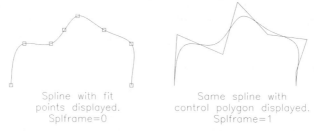

Spline with fit
points displayed.
Splframe=0

Same spline with
control polygon displayed.
Splframe=1

Figure 3.43

TRY IT! - DRAWING A 2D OPEN SPLINE

In this exercise you will draw a simple 2D spline to get a feel for how the SPLINE command works.

Command: SPLINE

Specify first point or [Object]: 0,0

Specify next point: 1,0

Specify next point or [Close/Fit Tolerance] <start tangent>: 2,1

Specify next point or [Close/Fit Tolerance] <start tangent>: 1,2

Specify next point or [Close/Fit Tolerance] <start tangent>: 1,3

Specify next point or [Close/Fit Tolerance] <start tangent>: 3,3

Specify next point or [Close/Fit Tolerance] <start tangent>: 3,2

Specify next point or [Close/Fit Tolerance] <start tangent>: *(Press ENTER.)*

Specify start tangent: @-1,0

Specify end tangent: @0,-1

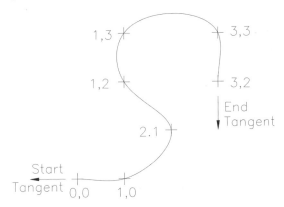

Figure 3.44

Notice that the spline goes through each entered point (if your spline did not, check to see if the tolerance is 0), and that the curve blends smoothly as it passes through them. We used relative coordinates to set the start and end tangent directions to show that they are accepted by the command. These particular directions force the start of the spline to be parallel to a horizontal line and the end of the spline to be parallel to a vertical line, although you'll have to zoom in extremely close to the endpoints to verify this.

When you are trying to understand the shape of a spline, it might help to imagine it as a flexible rod made of hard rubber or spring steel. The rod is attached to points in space at its interior fit points, but is free to swivel about those points as it assumes a natural shape. If an end tangent is not specified, the rod can also swivel about the endpoint. On the other hand, when an end tangent is specified, the rod acts as if its end is clamped and forced to point in the tangent direction.

When you use the LIST command on this object AutoCAD will report that the spline is:

1. Fourth-order—The SPLINE command always uses fourth-order (cubic) equations in creating splines. While you cannot change this default equation order, you can change the order of a specific spline through the SPLINEDIT command (discussed shortly).

2. Planar—The spline is confined to a single, flat plane.

3. Nonrational—All control points have equal weight.

4. Nonperiodic—The spline is open.

The LIST command will also show the coordinates of the control points and the fit points.

The fit points are the points you specified when drawing the spline, while the control points are for AutoCAD to mathematically pull the spline into position. A control point will be on the spline's first point and another will be on the spline's last point. Virtually all the interior control points, however, will be located off the spline. Finally, the LIST command will show the directions of the start and end tangents of the spline.

Drawing a 3D Closed Spline

Next you will use the SPLINE command to create a 3D, closed spline.

Command: SPLINE

Specify first point or [Object]: 0,1,0

Specify next point: 0,0,0

Specify next point or [Close/Fit Tolerance] <start tangent>: 2,-.5,-.5

Specify next point or [Close/Fit Tolerance] <start tangent>: 4,0,0

Specify next point or [Close/Fit Tolerance] <start tangent>: 4,1,0

Specify next point or [Close/Fit Tolerance] <start tangent>: 4,2,0

Specify next point or [Close/Fit Tolerance] <start tangent>: 2,2.5,-.5

Specify next point or [Close/Fit Tolerance] <start tangent>: 0,2,0

Specify next point or [Close/Fit Tolerance] <start tangent>:C

Specify tangent: @0,1

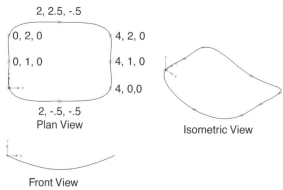

Figure 3.45

Except for the two points that have a negative Z coordinate, all of these points can be entered using a pointing device with a grid snap setting of 0.5. Notice that the end tangent points in the Y direction. Pointing it in any other direction would significantly change the spline's shape. This closed spline is in the file 3d_ch3_02.dwg on the CD-ROM that accompanies this book.

The resulting curve can be roughly described as rectangular, with rounded corners, and bowed down in its middle along the Y direction. There is a slight indentation in the left and right ends, even though the three points on each of these ends are in line. This is because AutoCAD always makes splines that have C2 continuity—every curve in the spline has both curvature and tangent continuity with the adjacent curve—which forces the curves into this dimpled shape. If we had used two interior points on the ends, rather than just one, the ends would have had a straight section.

The LIST command will report that this spline is fourth-order and nonrational, similar to the previous spline, but unlike the previous one, this spline is nonplanar and periodic.

Wireframe Model of a Display or Monitor Enclosure

In this exercise you will make the wireframe model of a shell, or case, that could be used as an enclosure for an electrical device, such as a small cathode ray tube. In Chapter 4 of this book you will use this wireframe as the basis of a surface model. Figure 1.4 in Chapter 1 shows a rendered view of the completed surface model, whereas Figure 1.5 shows a dimensioned orthographic drawing from the model.

Start the wireframe by drawing four rectangles having the dimensions and locations relative to the WCS shown in Figure 3.46. Since there are many equally good techniques to draw these rectangles, we will not give you any instructions as to how to draw them. Use lines rather than 2D polylines because they will be used as the boundaries of separate surfaces.

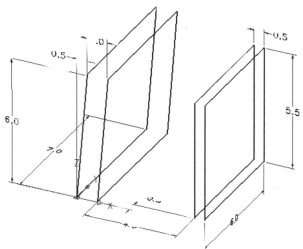

Figure 3.46

Then, fillet the rectangles using a one-unit radius for the eight upper fillets, and a one-half-unit radius for the lower eight fillets. Your model should now look like the one shown in Figure 3.47.

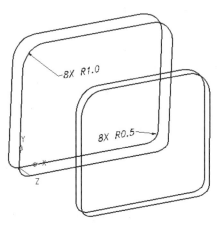

Figure 3.47

The enclosed end of the case will have ball-fillets on its corners. Each of the ball-fillets will have three arcs serving as its edges—one of which has just been made. Move the UCS to the center of an existing upper fillet's arc, as shown in Figure 3.48, and draw a one-unit radius arc for the lower edge of the ball-fillet. The arc for the third edge of the corner can either be drawn (after rotating the UCS 90° about the X axis), or copied from the existing vertical arc (using a polar array).

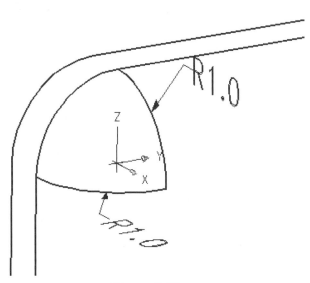

Figure 3.48

Make one of the lower ball-fillet corners next. To do this, first move the back lower edge and the two one-half unit arcs one-half-unit in the X direction. See Figure 3.49. This will adjust for the two different fillet radii, and will make the enclosed back end of the case vertical. Then draw two more one-half-unit arcs to serve as edges of the ball-fillet corner, using the same techniques you used to make the upper corner. The vertical straight line that is now hanging in space is no longer needed, so it should be erased.

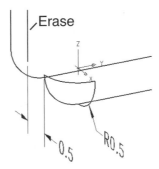

Figure 3.49

Zoom out so that you can see the entire model. From the WCS, mirror the arcs you have just drawn to make the edges for the opposite corners. Then connect the ends of the fillets to each other with lines as shown in Figure 3.50. Incidentally, the viewpoint in Figure 3.50 is 300° from the X axis, and 20° from the XY plane.

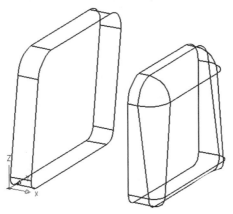

Figure 3.50

Notice that the wireframe is now in two halves. You will connect them with spline curves. In the surface model, these splines will be the edges of smooth transitions that will blend into the adjoining surfaces.

You will use three construction lines, as shown in Figure 3.51, for drawing each spline curve. Two of these construction lines extend 0.25 units from the lines that represent the edges of the rounded corners. Remember that these edges are in different planes, so you will have to move the UCS to draw them. Then object endpoint snaps to connect the two 0.25-length lines with a third construction line.

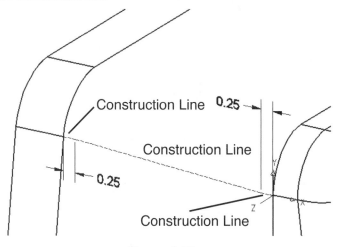

Figure 3.51

Now, draw the spline using the points labeled in Figure 3.52. All of these points are end-points, except for point D, which is a midpoint.

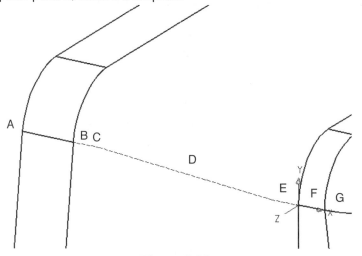

Figure 3.52

Command: SPLINE

Specify first point or [Object]: (*Select Point B.*)

Specify next point: (*Select Point C.*)

Specify next point or [Close/Fit Tolerance] <start tangent>: *(Select Point D.)*

Specify next point or [Close/Fit Tolerance] <start tangent>: *(Select Point E.)*

Specify next point or [Close/Fit Tolerance] <start tangent>: *(Select Point F.)*

Specify next point or [Close/Fit Tolerance] <start tangent>: *(Press ENTER.)*

Specify start tangent: *(Select Point A.)*

Specify end tangent: *(Select Point G.)*

After you erase the construction lines, your spline should look like the one in Figure 3.53. It is extremely important that the ends of all lines and curves meet exactly because they will be used as surface boundaries.

Figure 3.53

Use the same technique to draw the spline for the other boundary of this upper rounded edge, as well as for the top boundary of the lower rounded edge. The bottom boundary of the lower edge is planar, so it can be drawn with a 2D polyline. Then mirror the four curves to the other half of the wireframe. Your completed wireframe model should be similar to the one shown in Figure 3.54.

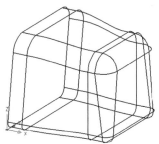

Figure 3.54

You can compare your wireframe with the one in file 3d_ch3_03.dwg on the CD-ROM that accompanies this book. In Chapter 4 you will add surfaces to this wireframe.

THE SPLINEDIT COMMAND

Several different techniques can be used to modify spline objects. As you would expect, object manipulation commands such as MOVE, COPY, ROTATE, and OFFSET all work with splines, as do the editing commands of TRIM, BREAK, SCALE, and STRETCH. You cannot, however, EXTEND, EXPLODE, or LENGTHEN splines.

You can also use grips to stretch and pull a spline into another shape. Normally, AutoCAD will display grips only on the spline's fit points, but if the system variable Splframe has been set to 1, grips will be located on the control points as well (see Figure 3.55). Although you can use the control point grips to modify the spline's shape, this will cause the spline to lose its fit points. When a spline has no fit points, grips will be located on the control points regardless of Splframe's value.

Figure 3.55

AutoCAD also has a special command for just for modifying splines: SPLINEDIT. (Notice that there is only one E in the command name.) The format for this command is:

Command: SPLINEDIT

Select spline: *(Select a spline.)*

Enter an option [Fit data/Close/Move vertex/Refine/rEverse/Undo]: *(Select an option or press ENTER.)*

You can edit only one spline at a time, so you cannot use group selection methods, such as windows or crossings. After you select the spline, its control points will be shown as grips. If the spline no longer has data points and fit data, the first option for the follow-up prompt will not be available or shown. Also, if the selected spline is closed, the Close option will be replaced by the Open option. After choosing an option, this prompt will be redisplayed when you have completed the option's action, and pressing ENTER will end the SPLINEDIT command.

FIT DATA

This option, which you can choose by typing in either an F or a D, permits you to add, move, and delete fit points. You may also close and open splines, change the tolerance setting, change the tangent direction, and even purge the spline of all fit data with this option. The spline's control points are no longer shown; instead, the fit points are shown as grips. The option's follow-up prompt is:

Enter a fit data option

[Add/Close/Delete/Move/Purge/Tangents/toLerance/eXit] <eXit>: *(Select an option or press ENTER.)*

The Close option will be replaced by an Open option if the spline is already closed.

- Add

- This option adds one or more fit points between two existing ones. Follow-up prompts to select a section of the spline for the new points along with the locations for new points are displayed:

Specify control point: *(Select a fit point.)*

Specify new point <exit>: *(Select a point, press u, or press ENTER.)*

- Despite the prompt, you will select a fit point rather than a control point. When you select a point AutoCAD will highlight its grip, along with the next point's grip. Direction is from the start of the spline toward the endpoint. Then AutoCAD prompts for new points between these existing points until you press ENTER. As you add points, AutoCAD adjusts the shape of the spline to fit each new point. Pressing the U key will undo the last added point (see Figure 3.56).

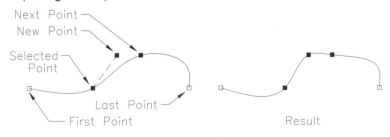

Figure 3.56

- If you select the spline's first point, AutoCAD issues prompts that allow you to position the new points either before or after the first point.

Specify new point or [After/Before] <exit>: *(Select a point, an option, or press ENTER.)*

Select new point <exit>: *(Select a point or press ENTER.)*

- The default response for the first prompt is to select a point before the first point. When you choose the After option, the spline's second point will be highlighted for you to add points between the two. In either case, the "Specify new point" prompt repeats until you press ENTER.

- If you select the last point of the spline, no other point is highlighted and you can only add points after the last point.

- Close

- Closes an open spline by drawing a segment from the last fit point to the first one. The last segment is shaped so that becomes tangent to the start of the spline. If the spline's last point and first point are in the same location, AutoCAD reshapes the last segment so that the spline has tangent continuity at that point (see Figure 3.57).

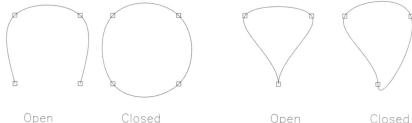

Open   Closed   Open   Closed

Figure 3.57

- Open

- Opens a closed spline by removing the last segment. If the spline was originally open, then closed, and then reopened; its resulting shape is not necessarily as it was originally, as shown in Figure 3.58.

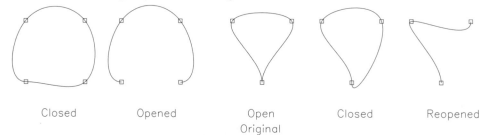

Closed   Opened   Open   Closed   Reopened
Original

Figure 3.58

- Delete

- Deletes a fit point from the spline. A follow-up prompt asks you to select the point to be deleted. The curve will readjust itself to fit the reduced set of points.

- Move

- Moves a fit point to a new location. AutoCAD highlights the grip on the first point and displays the follow-up prompt:

Specify new location or [Next/Previous/Select Point/exit] <N>: *(Select an option or press ENTER.)*

- This prompt is repeated until you select the eXit option to return to the Fit Data menu. You can use the Next and Previous options to move the highlighted grip to the point you want to move; or use the Select Point option to go directly to the point by picking it. Once the fit point is highlighted, it can be moved to a new location. As you move the point, AutoCAD will continually redraw the spline to show you the resulting shape (see Figure 3.59).

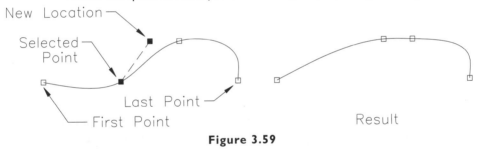

Figure 3.59

- Purge

- This option erases the spline's fit data, which includes the fit points, tolerance, and tangent directions. It cannot be recovered (except with Undo).

- Tangents

- Allows you to change the directions of the spline's start and end tangents. For open splines a follow-up prompt for the start is shown first.

Specify start tangent or [System default/]: (*Specify a point or press ENTER.*)

- AutoCAD will draw a rubberband line from the start point and show the new spline shape as you move your pointing device to a new point (see Figure 3.60). This point establishes the direction for the start of the spline. If you accept the system default, AutoCAD will set an appropriate tangent.

- After the tangent for the start has been established, AutoCAD displays a similar follow-up prompt for the end tangent.

Specify end tangent or [System default]: *(Specify a point or press ENTER.)*

- If the spline is closed, only the following follow-up prompt is used.

Specify tangent or [System default]: *(Specify a point or press ENTER.)*

- toLerance

- This option changes the spline's fit tolerance value. Its prompt is:

Enter fit tolerance <current>: *(Enter a value or press ENTER.)*

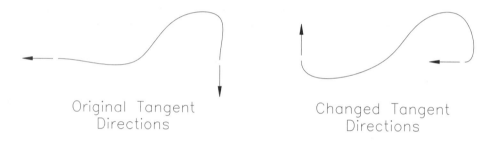

Original Tangent
Directions

Changed Tangent
Directions

Figure 3.60

- Pressing ENTER will leave the fit tolerance as it is. When the fit tolerance is zero, the spline will pass through each fit point. As the tolerance is increased, the spline will miss the fit points by a wider margin (see Figure 3.61).

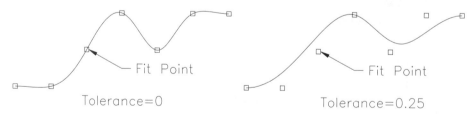

Fit Point

Fit Point

Tolerance=0

Tolerance=0.25

Figure 3.61

- eXit

- Exits the Fit Data menu and returns control to the main SPLINEDIT menu.

CLOSE

This option closes an open spline using control points rather than fit points. Notice in Figure 3.62 that the closed spline no longer goes through the original first and last points. If you wanted the closed spline to go through them, you would use the Close option in the Fit Data follow-up prompts rather than this Close option in the main prompt. This option will cause the spline to lose its fit points.

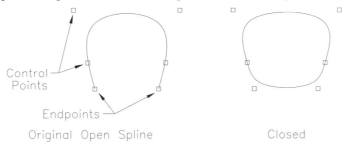

Figure 3.62

OPEN

Opens a closed spline between the first and last control points, rather than the first and last fit points, as the Close option in the Fit Data submenu does. This option will cause the spline to lose its fit points. Figure 3.63 shows the same spline shown on the left in Figure 3.58 so that you can see the difference between opening a spline through its fit points and opening a spline through its control points.

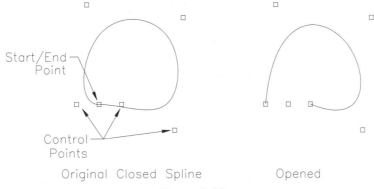

Figure 3.63

MOVE VERTEX

Moves a control point to a new location. AutoCAD highlights the grip on the spline's first point and displays the follow-up prompt:

Specify new location or [Next/Previous/Select Point/eXit] <N>: (*Select an option or press ENTER.*)

You can use either the Next or the Previous option to move the highlighted grip to the control point you want to move, or use the Select Point option to go directly to the point by picking it. Once the control point is highlighted, it can be moved to a new location. As you move the point, AutoCAD will continually redraw the spline to show you the resulting shape. The prompt is repeated, allowing you to move other control points, until you select the eXit option.

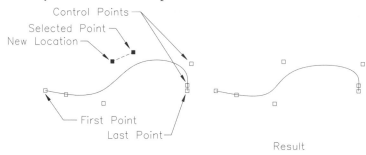

Figure 3.64

REFINE

This option is for adding a new control point in a specific location, changing the weight of selected control points, and increasing the order of the spline's equation. It uses the follow-up prompt:

Enter a refine option [Add control point/Elevate order/Weight/eXit] <eXit>: (*Select an option or press ENTER.*)

Each of these actions causes the spline to lose its fit points.

- Add Control point

- Adds a control point to a portion of the spline. The follow-up prompt is:

Specify a point on the spline <exit>: (*Select a point or press ENTER.*)

- This prompt repeats until you press ENTER; allowing you to add more than one control point. Notice that you are to pick a point on the spline, not a control point. If your selected point is not exactly on the spline, AutoCAD proceeds using a point on the spline near the one you picked. A new control point is added near the selected point, and the existing control point is moved to another location. The shape of the spline is not changed—there is just an additional control point for the spline (see Figure 3.65).

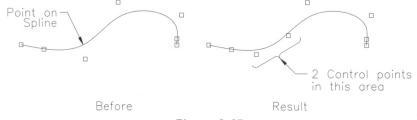

Figure 3.65

- Elevate Order
- AutoCAD always uses fourth-order (cubic) equations in drawing splines, but you can increase an existing spline's order to as high as 26 with this option. The follow-up prompt is:

Enter new order <current>: (*Enter an integer or press ENTER.*)

- AutoCAD shows the current order as a default value, which can be accepted by pressing ENTER. Although elevating the order does not change the shape of the spline, the number of control points increases dramatically. As shown in Figure 3.66, increasing a spline's order from four to five increases the number of control points from 7 to 13. Once a spline's order has been elevated, it cannot be reduced (except with Undo). Furthermore, elevating a spline's order erases all of its fit data.

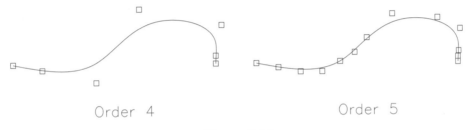

Order 4   Order 5

Figure 3.66

- Weight
- Normally, all control points are equally weighted and have an equal effect on the shape of the spline. This option allows you to change the weight of selected control points, thus changing their pull on the spline. A control point that has been given more weight than the others will pull the spline in closer, whereas one with less weight will allow the spline to recede. The prompt for this option is:

Spline is not rational. Will make it so.

Enter new weight (*current*) or [Next/Previous/Select Point/eXit] <N>: (*Enter number, select an option, or press ENTER.*)

- If a spline already has unequally weighted control points, the message preceding the prompt will not be shown. The grip on the spline's first point will be highlighted to show that it is the current point. Use the Next option to move to the next point toward the endpoint, the Previous option to move to the next point toward the first point, or the Select Point option to go directly to the control point whose weight you wish to change. Once you are on the point, type in a number for the new weight. The initial weight for all points is

1. Selecting the eXit option returns to the Refine prompt.

- In the example shown in Figure 3.67, the highlighted control point is given a weight of 4, while the other control point weights remain at 1. This additional weight pulls the spline very close to that control point. The list command will report that this spline is rational, and will give the weight of each control point. The spline will no longer have any fit data.

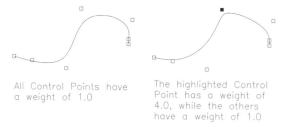

All Control Points have a weight of 1.0

The highlighted Control Point has a weight of 4.0, while the others have a weight of 1.0

Figure 3.67

REVERSE

Reverses the spline's direction. Although there is no change in the spline's appearance, the LIST command will show that order of both the control points and fit points has been reversed.

UNDO

Cancels the last SPLINEDIT operation.

 Tip: The SPLINEDIT command has more tools and options for modifying splines than you are ever likely to need; in most cases you should confine your editing to modifications of fit points.

A diagram of the numerous options of SPLINEDIT is shown in Figure 3.68.

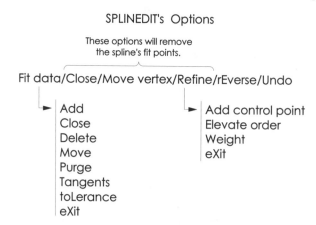

Figure 3.68

TRY IT! - EDITING A SPLINE

You will use SPLINEDIT to modify the closed spline you made earlier in this chapter. You should now open your file of the closed spline, or else open file 3d_ch3_02.dwg. You will recall that this spline had a dimple on two sides. To straighten these sides, you will add two new points to each side in line with the existing data points to force a straight line in the curve. In making these changes, it will probably be most convenient to work within the WCS and in the plan view because all of the work will be on the XY plane.

First, invoke the SPLINEDIT command, and select the spline. Then select the Fit Data option from the command line prompt.

Command: SPLINEDIT

Select spline: (*Select the spline.*)

Enter an option [Fit Data/Close/Move Vertex/Refine/rEverse/Undo]: F

AutoCAD will show the fit points as grips and display the Fit Data menu.

Enter a fit data option

[Add/Open/Delete/Move/Purge/Tangents/toLerance/eXit] <eXit>: A

Specify control point <exit>: 0,1

You can skip the Z coordinate when specifying point locations since you are on the XY plane. AutoCAD will highlight the selected point, along with the next point, and will ask for the location of the new point to be located between the two highlighted fit points.

Specify new point <exit>: 0,.25

Specify new point <exit>: (*Press ENTER.*)

Specify control point <exit>: 4,0

Specify new point <exit>: 4,.25

Specify new point <exit>: (*Press ENTER.*)

Specify control point <exit>: 4,1

Specify new point <exit>: 4,1.75

Specify new point <exit>: (*Press ENTER.*)

Specify control point <exit>: 0,2

Specify new point <exit>: 0,1.75

Specify new point <exit>: (*Press ENTER.*)

Specify control point: <exit>:(*Press ENTER.*)

Enter a fit data option

[Add/Open/Delete/Move/Purge/Tangents/toLerance/eXit] <eXit: (*Press ENTER.*)

Enter an option [Fit Data/Close/Move Vertex/Refine/rEverse/Undo/]: (*Press ENTER.*)

Now the dimples are gone, leaving the sides relatively straight. Each time you selected a new point, AutoCAD highlighted it and the following point, and as you added the new fit points AutoCAD adjusted the shape of the spline (see Figure 3.69).

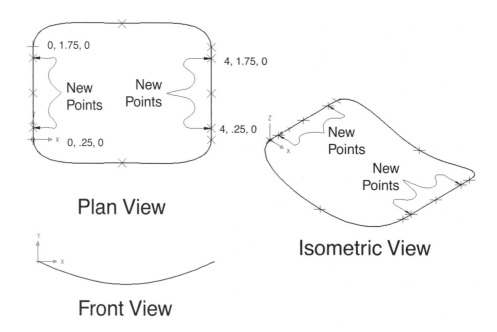

Figure 3.69

This edited version of the spline is in file 3d_ch3_04.dwg on the accompanying CD-ROM.

Put the tools you have acquired in this chapter to work by building two 3D wireframe models. You will construct a wireframe model of a sheet metal part in the first exercise and the wireframe for a boat hull in the second exercise. In Chapter 4, you will add a surface to the boat hull.

TRY IT! - SHEET METAL PART

The wireframe you will construct during this exercise is relatively simple, but it could be confusing if it is not built systematically. You will build the model one plane at a time. You will draw the edges representing just one side of the metal and then use copies of those edges for the other side. Initially, you will make all of the folded corners square. When you have all of the objects drawn and in place, you will use AutoCAD's FILLET command to simulate rounded bend corners. Start the part by drawing its left-hand flange on the WCS XY plane using the dimensions shown in Figure 3.70. This flange consists of three lines and three circles. The units for the dimensions are centimeters.

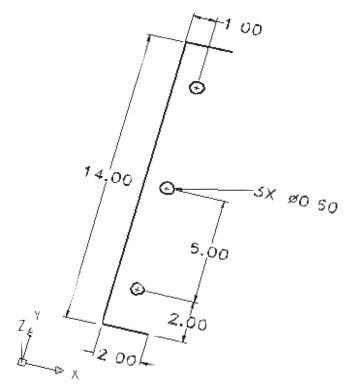

Figure 3.70

Next, rotate the UCS XY plane so that it is parallel to the WCS ZX plane and is 0.5 cm in front of the flange, as shown in Figure 3.71. The previously drawn objects are shown in dashed lines in this figure. Draw the four lines and one circle shown in the figure. (Actually, you do not need the two vertical lines, and you will erase them later, but for now they will help you stay oriented.)

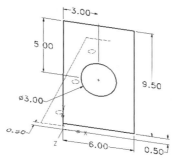

Figure 3.71

Rotate the UCS XY plane so that it is parallel to the WCS YZ plane and is at the inside edge of the flange. Draw the five lines representing one of the part's vertical sides, using the dimensions given in Figure 3.72. The objects you have already drawn are shown in dashed lines in this figure. The viewpoint for this figure is 220° in the XY plane from the X axis, and 30° from the XY plane.

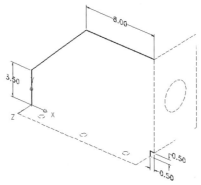

Figure 3.72

Restore the WCS and erase the two vertical lines on the front side of the part. The thickness of the part's metal is 0.20 cm. We have drawn all of the edges for one side of the metal, and we will use AutoCAD's COPY command to make the edges for the other side. Start with the part's horizontal flange. The copies for the top side of this flange can be conveniently made with the following command line input:

Command: COPY

Select objects: (*Select the three lines and three circles that are on the WCS XY plane.*)

Specify base point or displacement, or [Multiple]: 0,0,.2

Specify second point of displacement: or<use first point as displace-ment>: (*Press ENTER.*)

The copies will be located 0.20 cm in the Z direction from the original objects. Use the same method to make copies of the objects that are on a plane parallel with the WCS YZ plane, with a displacement of -.2,0,0 (in the minus X direction). The edges for the opposite metal side of the front of the part (the edges are in a plane parallel with the WCS ZX plane) can be made with a copy displacement of 0,-.2,0 (in the minus Y direction). Your part should look similar to the one shown in Figure 3.73.

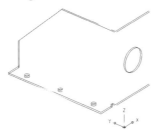

Figure 3.73

Use AutoCAD's MIRROR command to make the part's other vertical side and flange. Next, use the FILLET command to round the four bend corners of the part. The inside bend radius is 0.20, and the outside bend radius is 0.40 (see Figure 3.74). You will have to zoom in close to the bend edges that are near the relief cut-outs on the model before you can clearly identify the objects to be filleted.

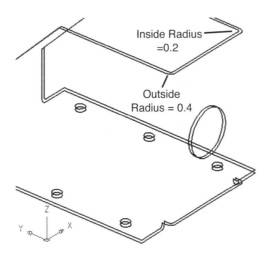

Figure 3.74

Finish your model by drawing lines between the tangent points of the bends, and by drawing lines at the sharp edges on the part. Your completed model should look similar to the one shown in Figure 3.75.

 A completed version of this model is in file 3d_ch3_05.dwg on the CD-ROM.

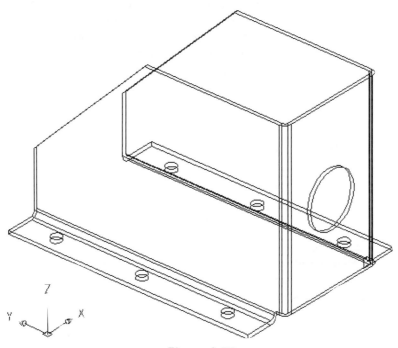

Figure 3.75

TRY IT! - BOAT HULL

You will construct a wireframe for the surface model of a boat hull during this exercise. You will add surfaces and complete the model in Chapter 4. The command you will use to add the surfaces (EDGESURF) requires four open wireframe objects for the surface's boundary. Although you have wide flexibility in the types and shapes of objects that comprise the boundary, there must be four of them and each must exactly touch its neighbor—the slightest gap or overlap is not allowed.

Since the boat hull is symmetrical, you will make just one side of it. After you have added the hull's surfaces in Chapter 4, you will use AutoCAD's MIRROR command to make the other half. The side you are going to make will have three separate surfaces; and the wireframe will have seven splines and two lines to be used as boundary objects for those surfaces. (Three splines will serve as boundaries for two surfaces.) The units for this model will be dimensionless—they can represent meters, feet, yards, or any other measurement units you like.

Start the wireframe by using the SPLINE command to draw the curve shown in Figure 3.76 on the WCS XY plane. If you work in AutoCAD's snap mode with a snap spacing of 0.25, you can draw this spline with your pointing device. The spline has only a start point (at the WCS origin), one interior fit point (at 0.75,0.75), and an endpoint (at 2.75,15). Be certain that you specify the start and end tangents as shown in the figure. Although a 2D polyline could be used for this curve, a spline gives more control over the endpoint tangent directions and more flexibility in making changes.

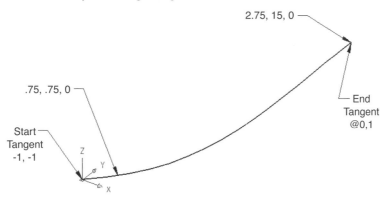

Figure 3.76

Position the UCS XY plane so that it lies in the WCS YZ plane, and its origin is at the WCS origin. Then draw the three spline curves shown in Figure 3.77, using the points and tangents given in the figure. These curves have only start points and endpoints, with their shape controlled by their start and end tangents. Although you could have draw this shape as a single object, you need three separate curves to make the surfaces.

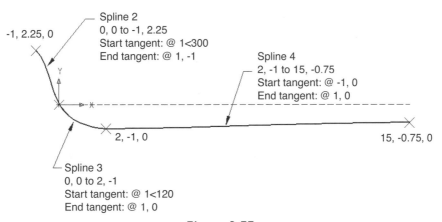

Figure 3.77

Move the UCS origin to the WCS 0,15,0 point and rotate it so that its XY plane is parallel to the WCS ZX plane. Then draw the spline curves numbered 5 and 6 in Figure 3.78. Notice that spline number 5 has one interior fit point, and spline 6 has only a start point and an endpoint. Finish this area of the hull by drawing the two lines shown in the figure. Previously drawn spline numbers 1 and 4 are shown in dashed lines in Figure 3.78.

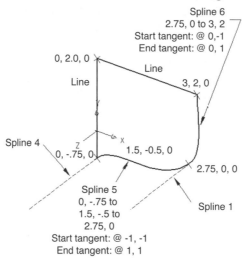

Figure 3.78

Restore the WCS and use the data shown in Figure 3.79 to draw the last object, spline number 7, in the wireframe. The objects you have already drawn are shown in dashed lines in this figure. You can use object endpoint snaps for the spline's start point and endpoint, but you will probably use a typed-in coordinate for the interior fit point. Your completed wireframe should look similar to the one shown in Figure 3.79. You should save your wireframe model so that you can add surfaces to it as you experiment with the EDGESURF command in Chapter 4.

 A completed version of this model is in file 3d_ch3_06.dwg on the CD-ROM.

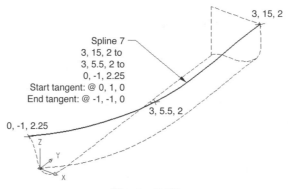

Figure 3.79

COMMAND REVIEW

3DPOLY

This command makes 3D polylines.

ALIGN

This command moves objects and rotates them using source and destination points rather than angles.

ARRAY

This command makes copies of objects using a rectangular grid in the **X** and **Y** directions; as well as around an axis perpendicular to the **XY** plane. It cannot, however, make copies in the **Z** direction or change the orientation of the rotational axis of polar arrays.

MIRROR

This command makes mirror image copies of objects through a mirror plane perpendicular to the current **XY** plane.

MIRROR3D

This command mirrors objects through a user-defined plane.

MOVE

This command is equivalent to using just the first pair of source and destination points with **ALIGN**.

PEDIT

This command edits both **2D** and **3D** polylines. One of the features allows 2D and 3D polylines to be transformed into spline fit curves. This command does not convert polylines into spline objects.

PLINE

This command constructs 2D polylines.

ROTATE

This command rotates objects around an axis perpendicular to the **XY** plane.

ROTATE3D

This command rotates objects about a user-defined axis. It does not make copies of objects.

SPLINE

This command creates non-uniform, rational, B-spline curves (**NURBS**). This type of curve can be edited with the **SPLINEDIT** command.

SPLINEDIT

Splines made by the **SPLINE** command can be modified by this command. Through it, you can add or delete fit points, change the spline's fit tolerance, and more.

STRETCH

This command moves spline data points and adjusts the curve to fit the new locations.

SYSTEM VARIABLE REVIEW

ANGBASE

Stores the base 0 angle relative to the **X** axis of the current **UCS**. The default setting is 0°.

ANGDIR

Determines whether positive rotation is clockwise or counterclockwise. The default setting is counterclockwise.

DELOBJ

This variable controls whether the original splined 2D or 3D polyline is retained or deleted when converted into a spline object by the **Object** option of **SPLINE**. When Delobj is set to 0, the original object is retained; when it is set to 1 (its default value), the original object is deleted.

SPLFRAME

This variable can be set either to 0 or 1. When Splframe is set to 1, AutoCAD connects the control points of spline curves with a frame of lines. (You must cause a regeneration after setting Splframe for it to go into effect.) With 3D polylines, this control point frame has the same appearance as the original, uncurved polyline, which might sometimes be useful for viewing the original polyline without actually decurving it.

SPLINETYPE

This variable sets the order of spline curves. When Splinetype is 5, spline-curved polylines are quadratic. When Splinetype is 6, the default value, they are cubic.

SPLINESEGS

This variable controls the relative number of vertices (and therefore, the number of straight line segments used to approximate the curve) in spline-curved polylines. The default setting is 8.

CHAPTER REVIEW

Directions: Circle the letter corresponding to the correct response in each of the following.

1. Arcs and circles can be drawn in any orientation relative to the current UCS.

   a. true

   b. false

2. Two lines can be filleted regardless of their orientation to the UCS, provided they are coplanar.

   a. true

   b. false

3. 3D objects, such as a wireframe model containing objects having Z coordinates, cannot be rotated with AutoCAD's ROTATE command; you must use the ROTATE3D command to rotate them because ROTATE can only rotate objects that are on the XY plane.

   a. true

   b. false

4. As you view them on your computer display, you cannot distinguish between ordinary lines, polylines, and 3D polylines.

   a. true

   b. false

5. The PEDIT command does not allow you to join objects with 3D polylines, as it does with 2D polylines.

   a. true

   b. false

6. When you increase a spline's order through the SPLINEDIT command, it will become more tightly curved.

   a. true

   b. false

Directions: Match the following items as indicated.

7. Match a project option of the TRIM and EXTEND commands in the left column with the appropriate result in the right column.

   _____a. None

   1. Boundary objects are projected perpendicularly, relative to the XY plane, onto the plane of the objects to be extended or trimmed.

   _____b. UCS

   2. Boundary objects must be in the same plane as the objects to be extended or trimmed.

   _____c. View

   3. Objects are trimmed if they appear to intersect the boundary object, and are extended if they look as if they would intersect the boundary.

8. Match the system variable from the list on the left with a function in the list on the right.

   _____ a. Delobj   1.   Controls the curve order for smoothed 3D polylines.

   _____ b. Splframe   2.   Controls the relative number of vertices in spline fit 3D polylines.

   _____ c. Splinesegs   3.   Displays spline control points and draws a connecting line between them.

   _____ d. Splinetype   4.   Erases spline-fit 3D polylines when they are converted to true spline objects.

Directions: Answer the following questions.

9.   What can 3DARRAY do that ARRAY cannot?

10.   What is the difference between a planar curve and a nonplanar curve?

11.   What does the acronym NURBS stand for?

12.   In what ways are AutoCAD spline objects similar to 3D polylines? In what ways are they different?

13.   What is the difference between a spline's fit points and its control points? Does a spline always have fit points? Does a spline always have control points?

14.   Does a spline curve always pass through its fit points?

15.   How does control point weight affect spline curves, and how can you change the weight of a control point?

Surface Models

LEARNING OBJECTIVES

In addition to defining the edges of a 3D object, surface models define surfaces between those edges. These surfaces enable you to create models that not only hide objects behind their surfaces but also can be used to make realistic renderings. When you have completed Chapter 4, you will:

- Understand the properties and characteristics of AutoCAD surface objects.

- Know where extruded surfaces are appropriate and how to make them.

- Know which AutoCAD command is best for surfacing a particular planar area and how to use it.

- Be able to create nonplanar surfaces with almost any shape.

- Know how to improve surface shapes through AutoCAD's commands for editing surface objects.

- Be able to use clipping planes to eliminate obstructions from surface models and create realistic perspective views.

SURFACE MODELING

Even though wireframes are a necessary component of 3D modeling, working with them is seldom satisfying since they are not realistic looking and, due to their very nature, generally have an unfinished feel and appearance. Surface models are a step up for creating realistic models. You can actually make houses that look like houses, gears that look like gears, and teapots that look like teapots (see Figure 4.1). Solid models, which we'll cover in the next part, can come even closer to representing some objects because they have mass in addition to surface; surface models are still the best choice for many objects, especially in architectural disciplines. Furthermore, you can make surface models of some objects that are impossible with solid modeling.

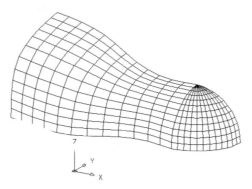

Figure 4.1

CHARACTERISTICS OF AUTOCAD SURFACES

As shown in Table 4.1, AutoCAD has seven different commands for making surfaces and three different database objects for surfaces, plus a couple of surface-like objects. Although these surface objects vary somewhat in their properties, they all have the following characteristics in common:

- They have 0 thickness. Surface models are just empty shells. What may look like a rectangular steel bar is actually an infinitely thin top surface, with infinitely thin sides, and another infinitely thin surface for the bottom. There is absolutely nothing in between. If you want a round hole through the bar, you'll have to simulate the hole by making a round opening in the top and bottom surfaces and represent the sides of the hole with a tube.

- They can hide objects, including other surfaces, that are behind them, although this property is apparent only when certain commands, such as HIDE and SHADEMODE, are invoked. At other times, surfaces are seen in a wireframe condition, which is completely transparent.

- In their wireframe condition, the edges of surfaces are shown (although in some cases you can hide surface edges); if the surface is curved or rounded, it is delineated by a pattern of lines. This pattern may be in the form of a rectangular grid, a triangular mesh, or a set of lines that may be roughly parallel to each other or radiate. Which pattern form applies depends largely upon the shape of the surface.

- In renderings, they can be given colors and material properties, and they respond to light. These model the physical laws of light and material for producing shaded realistic images of the 3D model. (Rendering will be discussed in Chapter 7.)

- AutoCAD surfaces are always flat, with curved and rounded surfaces approximated by small rectangular or triangular faces.

- Surfaces cannot be used as objects in other AutoCAD commands. For instance, you cannot select a surface as a cutting edge for use in the TRIM command. The same is true for the EXTEND command; surfaces cannot be used as boundary edges.

Table 4.1 AutoCAD Surfaces

Command	Object Type	Surface Description and Remarks
ELEV	Varies	An entity property in which wireframe objects have an extrusion thickness.
3DFACE	3D face	Planar surface with three or four vertices.
PFACE	Polyface mesh	Planar surface with unlimited number of vertices.
REGION	Region	Planar version of 3D solids.
RULESURF	Polygon mesh	Surface mesh between two boundary curves.
TABSURF	Polygon mesh	Surface mesh made by extruding a boundary curve.
REVSURF	Polygon mesh	Surface mesh made by revolving a boundary curve.
EDGESURF	Polygon mesh	Surface mesh between four boundary curves.
3DMESH	Polygon mesh	General surface mesh, defined on a point-by-point basis.
CIRCLE	Circle	Every ordinary AutoCAD circle is a disk-like surface.

EXTRUDED SURFACES

An extruded surface is a wireframe object—such as a line, arc, circle, or 2D polyline (see Figure 4.2)—that has thickness in addition to length. It is as if the object was stretched in the Z direction. An extruded line, for instance, will look like a wall, and a circle will look like a column. Even though extrusions are an old AutoCAD device for making surfaces—one that is no longer really needed—they are still often convenient to use.

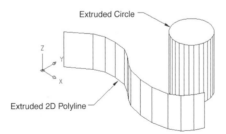

Figure 4.2

Extrusion Characteristics

Though extrusion thickness is an object property stored in AutoCAD's database (as are color and line type), it is a property that acts like a surface. Extruded surfaces can hide objects that are behind them, and they reflect light in renderings. The objects in Figure 4.2 are shown with HIDE on. For curving, ribbon-like surfaces they can even be superior to AutoCAD surface objects that approximate such curves using small rectangular faces. Curved extruded surfaces, on the other hand, will be just as accurate as the polyline or set of arcs used to make them.

Extrusion is always in the object's Z direction; that is, the direction of the Z axis at the time the object was created. This direction, which is often called extrusion direction, is also an object property stored in AutoCAD's database. Therefore, it remains fixed and unchanging relative to the world coordinate system.

Because extrusion thickness is an object property, it cannot vary within any single object. A line, for instance, cannot have a thickness of 3 at one end and 5 on the other end. AutoCAD adds object snap points to an extruded object to accommodate the added dimension. As shown in Figure 4.3, an extruded line, which in effect is a rectangle, will have endpoint snap locations at each corner and midpoint snap location in between. Including just one of the corners within the STRETCH command's crossing window effectively includes the other corner. AutoCAD displays grips on all of an extruded line's endpoints and the horizontal midpoints, but not on the vertical endpoints.

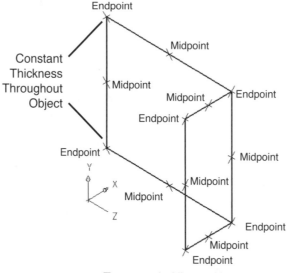

Two extruded lines with
object snap locations shown

Figure 4.3

Curved objects such as circles and arcs will have a series of parallel lines on their extruded surface. These are called tessellation lines, and their relative number is controlled by the VIEWRES command. Figure 4.4 shows the effect of VIEWRES. Although this figure shows two side-by-side curves with seemingly different VIEWRES settings, it is not possible for some objects in a drawing to have a different setting than other objects.

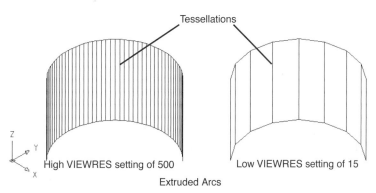

High VIEWRES setting of 500   Low VIEWRES setting of 15

Extruded Arcs

Figure 4.4

Making Extrusions

Extrusion thickness is controlled by the Thickness system variable, which has a default value of 0. Objects are the given a height equal to the Thickness value as they are drawn. If the value of Thickness is positive, extrusion is in the positive Z direction; if it is negative, extrusion is in the negative Z direction. Not all AutoCAD wireframe objects accept thickness. The wireframe object types that cannot are mline, xline, ray, spline, and 3D polyline. Text cannot have thickness initially, but it can be given thickness through one of the commands that change object properties. Mtext, however, cannot be given thickness, neither initially nor by changing its properties. In AutoCAD Releases 13 and earlier, hatches can have a thickness. This causes them to show up in renderings, and it is sometimes useful for creating patterns of roof shingles and concrete blocks. In Release 13, however, only nonassociative hatches (made with the HATCH command, rather than with BHATCH) can have thickness, and beginning with Release 14, not even nonassociative hatches can have thickness.

AutoCAD provides three different ways to set Thickness:

- You can set it through the SETVAR command or even more directly by typing in the variable's name, Thickness, on the AutoCAD command line and typing in a value as follows:

Command: THICKNESS

Enter new value for THICKNESS <current>: (*Enter a new value or press ENTER to accept the current value.*)

- You can use the PROPERTIES command. This brings up a two-column dialog box with object properties listed in the left column and their properties in the right column. Find thickness in this dialog box and enter a new value for it.

- You can use the ELEV command. AutoCAD will first prompt for an elevation, which you should always leave at 0. Then it will prompt for a new thickness.

Command: ELEV

Specify new default elevation <0.0000>: *(Press ENTER to leave elevation at zero.)*

Specify new default thickness <current>: *(Enter a new value or press ENTER to accept the current value.)*

The Thickness setting is stored in each drawing's database and therefore remains in effect when the drawing file is reopened. This can sometimes cause surprises, because AutoCAD does not have any built-in, on-screen flags or status bar signals to indicate that the current Thickness is not 0.

AutoCAD also provides several different ways to change an object's thickness after it has been drawn:

- You can use the CHANGE command:

Command: CHANGE

Select Objects: *(Use any object selection method.)*

Specify change point or [Properties]: P

Enter property to change [Color/Elev/LAyer/LType/ltScale/Thickness)]: T

Specify new thickness <current>: *(Enter a new thickness value, or press ENTER to accept the current value.)*

- You can use the CHPROP command which is similar to but more direct than CHANGE for changing object properties:

Command: CHPROP

Select Objects: *(Use any object selection method.)*

Enter property to change [Color/LAyer/LType/ltScale/Thickness)]: T

Specify new thickness <current>: *(Enter a new thickness value or press ENTER to accept the current value.)*

- You can use the PROPERTIES command. Select the object, find Thickness in the left column, and enter a new value in the right column.

 Tip: Although text does not respond to the Thickness system variable during input, you can give existing text thickness using any of the methods we've described. Sometimes this is useful because text with 0 thickness is not hidden during the HIDE command. Also, text must have thickness if it is to be shown in a rendering.

TRY IT! -

We will first demonstrate extruded surfaces with three lines. First, we will set Thickness to 0.50 using any of the three methods we described. Then, we will set VPOINT to 0,-1,0. From this view direction, looking at the edge of the XY plane, we will have to type in coordinates when drawing the lines

Command: LINE

Specify first point: 1,1,0

Specify next point or [Undo]: 2,1,0

Specify next point or [Undo]: 3,1,.5

Specify next point or [Undo/Close]: 4,1,0

Specify next point or [Undo/Close]: (*Press ENTER.*)

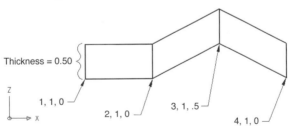

Thickness = 0.50

1, 1, 0
2, 1, 0
3, 1, .5
4, 1, 0

Figure 4.5

The extruded lines are shown in Figure 4.5. Notice that the lines that have one end off of the XY plane assume a parallelogram shape since extrusion thickness is always a distance in the Z direction.

Next, we will give a 2D polyline both width and thickness. We will leave Thickness at 0.5, but use the Rotate option of VPOINT to set a view direction 300° from the X axis and 30° from the XY plane. Then we will draw a 2D polyline.

Command: PLINE

Specify start point: 1,1

Current line-width is 0.0000

Specify next point or [Arc/Close/Halfwidth/Length/Undo/Width]: W

Specify starting width <0.0000>: 1

Specify ending width <1.0000>: 1

Specify next point or [Arc/Close/Halfwidth/Length/Undo/Width]: 2,1

Specify next point or [Arc/Close/Halfwidth/Length/Undo/Width]: A

Specify endpoint of arc or
[Angle/CEnter/CLose/Direction/Halfwidth/Line/Radius/Second
pt/Undo/Width/]: 3,2

Specify endpoint of arc or
[Angle/CEnter/CLose/Direction/Halfwidth/Line/Radius/Second
pt/Undo/Width/]: L

Specify next point or [Arc/Close/Halfwidth/Length/Undo/Width]: 3,3

Specify next point or [Arc/Close/Halfwidth/Length/Undo/Width]:
(*Press ENTER.*)

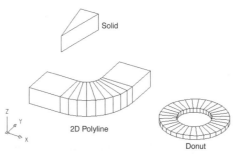

Solid

2D Polyline

Donut

Figure 4.6

The result, as shown in Figure 4.6, is almost an instant 3D model of a rectangular ventilation duct. Other AutoCAD objects that have width can also be given thickness to make 3D shapes. The SOLID command can make box- and wedge-shaped objects, while DOUGH-

NUT can make washers. Samples of these objects are shown in Figure 4.6.

PLANAR SURFACES

Although AutoCAD has several commands that can make planar surfaces—surfaces that are flat—it has two that are intended for nothing else: 3DFACE and PFACE. Another AutoCAD command, REGION, also makes objects that are planar and, similar to surface objects, have no thickness, are able to hide objects, and reflect light in renderings. Regions, however, are a special 2D form of solids, so we will discuss them in Chapter 5.

TRY IT! - SURFACE MODEL TABLE

As we go through the topics in surface modeling, you will build models to demonstrate each of AutoCAD's commands for making surfaces. These surface models will be used to build and furnish a room in a house, so you will want to keep them.

As the first model in this set, you will use extruded objects to build the table shown in Figure 4–9. We will describe each step in building this model and give some suggestions to make your work go smoothly. We will not, however, detail the use of each command because you should have a good working knowledge of them by now. The drawing setup of layers and so forth are not important, other than using inches as the unit of measure.

The first step is to build the supports for the table top. These will consist entirely of extruded lines. Use the UCS command and move the coordinate system origin 25 units in the Z direction. Then give Thickness a value of 4, and draw the supports using the LINE command, with the dimensions shown in Figure 4.7. Although this figure shows an isometric-type view, you will probably find working from the plan view to be more convenient at this stage of the model.

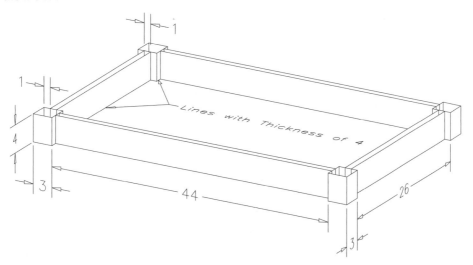

Figure 4.7

The corner posts are simply 3-by-3 squares; the long supports between the corner posts are two lines, each 44 units long and spaced 1 unit apart; and each of the short supports are lines 26 units long spaced 1 unit apart. Of course, you need to draw only one of each item and then use copies of it for the other pieces. Although the top and bottom of each of these pieces will be open, the table top will cover their top, and we do not intend to look at their lower sides, so we will leave them open. Building this part of the table is exactly like working in 2D (especially if you use the plan view as shown in Figure 4.8), except that the lines have height in addition to length.

Now, we are ready to draw the table top. We will use an extruded 2D polyline that is 36 units wide. First, move the UCS 4 units in the Z direction. The XY plane should now be 29 units above the WCS XY plane. If you want to check this, take a look at the Ucsorg system variable. It should read:

UCSORG= 0.0000,0.0000,29.0000 (*read only*)

Change the Thickness system variable from 4 to 1, and invoke the PLINE command. Start the polyline at the point shown in Figure 4.8, set its width to 36 and end it 54 units away in the X direction. That completes the table top. In plan view the polyline will be filled with its entity color. From all other viewpoints AutoCAD automatically turns fill off, but if it obscures objects while you are working in plan views you can turn it off through the FILL command.

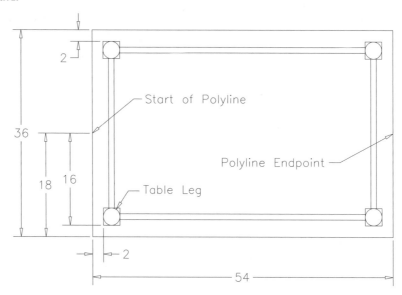

Figure 4.8

The table's legs are all that remain, and you can stay in the plan view while you draw them. Restore the WCS to place the drawing plane back on the WCS XY plane, and change

Thickness from one to 29. Then draw a circle centered within the 3-by-3 table supports and having a diameter of 3 units (see Figure 4.8). You can either copy or draw the other three legs, whichever you prefer. That finishes the table.

 You can compare your model with the one in file 3d_ch4_01.dwg on the CD-ROM that accompanies this book.

For a better look, switch to an isometric-type view direction. The viewpoint of Figure 4.9 is 300° in the XY plane from the X axis, and 20° from the XY plane. Use the HIDE command to cover objects that lie behind surfaces. There are no options with the HIDE command—you simply invoke the command, and it does its work. A screen regeneration will restore the hidden lines. We will describe HIDE more completely later in this chapter.

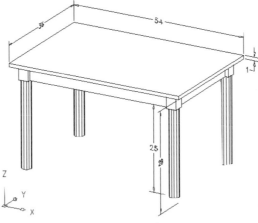

Figure 4.9

The table legs would look better if they were tapered, but extruded surfaces made from circles cannot be tapered. In a later section we will redo these table legs using another AutoCAD surface object that is able to make tapered, cylindrical surfaces.

THE 3DFACE COMMAND

This command makes 3D faces, which are a basic AutoCAD object type, just as lines and circles are. 3D faces are one of AutoCAD's first 3D surface objects (they were introduced with version 2.6 in 1987), but they are still often used for making planar surfaces that have three or four sides.

Their clean, unmeshed surfaces, along with their ability to hide edges, give them an advantage over AutoCAD's newer surface object types, which generally have mesh lines running across their surfaces. These mesh lines can quickly add confusing visual clutter to a 3D model.

3D faces always have either three or four straight sides, and for all practical purposes they are always flat. Although it is possible to position the corners of a four-sided 3D face so that the surface is warped, or bowed, this is seldom needed. Only the edges of 3D faces are shown; there are no mesh lines or fill to delineate the surface of the face. Nevertheless, they can hide objects, and they are colored in during rendering and shading. The edges of 3D faces can be made to be invisible, either during command input or by modifying the properties of an existing 3D face.

Release 14's pull-down menu and the button on the Surfaces toolbar for initiating the 3DFACE command are shown in Figure 4.10. Both the menu and the toolbar include AutoCAD's SOLID command, which makes 2D solids, but you are not likely to use that command for making surfaces. 2D solids are filled 2D objects similar to 2D polylines with width. Although you can use them as planar surface objects, any surface shape that the SOLID command can make can be just as easily made with the 3DFACE command. Moreover, you can control the visibility of 3D face edges, but you cannot control the visibility of 2D solid edges.

The command line format for 3DFACE is:

Command: 3DFACE

Specify first point or [Invisible]: *(Specify a point.)*

Specify second point: or [Invisible]: *(Specify a point.)*

Specify third point or [Invisible] <exit>: *(Enter an I, specify a point, or press ENTER.)*

Specify fourth point or [Invisible] < create three-sided face>: *(Enter an I, specify a point, or press ENTER.)*

Specify third point or [Invisible] exit>: *(Enter an I, specify a point, or press ENTER.)*

If you press ENTER when prompted for a fourth point, AutoCAD will draw an edge back to the first point and end the command. This results in a three-sided face. If you specify a fourth point, AutoCAD will draw an edge from it back to the first point, making a four-sided face, and prompt for another third point. If you press ENTER at this prompt, AutoCAD will end the command, leaving the four-sided face.

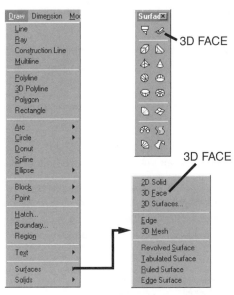

Figure 4.10

When you add points after the fourth point, the prompts alternate between fourth point and third point; with each pair of prompts AutoCAD makes another 3D face by drawing an edge from every new fourth point to the third point of the preceding pair. Pressing ENTER will end the command. There is no provision for undoing a point selection or for stepping back through the points.

Edges are drawn using the current color. You can make an edge invisible by typing in the letter i, or the word invisible, prior to specifying the first point of the edge. In fact, it is even possible to make every edge of a 3D face invisible. Such a 3D face would still hide objects, but could be seen only when it was rendered or shaded.

 Tip: Although you can cover some reasonably complex surfaces in one session of the 3DFACE command, you will probably do better by entering a maximum of four or five points and then repeating the command to add another face to the area you are surfacing. The lack of an undo option leaves no room for mistakes.

Edges are used to select 3D faces for moves, copies, and so on. Invisible edges, however, cannot be seen, even by AutoCAD. Therefore, you must pick on an edge that is visible, or make invisible edges temporarily visible by setting the Splframe system variable to a value of 1.

Making 3D faces is a straightforward process, but you have to be systematic and keep track of where you have added them. Their presence and position is sometimes not very obvious, especially if they have invisible edges. But there are several tools available to help you:

- Setting the Splframe system variable to 1 allows you to check on the location of invisible edges.

- The HIDE command can help you see which faces have been completed and if they are correct, but only if there is something behind the faces to hide.

- SHADEMODE, which fills surfaces with their object color, is also helpful for checking on your progress, especially if there is nothing behind a surface that can be hidden.

- Having more than one viewport on the screen, so you can simultaneously see the model from different viewpoints, is also helpful in keeping track of your progress.

Tip: It is considered good practice to build all surfaces on separate layers. This will allow you to select wireframe objects easier as you generate the surface. In the case of complex surface models, various layers need to be used for holding surface information.

TRY IT! - CREATING 3D FACES

These four examples will demonstrate how the 3DFACE command works. Create a new drawing from scratch and follow the next series of prompts that create various 3D Face objects. First, we will the simple, three-edged 3D face shown in Figure 4.11.

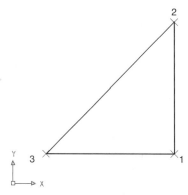

Figure 4.11

Command: 3DFACE

Specify first point or [Invisible]: *(Point 1.)*

Specify second point or [Invisible]: *(Point 2.)*

Specify third point or [Invisible] <exit>: *(Point 3.)*

Specify fourth point or [Invisible] <create three-sided face>: *(Press ENTER,)*

Specify third point or [Invisible] <exit>: *(Press ENTER.)*

Next, we will make the four-sided 3D face shown in Figure 4.12.

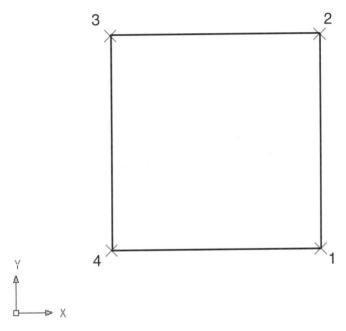

Figure 4.12

Command: 3DFACE

Specify first point or [Invisible]: *(Point 1.)*

Specify second point or [Invisible]: *(Point 2.)*

Specify third point or [Invisible] <exit>: *(Point 3.)*

Specify fourth point or [Invisible] <create three-sided face>: *(Point 4.)*

Specify third point or [Invisible] <exit>: *(Press ENTER.)*

Now we will make two four-sided 3D faces, as shown in Figure 4.13. Notice that you must reverse directions in the second face—we will go counterclockwise on the first face and clockwise on the second. The edge between the two faces will be visible.

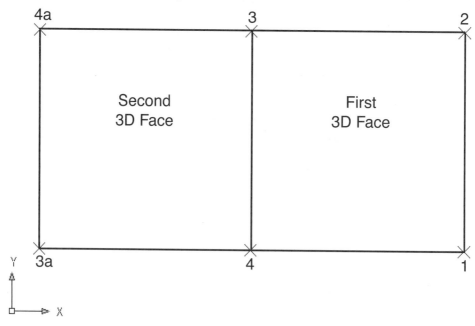

Figure 4.13

Command: 3DFACE

Specify first point or [Invisible]: *(Point 1.)*

Specify second point or [Invisible]: *(Point 2.)*

Specify third point or [Invisible] <exit>: *(Point 3.)*

Specify fourth point or [Invisible] <create three-sided face>: *(Point 4.)*

Specify third point or [Invisible] <exit>: *(Point 3a.)*

Specify fourth point or [Invisible] <create three-sided face>: *(Point 4a.)*

Specify third point or [Invisible] <exit>: *(Press ENTER.)*

The last application will make an L-shaped surface from two trapezoid-shaped 3D faces, with an invisible edge between the two faces, as shown in Figure 4.14. We will use this technique later in this chapter for making the walls of a room.

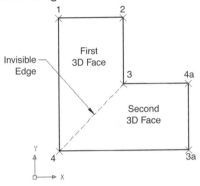

Figure 4.14

Command: 3DFACE

Specify first point or [Invisible]: (*Point 1.*)

Specify second point or [Invisible]: (*Point 2.*)

Specify third point or [Invisible] <exit> : i (*To make the edge from Point 3 to Point 4 invisible.*)

Specify third point or [Invisible] <exit> : (*Point 3.*)

Specify fourth point or [Invisible] <create three-sided face>: (*Point 4.*)

Specify third point or [Invisible] <exit>: (*Point 3a.*)

Specify fourth point or [Invisible] <create three-sided face>: (*Point 4a.*)

Specify third point or [Invisible] <exit>: (*Press ENTER.*)

TRY IT - SURFACING A WIREFRAME MODEL WITH 3D FACES

Now you will cover the wireframe model you made in Chapter 2 as an exercise in working in 3D. Retrieve and open your file of that wireframe, or open file 3d_ch2_01_dwg on the CD-ROM that accompanies this book. First, we will work on the model's right side.

Because this side has five corners, it will take two 3D faces—one with four edges, and one with three. Also, we will make the edge between the two faces invisible. Before you add these 3D faces, though, you should set up and make current a layer just for the 3D faces.

Command: 3DFACE

Specify first point or [Invisible]: *(Point 1.)*

Specify second point or [Invisible]: *(Point 2.)*

Specify third point or [Invisible] <exit>: i *(To make the edge from Point 3 to Point 4 invisible.)*

Specify third point or [Invisible] <exit>: *(Point 3.)*

Specify fourth point or [Invisible] <create three-sided face>: *(Point 4.)*

Specify third point or [Invisible] <exit>: *(Point 3a.)*

Specify fourth point or [Invisible] <create three-sided face>: *(Point 4a.) (Point is in same location as Point 3.)*

Specify third point or [Invisible] <exit>: *(Press ENTER.)*

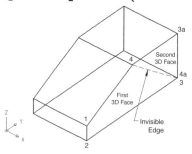

Figure 4.15

In Figure 4.15, the HIDE command has been used to hide the parts of the wireframe that are behind the 3D faces, and a dashed line is used to show the edge between the two 3D faces. Copies of these two 3D faces can be used to surface the left side of the wireframe. Next, you will cover the front and top surfaces of the wireframe (see Figure 4.16). Even though each surface is in a different plane, all three can be covered with one use of the 3DFACE command. When working on existing objects that are in different planes, as we will do here, running object snaps are useful for selecting points.

Command: 3DFACE

Specify first point or [Invisible]: *(Point 1.)*

Specify second point or [Invisible]: *(Point 2.)*

Specify third point or [Invisible] <exit> : *(Point 3.)*

Specify fourth point or [Invisible] <create three-sided face>: *(Point 4.)*

Specify third point or [Invisible] <exit>: *(Point 3a.)*

Specify fourth point or [Invisible] <create three-sided face>: *(Point 4a.)*

Specify third point or [Invisible] <exit>: *(Point 3b.)*

Specify fourth point or [Invisible] <create three-sided face>: *(Point 4b.)*

Specify third point or [Invisible] <exit>: *(Press ENTER.)*

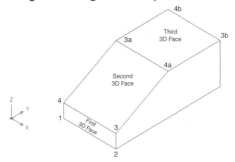

Figure 4.16

Now, when you use HIDE, the model finally begins to resemble a real object. We could have kept on going, adding 3D faces to cover the back and bottom sides of the wireframe. The model in file 3d_ch4_02.dwg on the accompanying CD-ROM does have 3D faces on all sides.

ADDING 3D FACES TO THE MONITOR ENCLOSURE

In this exercise, you will use the 3DFACE command to begin surfacing the wireframe model of an electrical device, or monitor, case that you made in Chapter 3 as an exercise in using the SPLINE command. Open your file of that model, or use file 3d_ch3_03.dwg on the CD-ROM. Constructing the wireframe was the hardest phase in building this model. Adding the surfaces will consist of simply picking wireframe objects and their endpoints. The eight flat, four-sided areas on the wireframe can be surfaced with 3D faces, as shown in Figure 4.17. These 3D faces should be easy for you to make using object endpoint snaps, so we will not go through the steps to add them. Be sure, however, to place them on a layer (assign a name such as SURF-01) that is separate from the wireframe's layer. Figure 4.17 shows the model after the SHADEMODE command has been used. At this stage of the surface model, HIDE will have little effect because there is very little to be hidden. There is one planar area on the wireframe, the end seen when looking from the positive end of the X axis, that will have some holes in it. Therefore, we will not surface that area with a 3D face.

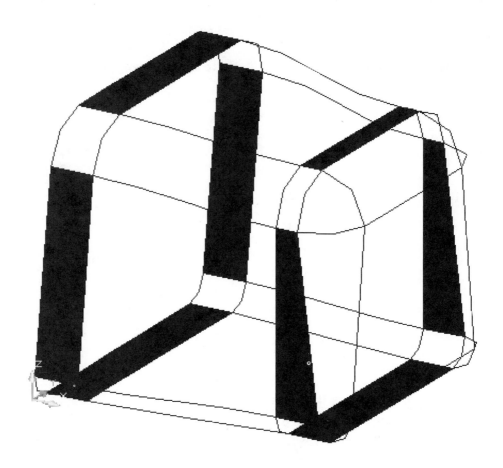

Figure 4.17

 File 3d_ch4_03.dwg on the accompanying CD-ROM has these eight 3D faces added to the wireframe.

CONSTRUCTING THE WALLS OF A ROOM

You will use the 3DFACE command to build a room in which you can place the table you made with extruded surfaces. In later exercises you will add more features and furnishings to this room as you work with AutoCAD's surface modeling commands.

The first step in building the room is to make a wireframe of it. Although most of this wireframe will be covered and no longer visible after the 3D faces have been added, it will help you locate the corners of the 3D faces. You should use a layer for this wireframe that is different than the 3D faces' layer. You should also set the drawing units to feet and inches.

Use the dimensions shown in Figure 4.18 when drawing the wireframe room. If you use AutoCAD's Snap mode with the snap distance set to 6 and position the UCS so that you can draw each side on the XY plane, this wireframe will be easy to draw. You may also find it helpful to work from an isometric-type viewpoint. Figure 4.18 uses VPOINT rotation angles of 300° in the XY plane from the X axis and 30° from the XY plane.

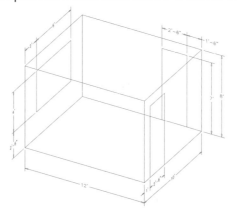

Figure 4.18

When the wireframe is finished, restore the WCS and switch to the layer you are going to use for the 3D faces. You will make the face for the large, unbroken wall on the WCS XZ plane first. To do this, rotate the UCS 90° about the X axis so that you can work in the XY plane, and then draw the four-cornered 3D face using the points shown and labeled in Figure 4.19.

Command: 3DFACE

Specify first point or [Invisible]: (*Point 1.*)

Specify second point or [Invisible]: (*Point 2.*)

Specify third point or [Invisible] <exit>: (*Point 3.*)

Specify fourth point or [Invisible] <create three-sided face>: (*Point 4.*)

Specify third point or [Invisible] <exit>: *(Press ENTER.)*

You could also have used endpoint object snaps to establish the 3D face corners. To draw the 3D faces on the wall with the window cutout, rotate the UCS about the Y axis 90°, or use object snaps to pick the 3D face corners. This wall will require at least four 3D faces because AutoCAD has no means for cutting a hole in a surface. If an area has a hole in it, you must build surfaces around the hole.

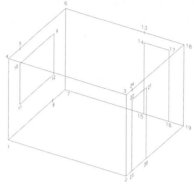

Figure 4.19

We will use two L-shaped pairs of 3D faces for this wall. The first pair of faces will use points 4, 5, 11, 12, 8, and 1; with the three edges from point 5 to 11, point 1 to 11, and 8 to 12 invisible.

Command: 3DFACE

Specify first point or [Invisible]: *(Point 4.)*

Specify second point or [Invisible]: i

Specify second point or [Invisible]: *(Point 5.)*

Specify third point or [Invisible] <exit>: i

Specify third point or [Invisible] <exit>: *(Point 11.)*

Specify fourth point or [Invisible] <create three-sided face>: *(Point 1.)*

Specify third point or [Invisible] <exit>: i

Specify third point or [Invisible] <exit>: *(Point 8.)*

Specify fourth point or [Invisible] <create three-sided face>: *(Point 12.)*

Specify third point or [Invisible] <exit>: *(Press ENTER.)*

If you forget to type in the letter I, for making an edge invisible, continue on with 3DFACE, and then use EDGE command (which we will describe shortly) to change the edge's visibility. The next L-shaped area we will surface will use points 7, 8, 9, 10, 5, and 6; with the edges from point 8 to 9, point 9 to 6, and 5 to 10 invisible.

Command: 3DFACE

Specify first point or [Invisible]: *(Point 7.)*

Specify second point or [Invisible]: i

Specify second point or [Invisible]: *(Point 8.)*

Specify third point or [Invisible] <exit>: i

Specify third point or [Invisible] <exit>: *(Point 9.)*

Specify fourth point or [Invisible] <create three-sided face>: *(Point 6.)*

Specify third point or [Invisible] <exit>: i

Specify third point or [Invisible] <exit>: *(Point 5.)*

Specify fourth point or [Invisible] <create three-sided face>: *(Point 10.)*

Specify third point or [Invisible] <exit>: *(Press ENTER.)*

That finishes the most complicated of the four walls. An alternative to the technique we used would be to implement 3DFACE four times to make four rectangular-shaped faces with appropriate invisible edges.

For the next wall, move the UCS origin to point 7 and rotate it about the Y axis minus 90°. (Or else, do not move the UCS and use object snaps to locate the 3DFACE points.) We will first make a simple, four-edged 3D face from points 6, 7, 15, and 13; with the edge from points 15 to 13 invisible.

Command: 3DFACE

Specify first point or [Invisible]: *(Point 6.)*

Specify second point or [Invisible]: *(Point 7.)*

Specify third point or [Invisible] <exit>: i

Specify third point or [Invisible] <exit>: (*Point 15.*)

Specify fourth point or [Invisible] <create three-sided face>: (*Point 13.*)

Specify third point or [Invisible] <exit>: (*Press ENTER.*)

For the next area of this wall, which is L-shaped, we will use points 19, 16, 17, 18, 13, and 14; with the edges from point 17 to 18 and 13 to 14 invisible.

Command: 3DFACE

Specify first point or [Invisible]: (*Point 19.*)

Specify second point or [Invisible]: (*Point 16.*)

Specify third point or [Invisible] <exit>: i

Specify third point or [Invisible] <exit>: (*Point 17.*)

Specify fourth point or [Invisible] <create three-sided face>: (*Point 18.*)

Specify third point or [Invisible] <exit>: i

Specify third point or [Invisible] <exit>: (*Point 13.*)

Specify fourth point or [Invisible] <create three-sided face>: (*Point 14.*)

Specify third point or [Invisible] <exit>: (*Press ENTER.*)

That finishes that wall, and you can move the UCS to the remaining wall for surfacing it also. This wall can also be surfaced with a rectangular 3D face from points 2, 3, 24, and 23, plus two faces in the L-shaped area between points 20, 19, 18, 21, 22, and 24. The command line
prompts and input for adding these 3D faces will be similar to those for the previous wall, so we will not go through them.

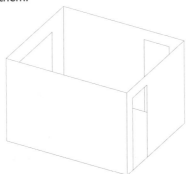

Figure 4.20

The completed walls are shown, with HIDE in effect, in Figure 4.20.

 The 3D room model walls are in file 3d_ch4_04.dwg on the accompanying CD-ROM.

THE PFACE COMMAND

The PFACE command makes three- and four-sided planar surfaces, as does the 3DFACE command. Also like 3DFACE, you can make any number of faces each time you use the command. Unlike 3DFACE, however, the faces made with PFACE are tied together as a single object and the edges between the faces are automatically invisible. Their object type is different as well. Whereas objects made by the 3DFACE command are 3D faces, which is a basic AutoCAD object type, objects made by PFACE are a polyline variation called a polyface mesh. This difference sometimes affects editing and modification operations.

Furthermore, the input for PFACE is entirely different than for 3DFACE. The command has two distinct phases. First, it prompts you for the vertices, the corners, of the surface you want to define. You pick points in space for the vertices, one at time, using any input method, and AutoCAD assigns a number to each vertex as it is selected. In the second phase, AutoCAD creates surface faces by having you enter the appropriate vertex numbers. (Although, at their core these polyfaces consist of three- and four-sided individual faces, in the context of this command the word face refers to a set of individual faces that are tied together.) The faces can lie in different planes. Accordingly, in this phase AutoCAD prompts for both a face number and a vertex number. AutoCAD does not include PFACE in any of its menus or toolbars. You must start the command from the command line.

Command: PFACE

Specify location for vertex 1: (*Specify a point.*)

Specify location for vertex 2 or <define faces>: (*Specify a point or press ENTER.*)

Specify location for vertex n or <define faces>: (*Specify a point or press ENTER.*)

You can enter as many point locations as you want. When you press ENTER, AutoCAD moves to the second phase of the command (Figure 4.21 shows the results):

Face 1, vertex 1:

Enter a vertex number or [Color/Layer] <next face>: (*Enter a C, L, or a vertex number.*)

Face 1, vertex 2:

Enter a vertex number or [Color/Layer] <next face>: (*Enter a C, L, a vertex number, or press ENTER.*)

Face 1, vertex n:

Enter a vertex number or [Color/Layer] <next face>: (*Enter a C, L, a vertex number, or press ENTER.*)

Face m, vertex n:

Enter a vertex number or [Color/Layer] <next face>: (*Enter a C, L, a vertex number, or press ENTER.*)

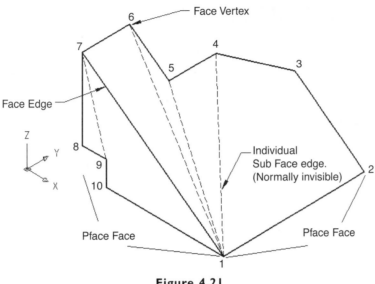

Figure 4.21

Whenever ENTER is pressed, AutoCAD moves to the next face, prompting for the number of the first vertex on this face. Pressing ENTER twice in succession ends the command. AutoCAD does not draw the faces until the command is ended. There is no undo for correcting an entry, and canceling the command cancels everything, even faces previously entered.

In the command's second phase you can enter the words color and layer, or the letters c and l, when prompted for a vertex number. AutoCAD will then prompt you for a new color or layer to be used for the current face, as well as for subsequent faces. Also, edges between faces will be shown, unless you precede the vertex numbers of the points between the faces with a minus sign.

Tips: The PFACE command is not often used due to its cumbersome input. Autodesk intended for the command to be used primarily by applications, such as AutoLISP programs, although Autodesk does not supply any such applications with AutoCAD 2002. Nevertheless, PFACE can sometimes be useful. Adding a surface to a multisided figure, such as a hexagon, is easier with PFACE than with 3DFACE, because you do not have to bother with making face edges invisible.

Polyface meshes are also useful in working with rendering material texture maps, because each object must have its own material. For example, if you constructed the walls of an interior room with three 3D faces, you would have to attach a texture map (such as for wallpaper) to each 3D face and adjust the scale of the texture map on each 3D face. On the other hand, if you used a single polyface mesh to construct the wall, you would have to attach just one texture map to the wall and adjust the scale of the texture map just once.

TRY IT! -CREATING A PFACE

We will demonstrate PFACE on the six-sided plane area shown in Figure 4.22. This is a shape that most people will find easier to surface with PFACE than with 3DFACE. In the figure, the point locations are shown on the left. Create a new drawing from scratch and follow the next series of prompts to create the Polyface mesh.

Command: PFACE

Specify location for vertex 1: (*p1.*)

Specify location for vertex 2 or <define faces>: (*p2.*)

Specify location for vertex 3 or <define faces>: (*p3.*)

Specify location for vertex 4 or <define faces>: (*p4.*)

Specify location for vertex 5 or <define faces>: (*p5.*)

Specify location for vertex 6 or <define faces>: (*p6.*)

Specify location for vertex 7 or <define faces>: (*Press ENTER.*)

Face 1, vertex 1:

Enter a vertex number or [Color/Layer]: 1

Face 1, vertex 2:

Enter a vertex number or [Color/Layer] <next face>: 2

Face 1, vertex 3:

Enter a vertex number or [Color/Layer] <next face>: 3

Face 1, vertex 4:

Enter a vertex number or [Color/Layer] <next face>: 4

Face 1, vertex 5:

Enter a vertex number or [Color/Layer] <next face>: 5

Face 1, vertex 6

Enter a vertex number or [Color/Layer] <next face>: 6

Face 1, vertex 7:

Enter a vertex number or [Color/Layer] <next face>: (*Press ENTER.*)

Face 2, vertex 1:

Enter a vertex number or [Color/Layer] <next face>: (*Press ENTER.*)

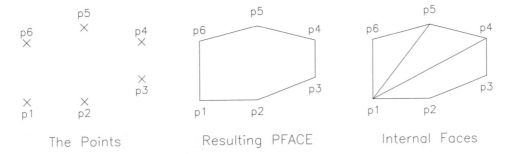

The Points   Resulting PFACE   Internal Faces

Figure 4.22

The resulting polyface mesh is shown in the center of Figure 4.22. On the right the polyface mesh is shown as it would appear when Splframe is set to 1. Notice that internally the polyface mesh consists of one four-edged face and two three-edged faces. Nonetheless, the three faces are all tied together, forming a single entity.

PFACE does not, however, handle internal holes or even edge cutouts or notches very well. Consequently, as shown in Figure 4.23, the results are not always as they first appear. If you were to use PFACE with the points shown, and input in the order shown, to surface the U-shaped area, the resulting surface edges would appear as they are in the upper left figure. However, the surface actually extends beyond the displayed polyface boundary as shown in the upper right figure. This would be apparent with SHADEMODE, or even with HIDE, provided there was an object behind the extra surface between points 4, 5, and 6.

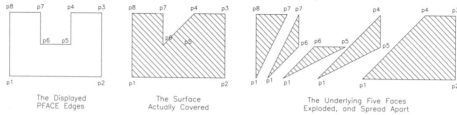

The Displayed PFACE Edges

The Surface Actually Covered

The Underlying Five Faces Exploded, and Spread Apart

Figure 4.23

Furthermore, the seemingly simple surface actually consists of five overlapping internal faces, as shown exploded and spread apart in the lower half of Figure 4.23. This surface comes out wrong because the PFACE input was wrong. The correct way to use PFACE in surfacing this area, is to divide it into two or more faces—such as using points 1, 2, 5, 6, 7, and 8 to define one face, and points 2, 3, 4, and 5 for the other.

TRY IT! - SURFACING A WIREFRAME WITH POLYFACE MESHES

In this exercise you will use PFACE to surface the wireframe you made in an exercise in Chapter 3 and surfaced with 3D faces in this chapter. This exercise will demonstrate making polyface meshes in several different planes. You will not, however, surface the entire wireframe. Open the file for your model, or open 3d_ch2_01.dwg on the CD-ROM. Set up a new layer for the polyface meshes; make it current, and freeze the layer containing the 3D faces. Using the points shown in Figure 4.24, the command line sequence of prompts and input will be:

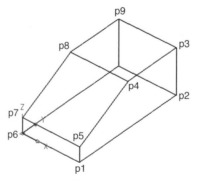

Figure 4.24

Command: PFACE

Specify location for vertex 1: (*p1.*)

Specify location for vertex 2 or <define faces>: (*p2.*)

Specify location for vertex 3 or <define faces>: (*p3.*)

Specify location for vertex 4 or <define faces>: (*p4.*)

Specify location for vertex 5 or <define faces>: (*p5.*)

Specify location for vertex 6 or <define faces>: (*p6.*)

Specify location for vertex 4 or <define faces>: (*p7.*)

Specify location for vertex 5 or <define faces>: (*p8.*)

Specify location for vertex 6 or <define faces>: (*p9.*)

Specify location for vertex 10 or <define faces>: (*Press ENTER.*)

Face 1, vertex 1:

Enter a vertex number or [Color/Layer]: 1

Face 1, vertex 2:

Enter a vertex number or [Color/Layer] <next face>: 5

Face 1, vertex 3:

Enter a vertex number or [Color/Layer] <next face>: 7

Face 1, vertex 4:

Enter a vertex number or [Color/Layer] <next face>: 6

Face 1, vertex 5:

Enter a vertex number or [Color/Layer] <next face>: (*Press ENTER.*)

Face 2, vertex 1

Enter a vertex number or [Color/Layer]: 1

Face 2, vertex 2:

Enter a vertex number or [Color/Layer] <next face>: 2

Face 2, vertex 3:

Enter a vertex number or [Color/Layer] <next face>: 3

Face 2, vertex 4:

Enter a vertex number or [Color/Layer] <next face>: 4

Face 2, vertex 5:

Enter a vertex number or [Color/Layer] <next face>: 5

Face 2, vertex 6:

Enter a vertex number or [Color/Layer] <next face>: (*Press ENTER.*)

Face 3, vertex 1

Enter a vertex number or [Color/Layer]: 5

Face 3, vertex 2:

Enter a vertex number or [Color/Layer] <next face>: 4

Face 3, vertex 3:

Enter a vertex number or [Color/Layer] <next face>: 8

Face 3, vertex 4:

Enter a vertex number or [Color/Layer] <next face>: 7

Face 3, vertex 5:

Enter a vertex number or [Color/Layer] <next face>: *(Press ENTER.)*

Face 4, vertex 1

Enter a vertex number or [Color/Layer]: 4

Face 4, vertex 2:

Enter a vertex number or [Color/Layer] <next face>: 3

Face 4, vertex 3:

Enter a vertex number or [Color/Layer] <next face>: 9

Face 4, vertex 4:

Enter a vertex number or [Color/Layer] <next face>: 8

Face 4, vertex 5:

Enter a vertex number or [Color/Layer] <next face>: *(Press ENTER.)*

Face 5, vertex 1:

Enter a vertex number or [Color/Layer] <next face>: *(Press ENTER.)*

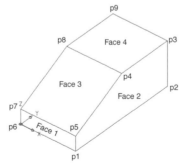

Figure 4.25

Notice that the order in which you assign point locations during the first phase of PFACE has no bearing on the order of vertex number input in the second phase. The completed polyface mesh, with HIDE in effect, is shown in Figure 4.25. The resulting surface, even though it has different faces in different planes, is a single entity and can be moved or copied as any other single entity.

Compare the polyface mesh on your model with the one in file 3d_ch4_05.dwg on the accompanying CD-ROM.

MODIFYING PLANAR SURFACES

As you would expect, objects made with both 3DFACE and PFACE can be moved, copied, rotated, erased, and stretched in the same way as any other AutoCAD object. All of the basic editing and modification commands apply to both 3D faces and polyface meshes. They cannot, however, be given an extrusion thickness, and their edges will always be continuous, regardless of what line type you assign them.

AutoCAD also recognizes the endpoints and midpoints of the visible edges of these objects, as well as intersection and perpendicular object snaps. Invisible edges, on the other hand, are not recognized. If you need to operate on an invisible edge, set the system variable Splframe to 1 to make the edge visible. When Splframe is returned to its default value of 0, the edge will again be invisible after a regen.

3DFACEs are a fundamental AutoCAD object type. The polyface mesh objects made by PFACE, on the other hand, are a 2D polyline variation, which will turn into 3D faces when they are exploded. Polyface meshes cannot, however, be edited with the PEDIT command, even though they are a polyline variation.

Both polyface meshes 3D faces can be modified by the PROPERTIES command. This command allows you to change the vertex locations of 3D faces, as well as the visibility of their edges. Although PROPERTIES allows you to change the vertex locations of polyface meshes, it does not allow you to change the visibility or the color of individual edges of their faces.

THE EDGE COMMAND

This command originates in an AutoLISP program, named EDGE.LSP, that is automatically loaded and functions like any built-in AutoCAD command. It changes the visibility of 3D face edges. It does not work on the polyface meshes made by PFACE. EDGE can change visible 3D face edges into invisible edges and invisible edges into visible edges. Unlike the temporary visibility settings of the Splframe system variable, changes made with EDGE are lasting.

Command: EDGE

Specify edge of 3dface to toggle visibility or [Display]: (Enter D, select an edge, or press ENTER.)

SELECT EDGE

The EDGE command applies to all 3D faces currently on the screen. You do not need to select a 3D face prior to selecting an edge. You must select a 3D face edge by picking on it—crossing and window selections are not allowed. If an invisible edge is selected, it is changed into a visible edge; if a visible edge is selected, it is changed into an invisible edge. The command line prompt is then repeated until ENTER is pressed to end the command.

If two or more 3D edges are on the pick point, all of them are selected and changed, even though they are edges of different 3D faces. Invisible edges cannot be selected unless the Display option of EDGE has been used to make them visible or Splframe has been set to a value of 1 (which is not recommended).

DISPLAY

This option causes invisible edges to be displayed as dashed lines so that they can be picked within the Select Edge option. The follow-up prompt is:

Enter selection method for display of hidden edges [Select/All]<All>: *(Enter S, A, or press ENTER.)*

All

Causes invisible edges in all 3D faces the screen to be displayed as dashed lines. AutoCAD will then repeat the "Specify edge" prompt.

Select

Allows you to select specific 3D faces to have their invisible edges displayed. Any standard selection method can be used to select the 3D faces. AutoCAD will then repeat the "Specify edge" prompt.

The menus and toolbar button for initiating the Edge command are shown in Figure 4.26.

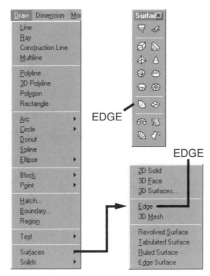

Figure 4.26

202

Tips: EDGE is a very useful command. You may even find it easier to ignore invisible edges when you are creating 3D faces, and then use EDGE to selectively change the edges you want to be invisible.

Do not display invisible edges with Splframe prior to using EDGE. This makes it difficult to distinguish visible edges from invisible edges. Use the display options of EDGE instead.

On the left side of Figure 4.27 are three 3D faces made during one call to the 3DFACE command. All of their edges are visible. To make the interior edges invisible, invoke EDGE and select the edges shown. The results are shown on the right.

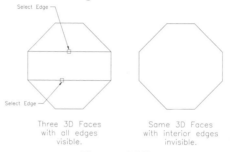

Figure 4.27

POLYGON MESH SURFACES

The most creative and interesting surface shapes produced by AutoCAD are usually made from polygon mesh surfaces. These surfaces seem analogous to a wire mesh because they have a flexible mesh framework that can be bent, pushed, and formed into almost any configuration you need. Moreover, on a computer screen, polygon mesh surfaces often look like a deformed wire mesh. Unlike a wire mesh, though, each grid in a polygon mesh consists of a surface (see Figure 4.28).

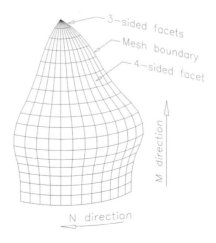

Figure 4.28

AutoCAD has five different commands for making polygon mesh surfaces—
RULESURF, TABSURF, REVSURF, EDGESURF, and 3DMESH. Which command you choose to make a particular surface will depend on the shape of the surface you want to create and on the available defining objects for the surface's shape and boundaries. All five of these commands make the same type of AutoCAD object, which has the following characteristics:

- The surface consists of a collection of flat faces. Most of these flat faces, or facets as they are often called, have four sides. Some, generally located in corners, have only three sides. Each face is similar to 3D face. In fact, when a polygon mesh is exploded, it becomes a set of 3D faces.

- These faces are organized into a matrix of rows and columns, though the rows and columns may bend and turn on the surface of the mesh. Some polygon mesh surfaces, however, will have only one row to go with their columns of faces, whereas others may have only a single column with several rows. A string, rather than a grid, of faces will delineate these surfaces.

- Each polygon mesh has surface directions, corresponding to the rows and columns of the surface matrix, for establishing the position of individual faces. AutoCAD labels one of these direction M, and the other direction N. Which direction is which depends on the AutoCAD command used to make the surface and occasionally on the steps taken to construct the surface.

- With one exception, all AutoCAD commands for making polygon mesh surfaces use the system variables Surftab1 and Surftab2 for setting the number of faces in the M and N direction. In the exception, which is 3DMESH, the number of faces in both directions is specified when the command is implemented.

- Surface meshes that wrap around with one edge joined with its opposite edge are called closed meshes. A mesh may be closed in either the M direction or the N direction, and in the case of a torus (a doughnut-shaped surface), closed in both directions.

- In AutoCAD's database, polygon mesh surfaces are classified as a 2D polyline variation. Consequently they can be edited, on a basic level, with the PEDIT command.

The menus and toolbar buttons for initiating the commands to make polygon mesh surfaces are shown in Figure 4.29.

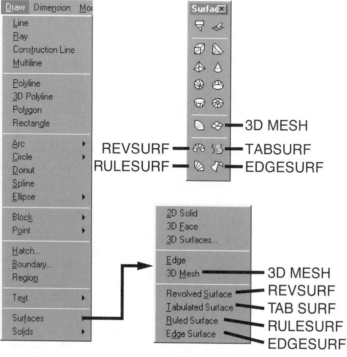

REVSURF
RULESURF
3D MESH
TABSURF
EDGESURF

3D MESH
REVSURF
TAB SURF
RULESURF
EDGESURF

Figure 4.29

THE RULESURF COMMAND

RULESURF makes a surface between two existing boundary objects. It will probably become one of your most frequently used commands for making polygon mesh surfaces. The command format is relatively simple:

Command: RULESURF

Current wire frame density: **SURFTAB1=6**

Select first defining curve: *(Select an object.)*

Select second defining curve: *(Select an object.)*

Each defining curve must be selected by picking a point on it—window and crossing selection methods do not work. The resulting surface will consist of a single row of faces between the two defining curves (see Figure 4.30). Although AutoCAD's reference manuals and online help often refer to the surface as a ruled surface, it is a polygon mesh surface.

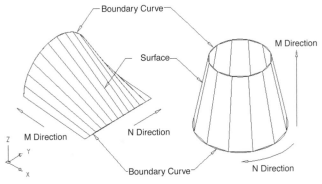

Figure 4.30

The command's prompts refer to the boundaries as curves, but they do not have to be curved. A boundary can be a line, 2D or 3D polyline, spline, arc, or even a point. The boundaries do not have to be the same object type. A line can serve as one boundary and an arc as the other. In fact, if one of the boundaries is a point, the other boundary must be some other object type. Closed objects, such as circles, closed 2D and 3D polylines, closed splines, polygons, doughnuts, and ellipses can also can also be used as boundaries, provided both boundaries are closed or the other boundary is a point.

The M direction on the surface is from one boundary curve to the other. In this direction there is only an edge at the beginning of the surface and another edge at its end. There is no further division, or mesh lines, in the M direction. The N direction is in line with the boundary curves, and the number of divisions in this direction is equal to the value in the Surftab1 system variable (see Figure 4.31). The default value of Surftab1 is 6.

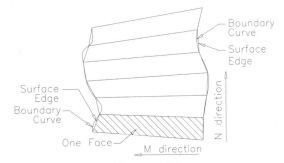

Figure 4.31

Because the edges of each individual face on the surface are straight, the ends of the surface consist of short, straight lines—they are not necessarily on the defining curves.

AutoCAD starts the edge of the first face on the surface at the beginning of the first defining curve and ends the edge of this first face at the beginning of the second defining curve. Moreover, the beginning end of an open defining curve is its end closest to the selection pick point. Therefore, it is important that you pick points on corresponding ends of both defining curves; otherwise the surface will cross over itself (see Figure 4.32).

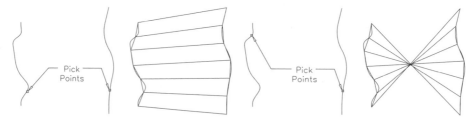

Pick Points

Pick Points

Figure 4.32

With closed boundary objects, AutoCAD starts the surface at a predetermined location, rather than the object selection point. If the boundary is a circle, the surface will start at the circle's 0° quadrant and proceed clockwise around the circle. If the boundary is a closed polyline, the surface starts at the last vertex and proceeds toward the first vertex. If the boundary object is a spline, the surface will start at the first data point and proceed toward the last data point.

Tips: The value you assign to Surftab1 will depend on the length of the surface boundaries and their curvature. The longer the boundaries and the tighter their curves, the larger the Surftab1 value.

It is extremely difficult to match a circle boundary with a closed polyline boundary because of their differences in the surface starting point and the direction in which the surface is built. An alternative is to use two 180° arcs in place of the circle and break the polyline into two pieces.

Using a layer for the surface that is different from the defining boundary object's layer is good practice.

TRY IT! - ADDING RULED SURFACES TO THE MONITOR ENCLOSURE

You will use the RULESURF command to add more surfaces to the wireframe model of an electrical device enclosure. You constructed the wireframe in Chapter 3 and added eight planar surfaces as an exercise for the 3DFACE command earlier in this chapter. Now you will begin adding nonplanar surfaces. Sixteen (out of a total of 32) surfaces will be made with the RULESURF command. All of the surfaces made with RULESURF will be similar in that the two defining boundaries will be curved, whereas the other two edges of the surface will be straight. Open your file of the model, or open file 3d_ch4_03.dwg on the CD-ROM.

File 3d_ch4_03.dwg has the eight 3D faces that were added to the wireframe during the previous exercise.

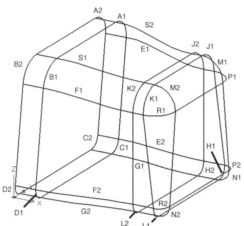

Figure 4.33

Adding the ruled surfaces will be done by simply selecting the pairs of boundary-defining curves. These pairs of RULESURF boundaries are labeled in Figure 4.33 as A1 and A2, B1 and B2, and so forth. (The letters i, o, and q were not used.) Use a separate layer, such as SURF-02, for the ruled surfaces. Then you can turn off the layer used for the 3D faces to get them out of the way. The default setting of Surftab1, which is 6, will be satisfactory for all of the surfaces made with RULESURF.

Your model should look similar to that shown in Figure 4.34 after you add these surfaces with RULESURF. The mesh lines of all 16 of the ruled surfaces are shown on the left in this figure, while the model with HIDE in effect is shown on the right.

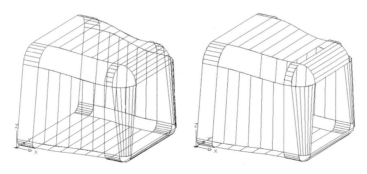

Figure 4.34

File 3d_ch4_06.dwg on the accompanying CD-ROM has these ruled surfaces added to the wireframe model.

Nine sections remain to be surfaced—the four ball-fillet corners, the rounded edges between the spline curves, and the back section with the cutouts. We will use REVSURF to make the corners and EDGESURF to make the surface between the spline curves.

RULED SURFACE TABLE LEGS

Open your drawing that contains the table you built in an exercise for extruded surfaces, or else open file 3d_ch4_01.dwg from the CD-ROM. In that exercise, you used extruded circles for the table's legs, which were not particularly attractive. Therefore you will replace them with tapered, polygon mesh legs made with the RULESURF command in this exercise.

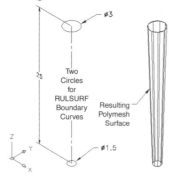

Figure 4.35

Erase the extruded table legs and then draw two circles: one 1.5 units in diameter on the XY plane, and the other located 25 units directly above, 3.0 units in diameter. Then set Surftab1 to 8, invoke RULESURF, and pick the two circles as defining curves. It does not matter which one you select first or where your pick point is on them (see Figure 4.35). Then copy the leg to the other corners of the table, so that it will look like the one shown in Figure 4.36.

 File 3d_ch4_07.dwg on the accompanying CD-ROM has the 3D table with these tapered legs.

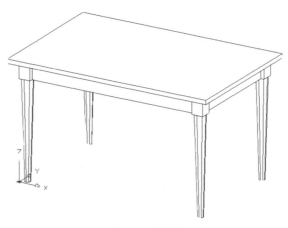

Figure 4.36

THE TABSURF COMMAND

THE TABSURF command makes a polygon mesh surface by extruding a defining curve in a direction specified by an existing object. The command line format is:

Command: TABSURF

Select object for path curve: (*Select an object.*)

Select object for direction vector: (*Select an object.*)

The first object selected, the path curve, controls the cross-section shape of the resulting polygon mesh surface. Only one path curve object is allowed, and it must be selected by picking a point on the object. Window and crossing selections are not permitted. The path curve can be a line, circle, arc, 2D polyline, any member of the 2D polyline family, 3D polyline, ellipse, or spline.

The second object selected, the direction vector, must be a line, an open 2D polyline, or an open 3D polyline. Spline objects are not accepted, although spline-fit polylines are. The direction vector can be located anywhere in 3D space; it doesn't have to be on or even close to the path curve.

The length of the resulting polygon mesh surface will be equal to the length of the direction vector (see Figure 4.37). If the direction vector is a polyline composed of segments that are not in a straight line or has arc segments, the length will be the distance from the polyline's first point to its last point—not the stretched-out length of the polyline.

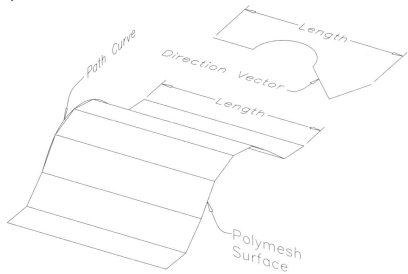

Figure 4.37

AutoCAD uses the pick point location on the direction vector for establishing the direction in which the path curve is extruded. The vector's end closest to the pick point is the start point. The path curve is then extruded, or pushed, toward the vector's opposite endpoint. If the direction vector is a crooked polyline, the direction of extrusion is from the polyline's end closest to the pick point to the polyline's opposite end (see Figure 4.38).

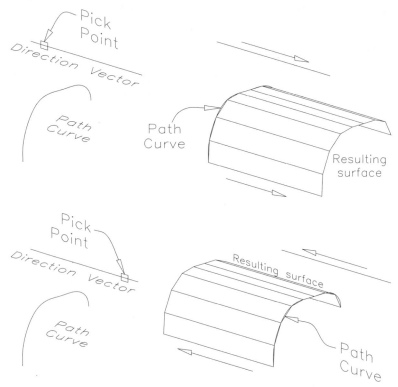

Figure 4.38

The resulting surface's M direction is the same as its extrusion direction, and similar to RULESURF, there is only one face in the surface's M direction. Also similar to RULESURF, the N direction on surfaces made with TABSURF is along the path curve, and the number of faces in the N direction is determined by the value in the Surftab1 system variable. However, how TABSURF uses Surftab1 to divide the path curve is different from RULESURF. RULESURF, you'll recall, simply made the number of faces in the N direction equal to the value of Surftab1. TABSURF, on the other hand, does this only with path curves that are lines, arcs, circles, ellipses, splines, or spline-fit polylines. When the path curve is a 2D or 3D polyline that has not been spline-fit, TABSURF makes a separate face from each straight segment in the polyline. Only arc segments of 2D polylines are affected by Surftab1, with each arc segment being divided into the number of faces equal to the value of Surftab1 (see Figure 4.39).

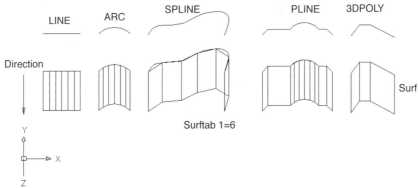

Figure 4.39

Tip: At first glance, TABSURF appears to be capable of doing nothing that the RULESURF command could not do. However, its unique division of faces in the N direction is definitely useful in some situations.

TRY IT! - EXPERIMENTING WITH TABSURF

Draw and make three copies of the planar path curve shown on the left side of Figure 4–40. Its size and actual dimensions are not critical. Then make four direction vectors similar to those shown, and use TABSURF to make four polygon mesh surfaces. You will probably need to view the objects simultaneously from two or more viewpoints (using several viewports) to see the differences between the direction vectors, as well as between the resulting surfaces.

These examples are in file 3d_ch4_08.dwg on the accompanying CD-ROM.

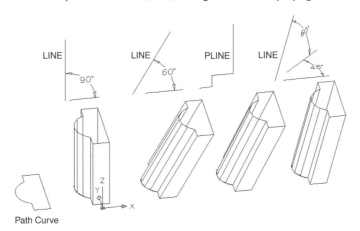

Figure 4.40

WINDOW CASING

In this exercise, you will use TABSURF to add casing around the window in the 3D room we have been working on. Start a new drawing for this casing, which will be inserted as a block into the main 3D room computer file. Set the drawing units to feet and inches. Since the window is in a plane parallel to the YZ plane, you may want to position the UCS in the same way to prevent alignment problems when the window is inserted in the room. Draw a wireframe outline that is 72 units by 48 units to serve as a framework for the casing (see Figure 4.41). Then, set the Thickness system variable to 1, draw a 3-unit-wide 2D polyline, 3 units long outside one of the corners of this frame, as shown in Figure 4.41. You will use this extruded surface as a corner of the casing because you cannot miter TABSURF surfaces.

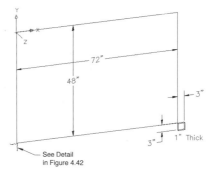

Figure 4.41

Next, at one of the other corners of the window frame, draw the path curve for the casing using the dimensions given in Figure 4.42, and make a 90° rotated copy of the path curve. Use a 2D polyline for these path curves, having first returned Thickness, as well as pline width, to 0. Although you could easily draw a more elaborate path curve, the details would not show up very well in views of the entire room.

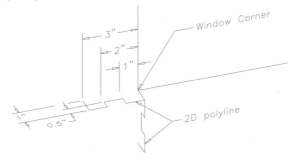

Figure 4.42

Now it is an easy matter to invoke TABSURF and use these path curves with the wireframe outline of the window serving as direction vectors, to make both a vertical and a horizontal strip of casing. Copy these polygon mesh surfaces to the opposite sides of the window frame, and also copy the extruded end piece to the other three corners of the window.

Your window should now look similar to the one shown in Figure 4.43.

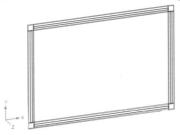

Figure 4.43

If you have time, you could add some rails and panes to these windows, plus any other embellishments you think of. These can all be done with the surface modeling tools you now know how to use—extruded surfaces, 3D faces, and polygon mesh surfaces made with RULESURF and TABSURF.

File 3d_ch4_09.dwg on the accompanying CD-ROM has these extras included with the window casing.

THE REVSURF COMMAND

REVSURF makes a polygon mesh surface by revolving a profile object about an axis. The command line sequence of prompts and input is:

Command: REVSURF

Current wireframe density: SURFTAB1=6 SURFTAB2=6

Select object to revolve: (*Select an object.*)

Select object that defines the axis of revolution: (*Select an object.*)

Specify start angle <0>: (*Enter a value or press ENTER.*)

Specify Included angle (+=ccw, -=cw) <full circle>: (*Enter a value or press ENTER.*)

The object to revolve establishes the cross-section shape of the resulting polygon mesh surface. Only one object is allowed, and it must be selected by picking a point on the object. Window and crossing selections are not permitted. The object can be a line, circle, arc, 2D polyline, any member of the 2D polyline family, 3D polyline, ellipse, or spline. If the object is closed, the surface will be ring- or doughnut-shaped.

The object selected as an axis must be a line, an open 2D polyline, or an open 3D polyline. Multi-segmented, crooked, and even spline-fit polylines may be used, although there is seldom any reason to use anything but a straight object. Spline objects, however, are not accepted, even if they are straight. Usually you will have the axis offset from the profile object, but it does not have to be. If the profile object crosses over the axis, AutoCAD will make an intersecting surface without complaint.

The axis of revolution has direction, which in turn determines the direction of revolution. The positive end of the axis is the end furthest from the object selection point.

The last two prompts of the command refer to the rotation angle of the profile object about the axis. The first of these, the "Start angle:" prompt, allows you to control where the revolved surface is to begin, relative to the profile object. The default value of 0 will begin the surface at the profile object. If you specify an angle, AutoCAD will not begin the surface until that angle from the profile object is reached. For example, if you specified a start angle of 90°, the surface would begin 90° (one-fourth of a revolution) from the profile object.

The final prompt, for included angle, controls both the size and the direction of the profile object's rotational angle. The default angle revolves the profile object a full circle about the axis. Specifying a lesser angle allows you to stop the surface before a complete revolution around the axis is made. Thus, entering 180 degrees at this prompt would revolve the profile object halfway around the axis. Rotation direction follows the right-hand rule, but you can reverse the direction by typing in a negative angle.

Direction of revolution has no importance if the angle of revolution is a full circle, but it becomes important when you are making surfaces that are only partially revolved around the axis. An easy way to visualize rotation direction is to mentally grasp the axis with your right hand, so that your thumb is pointed toward the positive end of the axis. Your fingers will then be curled in the positive rotation direction. Furthermore, if you look at the axis of revolution so that its positive end is pointed directly toward you, positive rotation will be counterclockwise (see Figure 4.44).

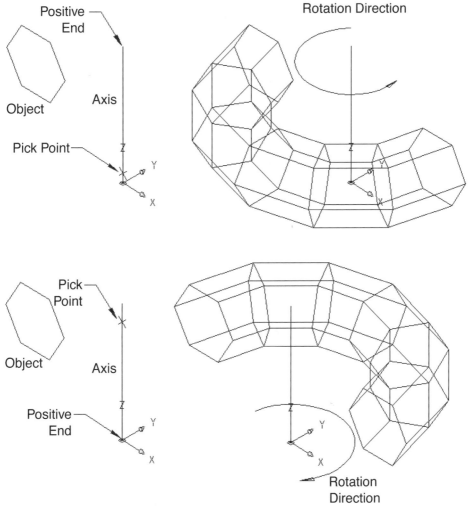

Figure 4.44

The M direction on revolved surfaces is in the direction of revolution around the axis, whereas the N direction is along the profile object. In a departure from RULESURF and TABSURF, in which the number of faces in the N direction is set by the

Surftab1 system variable, REVSURF uses Surftab1 to control the number of faces in the M direction. Surftab2, therefore, determines the number of faces in the N direction. This difference can be confusing, so you may prefer to ignore the M and N designations and just remember that the number of mesh faces along the path curve is set by Surftab2, and the number of mesh faces around the path of revolution is set by Surftab1 (see Figure 4.45).

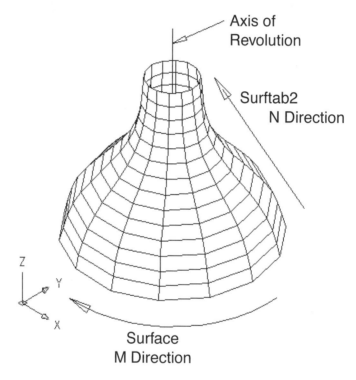

Axis of Revolution

Surftab2 N Direction

Surface M Direction

Figure 4.45

When the profile object is a line, arc, circle, ellipse, spline, or spline-fit polyline, REVSURF divides the boundary along the profile object into Surftab2 faces. However, when the profile object is a 2D or 3D polyline that has not been spline-fit, REVSURF makes a separate face from each straight segment in the polyline. Only arc segments of 2D polylines are affected by Surftab2, with each arc segment being divided into the number of faces equal to the value of Surftab2.

Tip: Of all the polygon mesh surfaces, revolved surfaces are probably the easiest to make. Notice, however, that in selecting the profile object and the axis of revolution you must be certain that your pick point is on an object. Otherwise, AutoCAD ends the command without a message and without giving you a second chance to make a selection.

TRY IT! -

Create a new drawing from scratch and draw a closed 2D polyline similar to the one shown in Figure 4.46 for a profile object. Then draw the two lines shown as Axis A and Axis B to serve as axes of revolution. Axis A is in line with the straight side of the polyline. When you revolve the profile object completely around it, the resulting revolved surface is shaped somewhat like a dumbbell (Figure 4.47). On the other hand, when you revolve the same profile object 180 degrees about Axis B, which is offset from the profile object and turned 90° from Axis A, the resulting surface is an arch (Figure 4.48).

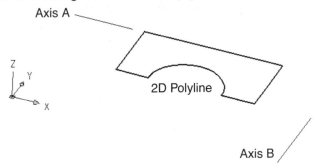

Figure 4.46

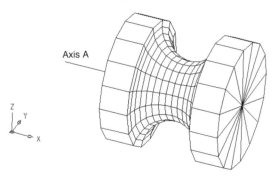

Figure 4.47

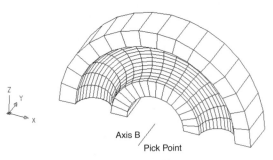

Figure 4.48

TRY IT! - ADDING REVOLVED SURFACES TO THE MONITOR ENCLOSURE

In this exercise, you will use REVSURF made the four ball-fillet corners on the surface model of the monitor enclosure you have been working on periodically. Open your most current file of the model, or else open file 3d_ch4_06.dwg on the CD-ROM. Create and make current a new layer to use for the revolved surfaces (with a name such as SURF-03) and freeze the layers you have used for the 3D faces and ruled surfaces. Draw the temporary construction lines shown in Figure 4.49 to serve as rotational axes for the ball-fillet corners. The default value of 6 for both Surftab1 and Surftab2 will be satisfactory for the surfaces. Then, use REVSURF twice, selecting the indicated arcs as profile objects and the lower part of each construction line for an axis, with an included angle of 90°.

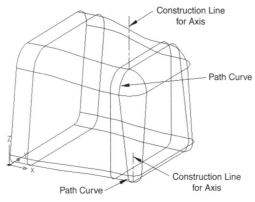

Figure 4.49

You can make the other two ball-fillet corners using the same technique, or else mirror the two existing revolved surfaces. Then your model should look like the one in Figure 4.50 when all of the layers used for the surfaces are on and thawed and HIDE is in effect.

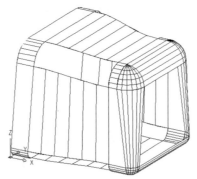

Figure 4.50

These revolved surfaces are on the model in file 3d_ch4_10.dwg on the accompanying CD-ROM.

LAMP AND TEAPOT

In this exercise you will make two more items to place in the 3D room model you have been working on. As with the window frame casing you made with TABSURF, these objects should be in their own computer file, to be inserted as blocks later in the file of the room. Use inches as the drawing's units. First, you will make a table lamp. Draw the body of the lamp as a single 2D polyline, using the dimensions shown in Figure 4.51.

Drawing profiles like this is more subjective than most drafting; and you usually end up drawing them so that they "look good." Consequently, we have not shown the dimensions of the arcs in this polyline. Make sure, however, that the center line of the profile is pointed in the world coordinate system's Z axis's positive direction. Use the ROTATE3D command to accomplish this.

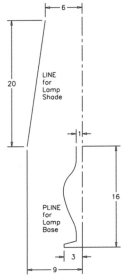

Figure 4.51

Figure 4.52

The profile for the lampshade will be the single line shown in Figure 4.51. We will not bother making a light socket or bulb since our objective is to model a room—not design a lamp.

Prior to making the revolved surfaces, set Surftab1 to a relatively large number, 16 for instance, to obtain a well-rounded appearance. But, set Surftab2 to a low number, such as 4, because the polyline profile contains several short curves. With those settings, your revolved surfaces should be similar to those shown in Figure 4.52.

 This lamp model is in file 3d_ch4_11.dwg on the accompanying CD-ROM.

The next object you will make for the room will be a teapot. The profile for this teapot is shown in Figure 4.53. The dimensions are rather loose; as with the lamp base your profile will probably be slightly different. Make sure, though, that there is a relatively long, flat spot in the profile. That is where the spout and handle will be placed. The lid profile is also shown in Figure 4.53.

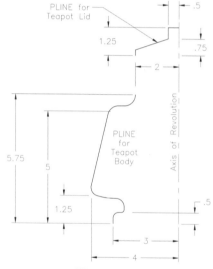

Figure 4.53

Set Surftab1 to 6. This will result in a hexagonal-shaped teapot, which is the shape we want for this particular teapot. Because the profile has several short radius curves, Surftab2 should also be set to a low number, such as 4. After REVSURF, with the lid placed on the teapot, your polygon mesh surfaces should look like those in Figure 4.54.

Figure 4.54

Compare your model teapot with the one in file 3d_ch4_12.dwg on the accompanying CD-ROM.

You will add the teapot's spout and handle later, using EDGESURF, another polygon mesh surfacing command.

THE EDGESURF COMMAND

EDGESURF uses four boundary curves to make a surface that is a blend of all four boundaries. AutoCAD's manuals often refer to these surfaces as a Coons surface patch. This term relates to the methods used to compute the surface, rather than to the surface itself. EDGESURF makes a polygon mesh surface that is no different from surfaces made with REVSURF, TABSURF, and so forth. The command line input for the command is:

Command: EDGESURF

Current wireframe density: SURFTAB1=6 SURFTAB2=6

Select object 1 for surface edge: (*Select an object.*)

Select object 2 for surface edge: (*Select an object.*)

Select object 3 for surface edge: (*Select an object.*)

Select object 4 for surface edge: (*Select an object.*)

Objects selected for surface edges can be lines, arcs, open 2D or 3D polylines, or splines. They can all be the same object type, or they can be a mixture; but there must be four of them, and their ends must touch. There cannot be the slightest gap between the edges or any overlapping intersections.

As is true with the other commands that make polygon meshes, the boundary objects must be selected by picking a point on them. Crossing and window selection methods are not allowed. If you do miss when picking an edge, however, AutoCAD will repeat the prompt rather than end the command.

The first of the four edges selected establishes the surface M and N directions (see Figure 4.55). The M direction on the surface is along the edge of the first object selected, and the Surftab1 system variable sets the number of faces in the M direction. Consequently, the surface's N direction is from the first edge selected toward its opposite edge, and Surftab2 sets the number of faces in this direction. After the first edge is selected, the other three edges may be selected in any order.

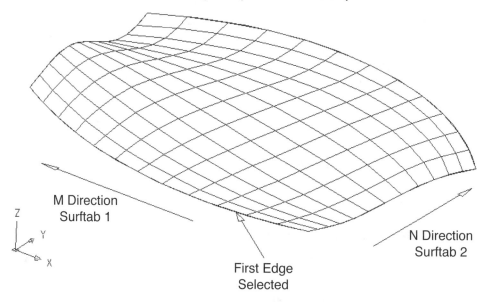

M Direction
Surftab 1

N Direction
Surftab 2

First Edge
Selected

Figure 4.55

Tip: If a wireframe object, such as a line or polyline, already is the edge of a surface, it is sometimes difficult to pick it as the edge of another surface. When you have such stacked objects, hold the computer's CTRL key down as you pick. AutoCAD will then step through the stack of objects each time you press the pick button, highlighting each object as it is selected. Press ENTER when the object you want selected is highlighted.

TRY IT! - USING EDGESURF ON THE MONITOR ENCLOSURE

In this exercise you will use EDGESURF to make the rounded-edge surfaces on the electrical device, or monitor, enclosure you have been working on. Open your most current version of that model or else open file 3d_ch4_10.dwg. Create and make current a new layer (with a name such as SURF-04) for the polymesh surfaces that are to be made with EDGESURF. You may want to freeze the layers for the other surfaces to make it easier to pick the wireframe objects of the model. The default value of 6 for both Surftab1 and Surftab2 will work satisfactorily for making the new surfaces.

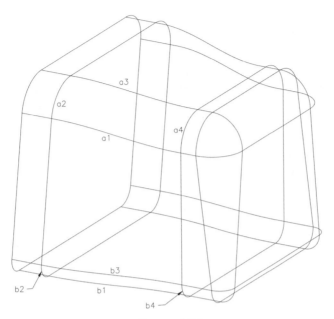

Figure 4.56

Invoke EDGESURF and pick the four boundary curves labeled a1, a2, a3, and a4 in Figure 4.56 to make the surface of the upper rounded edge. The order in which you pick the curves is of no importance because Surftab1 and Surftab2 contain equal values. Then invoke EDGESURF again and select the curves labeled b1, b2, b3, and b4 for the surface of the lower rounded edge.

The two rounded-edge surfaces on the opposite side of the model can also be easily made with EDGESURF by picking the appropriate boundary curves. Now your model should look similar to the one shown on the left in Figure 4.57 when HIDE is on.

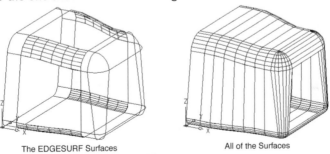

The EDGESURF Surfaces
All of the Surfaces

Figure 457

After all of the layers used for surfaces are thawed and HIDE is on, your model should look similar to the one on the right in Figure 4.57.

File 3d_ch4_13.dwg on the CD-ROM that comes with this book has these four surfaces added to the model.

224

The remaining flat surface on the closed end of the enclosure model will be made with the REGION command in Chapter 6.

BOAT HULL

You will complete the surface model of a boat hull you started as a wireframe exercise at the end of Chapter 3. You will add three surfaces to it using the EDGESURF command, and then make a mirror image copy with AutoCAD's MIRROR command of those surfaces for the other half of the hull.

Open your AutoCAD drawing file that contains the wireframe boat hull, or file 3d_ch3_06.dwg on the CD-ROM. Create three layers for the surfaces and make one of them the current layer, or if you are using the wireframe in 3d_ch3_06.dwg, make layer SURF-01 the current layer.

The first of the three surfaces you will make will be the planar surface on the back of the hull as shown in the Figure 4.58. The viewpoint for this figure is 275° in the XY plane from the X axis, and 20° from the XY plane. Set the value of system variable Surftab1 to 8, and set Surftab2 to 6. Then start the EDGESURF command and pick the two lines numbered 1 and 4, and the two curves numbered 2 and 3 as surface edges. Be certain that you pick line 1 or curve 3 first. The order of the other edge selections is not important. Your surface should look similar to the one on the right in Figure. 4.58.

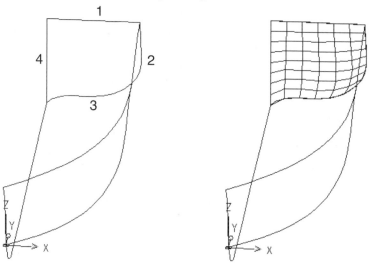

Figure 4.58

Next, you will surface the bottom of the hull. Switch to another layer you have established for the surfaces, such as SURF02, and turn off the previous layer. This will make it easier to pick the surface edges. Leave Surftab1 set to a value of 8, but change Surftab2 from 6 to 18. You will recall from the exercise in Chapter 3 that what appears to be a single curve down the middle of the hull is actually three separate curves. You will use two of those three curves as edges for this surface. Start the EDGESURF command again, and select curves 1,

2, 3, and 4 shown in the upper part of Figure 4.59, for edges. Pick curve 1 or 3 as the first edge. Your surface should look similar to the one in the lower part of Figure 4.59.

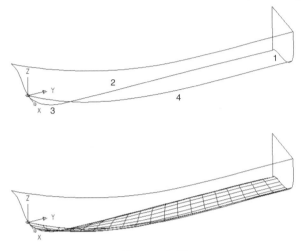

Figure 4.59

The third, and final, surface for hull will be its side. Turn off the layer you used for surface number two, and make another layer you have reserved for surfaces (such as SURF-03) the current layer. You do not need to change the Surftab1 and Surftab2 system variables. Start the EDGESURF command, and pick the curves numbered 1 through 4 in the upper part of Figure 4.60 as edges, with curve number 1 or 3 being selected first. Your surface should be similar to the one shown in the lower part of Figure 4.60.

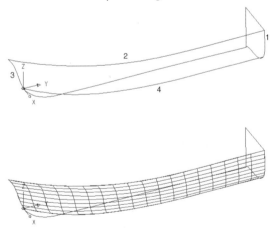

Figure 4.60

Turn on the layers for the first two surfaces, and turn off the layer (or layers) you used for the wireframe. Finish the hull by using AutoCAD's MIRROR command on the three surfaces. Pick two points along the WCS Y axis for the mirror plane, as shown on the left side of Figure 4.61. Your completed hull should look similar to the one on the right side of the figure. The viewpoint for Figure 4.61 is 270° in the XY plane from the X axis, and 30° from the XY plane.

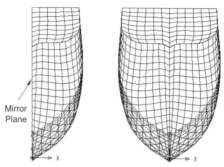

Figure 4.61

When you change to a viewpoint that is 310° in the XY plane and -20° from the XY plane, and invoke the HIDE command, to look at the lower side (the water side) of your boat hull, it should look similar to the one shown in Figure 4.62.

The completed surface model is on the CD-ROM that comes with this book, in file 3d_ch3_17.DWG.

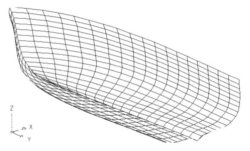

Figure 4.62

TEAPOT SPOUT AND HANDLE

In this exercise, you will add a spout and a handle to the teapot you started with the EDGESURF command. A low Surftab1 value was deliberately used when the revolved the teapot's profile object, so there would be flat faces on the teapot's surface. Because AutoCAD cannot trim intersecting surfaces, you must place the teapot's spout and handle on top of the teapot surface, and this is much easier to do if the matching areas are flat.

You will make just half of the teapot spout, and then mirror it to make the other half. Use the 3point option of the UCS command to place the XY plane on one of the six large faces on the teapot, as shown in Figure 4.63. Then draw half an ellipse in the middle of the face, using the given dimensions. Next, draw an arc and a 2D polyline from the ends of this half-ellipse using the dimensions in the profile view shown in Figure 4.64. The radii of the two arcs of the polyline and the radius of the arc are not critical. Draw these objects so that they are shaped approximately as shown in Figure 4.64. Lastly, draw the half-ellipse end of the spout as shown in Figure 4.63. Of course, you will have to move and reorient the UCS several times to draw these last three objects.

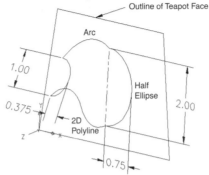

Figure 4.63

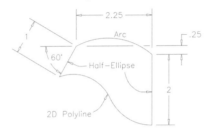

Figure 4.64

Once the wireframe edges are drawn, adding the surface is almost trivial. Just set Surftab1 and Surftab2 to 12 and 10, respectively (these values are not critical), invoke EDGESURF, and pick either the arc or the 2D polyline as the first edge. Your surface for one side of the teapot spout should look similar to the one in Figure 4.65. Then, mirror it to make the other half.

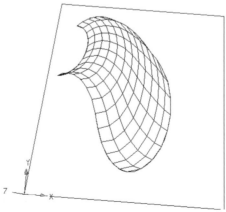

Figure 4.65

The steps to make the teapot's handle are similar, so we will not go through the details of making it. It will be located on the flat face of the teapot that is opposite the spout. Notice in Figure 4.66 that the four edges consist of two half-ellipses and two 2D polylines. Your wireframe for the handle should resemble the one shown on the left side of Figure 4.66 while the completed handle should be something like the one shown on the right side. The finished teapot surface model is shown in Figure 4.67.

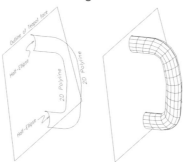

Figure 4.66

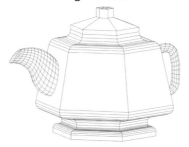

Figure 4.67

 File 3d_ch4_12.dwg on the accompanying CD-ROM has details of the spout and handle, along with the completed teapot.

THE 3DMESH COMMAND

This is the polygon mesh surface command you will use when none of the others will do the job. The input for 3DMESH specifies the location of each node in the mesh. This means you have complete control over shaping the mesh, but it also means that tedious and error-prone manual input is required. You'll use 3DMESH only in special circumstances and when a relatively small number of vertices are used. The command line sequence of prompts and input is:

Command: 3DMESH

Enter size of mesh in M direction: (*Enter a number from 2 to 256.*)

Enter size of mesh in N direction: (*Enter a number from 2 to 256.*)

Specify location for vertex (0,0): (*Specify a point.*)

Specify location for vertex (0,1): (*Specify a point.*)

Specify location for vertex (m,n): (*Specify a point.*)

The command begins by asking for the total number of nodes, or vertices, in the M direction, and then for the total number in the N direction. Then, after asking for the point location of the very first vertex on the surface—vertex (0,0)—AutoCAD begins a series of prompts for the location of every vertex on the surface. AutoCAD then constructs the polygon mesh surface using the vertices as corners of four-edged faces.

The first of the two numbers used to identify each vertex represents the M position, and the second represents the N position. You may find it helpful to think of these as matrix rows and columns, even though which direction represents rows and which represents columns is arbitrary. With each vertex prompt, AutoCAD increments the N index until the end of the row is reached, and then AutoCAD increments the M index and begins to step through the columns on the next row. Since AutoCAD starts numbering with zero, the maximum vertex index numbers will be one less than the M and N size.

There is no provision for undoing a point location or stepping back through the input.

TRY IT! - USING THE 3DMESH COMMAND

In this exercise, you will use 3DMESH to add a surface to the wireframe shown in Figure 4.68. This wireframe can be easily and quickly drawn from the plan view, as shown in Figure 4.69, since all but four of its points are on the XY plane. Those four points—the ones labeled 2,2; 2,3; 3,2; and 3,3 in Figure 4.69—are on a plane below and parallel to the XY plane. The actual point location of the nodes and the size of the wireframe are not important as long as the relative proportions are maintained. For easy point input during 3DMESH, you should use the Snap mode while drawing the wireframe objects that lie on the XY plane. The purpose of this wireframe is to help you keep oriented while executing the 3DMESH command.

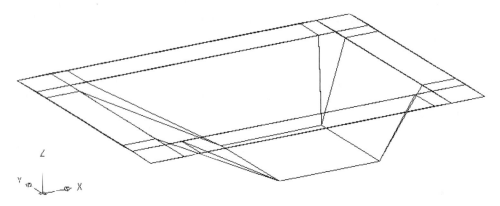

Figure 4.68

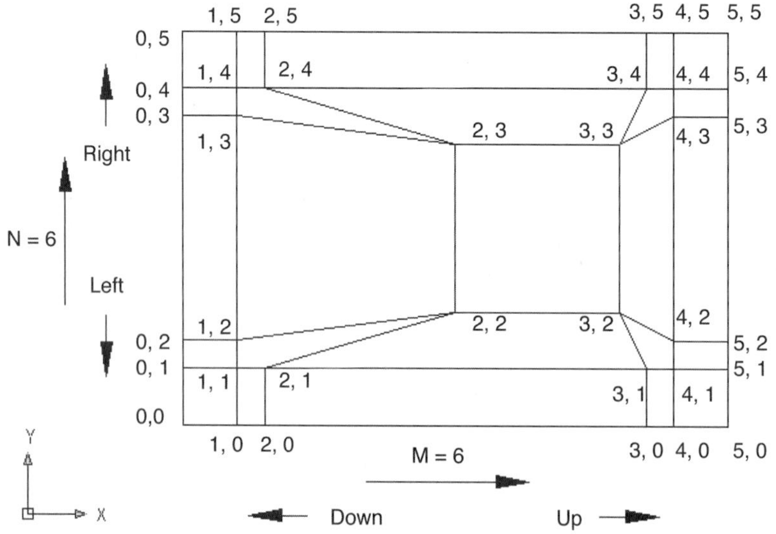

Figure 4.69

After you have drawn the wireframe, create and make current a layer for surface objects. Then start 3DMESH and enter 6 as the mesh size for both the M and N directions. Figure 4.69 shows the M and N directions we will use, along with the vertex index numbers AutoCAD will display in its prompts. The directions we selected, with M pointing to the right and N pointing up, are arbitrary; we could just as easily say instead that M points up and N points to the left. Any direction is fine, as long as you are consistent once you start specifying point locations to AutoCAD's vertex index number prompts.

When AutoCAD asks for the location of vertex (0,0), pick the point labeled 0,0 in Figure 4.69; when AutoCAD asks for the location of vertex (0,1), pick the point labeled 0,1; and so forth. After the input for vertex (0,5), AutoCAD will move to the next column, prompting for the location of vertex (1,0); then step through that column; and continue on with each row in each column until the locations of all 36 points have been specified.

The completed surface model, shown in Figure 4.70, looks exactly like the wireframe, except when HIDE or SHADE are used. You should keep this surface model. Later, when we discuss editing polymesh surfaces, we will smooth its rough edges so that such a surface model could be used for designing a product package, or even an oil pan.

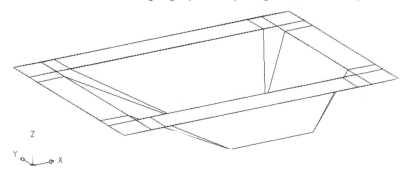

Figure 4.70

The model is in file 3d_ch4_14.dwg on the CD-ROM that accompanies this book.

THE 3D COMMAND

This command, named simply 3D, draws polygon mesh surface objects in any of nine different ready-made shapes. The command originates in an AutoLISP program, named 3D.LSP, that is automatically loaded and ready to be called at any time. Even though the polygon mesh commands we have just described can make any of the surface shapes that 3D can make, you will find 3D to be more convenient for making basic shapes. (In fact, 3D.LSP uses these built-in AutoCAD commands.) The menus and toolbar buttons for these ready-made shapes are shown in Figure 4.71. The Draw|Surfaces|3D Surfaces menu leads to a dialog box titled "3D Objects" that has image tiles showing all nine shapes. This dialog box is shown in Figure 4.72. The command line input and resulting prompt for 3D is:

Command: 3D

Box/Cone/DIsh/DOme/Mesh/Pyramid/Sphere/Torus/Wedge: (*Enter an option.*)

Notice that there is no default option. Therefore, pressing ENTER ends the command.

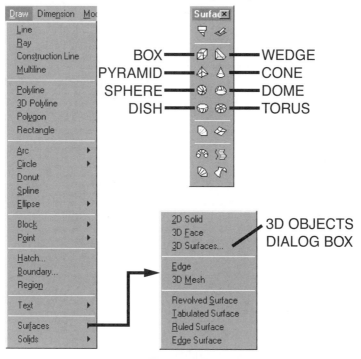

Figure 4.71

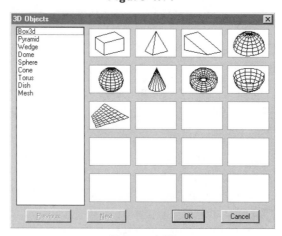

Figure 4.72

Tip: Although these basic shapes are easy to make, they are of limited use because AutoCAD has no means to combine them with other surface objects for making complex surface geometries.

MODIFYING POLYGON MESH SURFACES

Often there seem to be two phases involved when working with any AutoCAD object, whether it is a 2D or 3D object. Creating the object is the first phase, and the second is modifying it. Sometimes, of course, modification is necessary to correct a mistake, but some modifications are necessary because it is not convenient or perhaps not even possible, to create the object in the shape or form you want. For polygon mesh surfaces this may mean smoothing the mesh, stretching sections of the mesh, or moving individual vertices of the mesh. For instance, modifying a surface edge section can create a beaker's pouring lip on a straight polygon mesh tube, as shown in Figure 4.73.

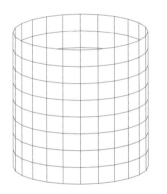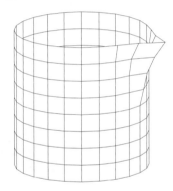

Figure 4.73

Modifications to polygon mesh surfaces are generally limited to operations that move vertices. You cannot add or remove vertices on a polygon mesh surface, and you cannot combine or join surfaces. Nor can you use the FILLET, CHAMFER, BREAK, EXTEND, TRIM, or LENGTHEN commands. The STRETCH command, however, does work on polygon mesh surfaces.

Grips can be used to copy, rotate, scale, mirror, and move entire polygon mesh surface objects, and they can also be used to move individual vertices. Just select a grip—there is one on each vertex—and use the Stretch option to move it to a new location.

THE PEDIT COMMAND

Since polygon mesh surfaces are a variation of 2D polylines, they can be edited through the PEDIT command. When an object made with the RULESURF, TABSURF, REVSURF, EDGESURF, and 3DMESH commands is selected for editing with PEDIT, AutoCAD brings up a special command line menu that allows you to move any vertex of the mesh to a new location, to smooth out bumps and sharp angles on the mesh surface, and to close an open mesh or open a closed one. The command line format for PEDIT with polygon mesh objects is:

Command: PEDIT

Select polyline: (*Select a polygon mesh object.*)

Select an option [Edit vertex/Smooth surface/Desmooth/Mclose/Nclose/Undo]: (*Select an option or press ENTER.*)

If the mesh is closed in the M direction, the Mclose option will be replaced by Mopen; if it is closed in the N direction, Nclose will be replaced by Nopen. This menu remains active until you exit by pressing ENTER.

EDIT VERTEX

This option is for moving individual vertices of a polygon mesh. It uses the follow-up prompt:

Vertex (m,n):

Enter an option [Next/Previous/Left/Right/Up/Down/Move/REgen/eXit] <N>: (*Select an option or press ENTER.*)

Notice that there are no options for inserting new vertices or for deleting existing ones, as there are for vertex editing on 2D and 3D polylines. The only vertex editing you can do with polygon mesh surfaces is to move them.

Moving vertices is straightforward; you simply go to the vertex that you wish to move, select Move from the menu, and specify a new location for the vertex. The Next, Previous, Left, Right, Up, and Down options are all used for going to the vertex you intend to move.

The current vertex is identified on the surface by a crosshair X and the vertex's position number is shown in parentheses on the left-hand side of the menu. Two numbers, separated by a comma, are used to identify vertex position—the first number is the M position and the second number is the N position. Zero is used as the first number in both directions.

M and N directions on a surface depend on how the surface was created, not on how it is oriented in space. AutoCAD uses the words up, down, right, and left in describing directions, but those directions are not likely to correspond to those same directions on your computer monitor. Figure 4.74 shows the vertex position numbers and surface directions for the object we made when we discussed the 3DMESH command.

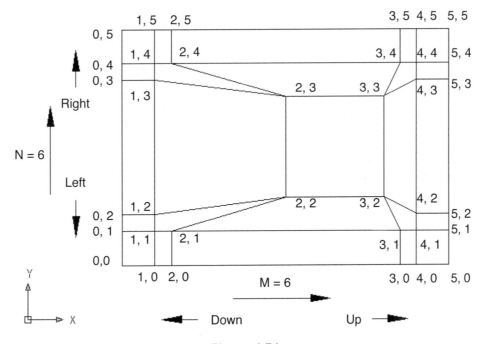

Figure 4.74

Next

Moves the vertex marker to the next vertex in the **N** direction. When the marker reaches the last vertex in the **N** direction it will then move one place in the **M** direction and proceed again in the **N** direction. If, for example, the marker is on vertex (1,3), the next vertex is (1,4). If you continued to select the **Next** option, the marker would eventually reach the end of the **N** direction, and the marker would go to vertex (2,0).

•Previous

Moves the vertex marker to the preceding vertex.

Left

Moves the marker one vertex in the minus **N** direction. Thus, if the marker is on vertex (3,4), the left vertex is (3,3).

Right

Moves the marker one vertex in the positive **N** direction.

Up

Moves the marker one vertex in the positive **M** direction. For example, if the marker is on vertex (3,4), the up vertex is (4,4).

•Down

Moves the marker one vertex in the negative **M** direction.

Move

This option signals that you want to move the vertex that currently has the marker. AutoCAD will display the prompt:

Specify new location for marked vertex: (*Specify a point.*)

REgen

Regenerates the polygon mesh.

eXit

Returns to the main **PEDIT** prompt.

SMOOTH SURFACE

Transforms the surface mesh into a 3D B-spline or Bézier curve, depending on the value in the Surftype system variable. We will discuss the effects of this option later.

DESMOOTH SURFACE

Returns a smoothed surface mesh into its original form. This option applies even to polygon mesh surfaces that were smoothed during previous AutoCAD editing sessions.

MCLOSE/MOPEN

Which option is displayed depends on whether the mesh is open or closed in the M direction. If the mesh is open in the M direction, AutoCAD will display the Mclose option, which will add a section to the mesh, thus closing it in the M direction. If the mesh is closed in the M direction, the menu will list the Mopen option, which will remove a section from the mesh to open it in the M direction.

NCLOSE/NOPEN

If the surface mesh is open in the N direction, AutoCAD will display the Nclose option, which adds a section to the mesh for closing it in the N direction. If the mesh is closed in the N direction, the menu will show the Nopen option for removing the last section from the surface mesh to open it in the N direction.

UNDO

Cancels the last operation.

When AutoCAD smoothes a 3D surface, it replaces the original surface mesh with a new one. This new surface, however, is still a mesh made of flat faces, so it is not perfectly smooth. The number of rectangular faces on this new mesh is controlled by two system variables—Surfu and Surfv. Surfu sets the number of faces in the M direction, while Surfv sets the number of faces in the N direction.

Figure 4.75 shows the effects of Surfu and Surfv. The original surface, shown at the top in this figure, has nine faces in the M direction (M = 10) and three faces in the N direction (N = 4). The three surfaces shown below the original surface were smoothed using the same curve type (cubic B-spline), but with different Surfu and Surfv settings. Because the surface has no curvature in the N direction, Surfv has no real effect, but larger values of Surfu create smoother curves in the M direction. The maximum allowed value for both Surfu and Surfv is 200, and the minimum is 2. The default setting for each is 6.

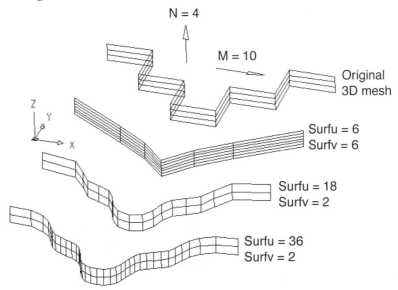

Figure 4.75

AutoCAD retains its information about the defining mesh of a smoothed surface, and the original surface can be restored by selecting the Desmooth option from PEDIT prompt. Moreover, you can see the defining surface mesh by setting the Splframe system variable to 1. When Splframe is set to 0, which is the default, the smooth curved surface is shown, but when it is set to 1, the original defining mesh is displayed. You may recall that for curve- and spline-fit polylines, the smoothed and original polyline are both displayed at the same time when Splframe is set to 1. On polygon meshes, however, just the original mesh is shown—the curve display would be too cluttered if both were shown at once.

Using the Edit Vertex option of PEDIT on a smoothed mesh can be confusing with smoothed curves, since AutoCAD still uses the vertices of the original mesh. You can only move the vertices of the defining surface mesh, not those of the smoothed mesh. Even though the smoothed mesh may be the one shown, the vertex position marker will always be located on one of the vertices of the defining mesh.

The type of curve AutoCAD uses in the Smooth Surface option of the PEDIT command's prompt for polygon mesh surfaces is controlled by the Surftype system variable, as shown in the following table:

SURFTYPE Value	Type of Curve	Minimum Number of Mesh Vertices in Either Direction	Maximum Number of Mesh Vertices in Either Direction
5	Quadratic B-spline	3	None
6	Cubic B-spline	4	None
8	Bézier curve	2	11

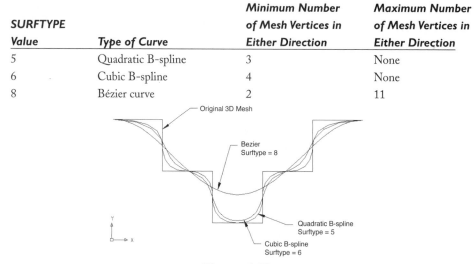

Figure 4.76

Figure 4.76 illustrates the differences between the three curve types that PEDIT can make. This figure shows a front view of the stairstep-shaped 3D mesh shown in the previous figure, along with copies that have been smoothed into quadratic and cubic B-spline curves and into a Bézier curve. Surfu was set to 36 for all three smoothed surfaces. Some characteristics of these surfaces are:

- The only vertices on the smoothed surfaces that match those of the defining surface are the end vertices.

- Although the interior vertices of the smoothed surfaces do not match those of the defining surface, they are controlled by them. The smoothed surface is pulled into shape according to the location of the vertices on the defining surface. Precisely how the curves are affected by the original vertices is controlled by the equations that make the new surface.

- The radii of the smoothed surface's curves are largest on the Bézier (pronounced BAY-zee-A) surface, and smallest on the quadratic B-spline surface. This makes the Bézier surface the smoothest of the three surface types, but it also has the greatest deviation from the defining surface. The quadratic B-spline surface is the closest match to the defining surface.

 Tip: Using grips to move surface vertices is more direct than using the PEDIT command's Move Vertex option, but with some large, complicated surfaces it is sometimes easier to find a particular vertex with the PEDIT vertex marker. AutoCAD places the marker on the first vertex shown on the screen. Therefore, you can zoom in on the particular area of the mesh you are interested in rather than start from the 0,0 vertex.

TRY IT! - SMOOTHING A 3D MESH

In this exercise, you smooth the polygon mesh surface you made earlier in this chapter with the 3DMESH command. Open your file that contains that surface, or else use the file 3d_ch4_14.dwg on the CD-ROM. Before smoothing the mesh, set both Surfu and Surfv to 24 and Surftype to 5 for a quadratic-curved surface. Then use PEDIT to smooth the surface. The result should be similar to the surface shown in Figure 4.77. The viewing angle in this figure is 285° in the XY plane from the X axis and 15° from the XY plane. A surface shaped somewhat like this one could be used to model a plastic cover or canopy, or even an oil pan.

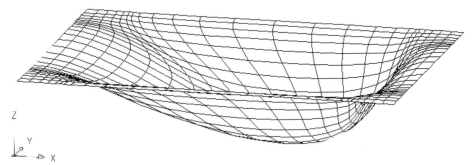

Figure 4.77

 File 3d_ch4_15.dwg on the accompanying CD-ROM contains this smoothed 3D mesh surface model.

VIEWING 3D SURFACES

The trouble with 3D surfaces is that most of the time they don't really look like surfaces on your computer screen. You can see right through them. Furthermore, they have mesh lines and edge lines cluttering up the model and making viewing difficult and confusing. Visualization—trying to understand what you are looking at—is possibly the biggest problem you will face in 3D modeling; until some practical 3D viewing system is developed, you must continue to view 3D objects on a 2D screen.

AutoCAD has some tools to assist in visualization. We've already discussed some—3D viewpoints and multiple viewports. One of the best tools for visualization is rendering, which deserves an entire chapter in this book, Chapter 9. Although rendering is sometimes helpful in visualizing a partially completed model, most of the time you will not render your model until it is finished. As you are working on your model, the visualization tools you will most often use are those we'll discuss next. The commands are HIDE, SHADEMODE, DVIEW, and a variety of commands associated with 3DORBIT. The menus and toolbar buttons for initiating these commands are shown in Figure 4.78.

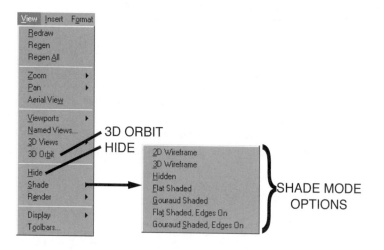

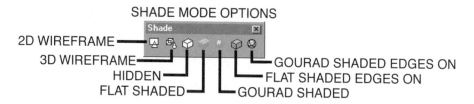

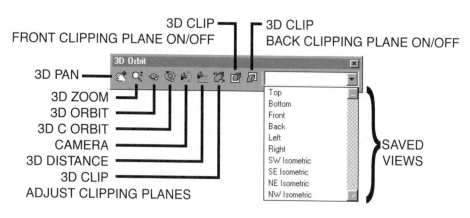

Figure 4.78

THE HIDE COMMAND

HIDE is an extremely simple command to use. There are no options and no selections. You simply invoke the command, and it does its job. Objects and parts of objects in the current viewport that are behind objects with surfaces, relative to the current view direction, are no longer shown—they are hidden (see Figure 4.79).

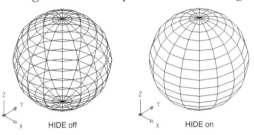

HIDE off   HIDE on

Figure 4.79

Some of the object types that are opaque during HIDE are: 3D faces, polygon meshes, polyface meshes, circles, traces, 2D solids, wide 2D polylines (which includes doughnuts), and edges of wireframe objects that have thickness. Objects having width—traces, 2D solids, and wide 2D polylines—are opaque even if Fill is turned off. Objects that have both width and thickness are opaque in both width and thickness. 3D solids and regions, which we'll cover in Chapter 5, also hide objects.

HIDE remains in effect until the viewport is regenerated. REDRAW and REDRAWALL, however, do not affect hidden line views. The ZOOM and PAN commands cause the view to return to the wireframe viewing mode, though real-time zooms and pans are not allowed.

Opaque objects in a frozen layer will not hide objects; however, opaque objects in a layer that is turned off continue to hide objects, even though they are invisible. Occasionally, this characteristic can be helpful. If you have some surfaces with a dense, confusing web of mesh lines, you can turn off their layer to eliminate the mesh lines, but they will continue to be opaque with HIDE. This may, however, cause some ghostly holes in your model; furthermore, it only works on your computer monitor. Objects in a turned-off layer will not hide objects when they are printed or plotted.

Text, whether it is in the form of text, mtext, or an attribute, is an exception to hidden objects. AutoCAD normally ignores text objects, and consequently they are not hidden. However, if you give text and attributes a thickness, no matter how small, they will be hidden by opaque objects. When text and attributes are created, they do not accept thickness, but existing text and attributes can be given thickness by modifying their properties. Mtext cannot be given a thickness, and hence, cannot be hidden.

Two objects in the same plane cannot hide each other. Thus, the tops of the walls of the 3D surface model house (made from 3D faces) in Figure 4.80 show through the roof when HIDE is on.

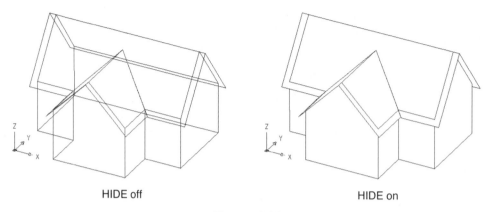

HIDE off   HIDE on

Figure 4.80

Even if an object is in a plane that is very close to the plane of an opaque object, it may not be hidden. Figure 4.81 shows three 3D faces and three hatched circles, each seen in a separate view. In the view on the left, the objects are in the same plane. Therefore, nothing is hidden when HIDE is on. In the center view, the 3D face is 0.0001 units above the hatch, but is still unable to hide it. In the right-hand view, the 3D face is 0.01 units above the hatch and is now able to hide it.

HIDE is on in all 3 views

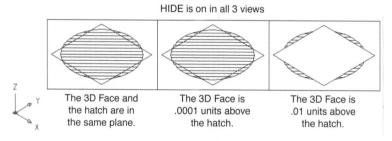

The 3D Face and   The 3D Face is   The 3D Face is
the hatch are in   .0001 units above   .01 units above
the same plane.   the hatch.   the hatch.

Figure 4.81

These characteristics are somewhat dependent on the viewpoint, and if you looked at the center view in Figure 4.81 from a plan view, the 3D face would properly hide the hatch. The characteristics also depend on the calculation techniques AutoCAD uses to determine what is to be hidden. By default, AutoCAD uses single-precision arithmetic during hidden-line calculations, but starting with Release 14 you can have AutoCAD perform the calculations with double-precision arithmetic. AutoCAD will correctly hide the hatch under the 3D face in the center view of Figure 4–81 when it uses double-precision arithmetic. In practice, however, you will seldom have objects that are so close together that you need to have AutoCAD use double-precision arithmetic during hidden-line calculations. Moreover, double-precision arithmetic may significantly increase the time required for performing hidden-line calculations.

Whether AutoCAD uses single- or double-precision arithmetic is controlled by the system variable Hideprecision. When Hideprecision is set to its default value of 0, single-precision arithmetic is used in hidden-line calculations; when it is set to 1, double-precision arithmetic is used

THE SHADEMODE COMMAND

THE SHADEMODE command is used for creating shaded images of solid and surface models. Shaded views differ from hidden line views in that surfaces are colored in with their object color and are illuminated as if a light was aimed over the viewer's shoulder toward the model. Surfaces that are perpendicular to the incoming light are the brightest, and surfaces almost parallel to the incoming light are the dimmest. Unlike rendered views, wireframe objects continue to be displayed in SHADEMODE's shaded images.

The command also has options for creating special wireframe and hidden line viewing modes. These viewing modes, as well as those for shaded views, use a 3D form of the UCS icon. This icon shows all three of the UCS axes as colored cone-shaped arrows on round shafts extending a short distance from the origin.

All of the SHADEMODE viewing modes are full working modes, in which you can freely draw and select objects. Unlike the HIDE command, screen regenerations do not affect the viewing modes. Also, real-time zooms and pans, as well as 3DORBIT and its options, work well with the SHADEMODE viewing modes.

The command line format for SHADEMODE is:

Command: SHADEMODE

Current mode: 2D wireframe

Enter option [2D wireframe/3D wireframe/Hidden/Flat/Gouraud/fLat+edges/gOuraud+edges] <2D wireframe>: (*Enter an option or press ENTER.*)

2D WIREFRAME

AutoCAD's normal wireframe viewing mode, along with the flat X and Y axes form of the UCS icon, is restored by this option (see Figure 4.82).

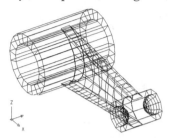

Figure 4.82

3D WIREFRAME

This option causes surfaces to be transparent, so that only wireframe objects, surface meshes, and solid model edges and isolines are displayed (see Figure 4.83). (Isolines will be described in Chapter 6.) However, fills, linetypes, lineweights, and raster images are not displayed. If the system variable Compass is set to a value of 1, the compass image will be displayed. (This image was described during the discussion of the 3DORBIT command in Chapter 2.)

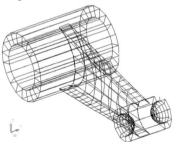

Figure 4.83

HIDDEN

Surfaces become opaque when this option is selected (see Figure 4.84). Even zero thickness text is hidden.

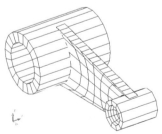

Figure 4.84

FLAT

Each face on surface models is colored with the object color and shaded relative to its orientation to the view direction. The surfaces of solid models are displayed as facets (see Figure 4.85).

Figure 4.85

GOURAUD

Faces on both surface and solid models are colored and shaded with the object color, but the faces are blended so that there are no distinct boundaries between faces (see Figure 4.86). Rounded and curved surfaces appear smooth.

Figure 4.86

FLAT+EDGES

This option displays surfaces as flat, shaded faces and also displays the edges of surface model faces in their object color (see Figure 4.87). The results are similar to combining flat shading with a hidden line view. Although the boundaries of facets on curved and rounded surfaces of solid models are not displayed, their edges and iso-lines are.

Figure 4.87

GOURAUD+EDGES

The Gouraud and Hidden options are combined with this option. Results are shown in Figure 4.88.

Figure 4.88

Tips: Shaded views are often better for checking on the existence and extent of surfaces than hidden line views because surfaces are apparent in hidden line views only when there is something behind them to be hidden. Shaded views, on the other hand, display surfaces even if there are no objects behind them.

If you prefer the SHADEMODE version of the UCS icon, you will work in the 3D wireframe mode rather than in the 2D. That mode's inability to display noncontinuous linetypes, however, can be a disadvantage.

If AutoCAD's 2D wireframe viewing mode is current, 3DORBIT automatically changes to SHADEMODE's 3D wireframe mode, and reverts to the 2D wireframe mode at the conclusion of 3DORBIT. The shaded and hidden line viewing modes of SHADEMODE are unchanged during 3DORBIT.

If you expect to perform numerous pans, zooms, and viewpoint changes, you may prefer SHADEMODE's hidden line removal over that of the HIDE command. The effects of the HIDE command, though, are easier to turn off, so you will HIDE use HIDE when you want to just briefly examine a model.

Gouraud shading gets it name from the person who first developed the method for smoothly shading and blending faces. Although it results in realistic images, you are seldom interested in realistic images as you construct a model, so you will probably spend most of your time in wireframe viewing modes and use shaded modes only when you have problems visualizing a model.

The SHADE command can be used as a shortcut to any one of four SHADEMODE options. You simply invoke the command, and the view is created.

THE DVIEW COMMAND

If you have used the HIDE or SHADEMODE command with the 3D room we have been working with, you have probably noticed that the room's walls get in the way. You cannot see inside the room except from view directions with very steep angles from the XY plane. Zooming in does not help because VPOINT and 3DORBIT set view directions, not a specific viewpoint location or distance.

AutoCAD has another viewing command that is more versatile than VPOINT, and even 3DORBIT. It is DVIEW. When you use DVIEW to look at the 3D room we've been making, you can:

- Set a viewpoint location anywhere in 3D space. It can even be inside the 3D room.

- Make clipping planes that slice through the walls, allowing you to see inside the room even when the viewpoint is outside the room.

- Make perspective views in which distant objects appear smaller than closer objects (see Figure 4.89).

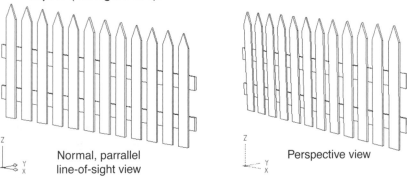

Normal, parrallel line-of-sight view   Perspective view

Figure 4.89

Although AutoCAD 2002 has some specialized commands—3DCLIP, 3DDIS-TANCE, and CAMERA, plus some options of 3DORBIT —that perform the same functions as some of DVIEW's options, you may prefer to use DVIEW because you can perform multiple tasks with one command. Also, the DVIEW techniques for setting perspective views and clipping planes work differently than those of the new commands, and you may still prefer DVIEW's methods.

DVIEW uses the term camera for the point in space from which you are looking, and target as the point in space toward which the camera is pointed. A line from the target to the camera represents the viewing direction, or line of sight. You can set only viewing direction with the VPOINT and 3DORBIT commands, but you can set any of these three parameters with DVIEW.

Unlike VPOINT, in which you must set a viewing direction by typing in numbers without the benefit of a preview, DVIEW (which stands for dynamic view) displays the model as you rotate the viewing direction about in 3D space. Because of this dynamic, real-time preview of the model (it also works for other DVIEW operations, including zooms and pans), the command begins by prompting you to select the objects to be previewed. You can select your entire model or a representative part of it. If you select just part, the remainder will temporarily disappear. When you are satisfied with your view and exit DVIEW, the rest of the model will reappear with the view characteristics established within DVIEW.

If, when prompted for an image preview selection set, you press ENTER without selecting anything, AutoCAD will use a block named DVIEWBLOCK for a DVIEW preview image. If there is no such block, then AutoCAD immediately creates one of a simple one-room 3D house (see Figure 4.90).

Figure 4.90

You will prefer to use the DVIEWBLOCK model if your own model is unsuitable for setting a view, and you will have to use it if you have invoked DVIEW prior to drawing anything. If you would rather use an image of your own, you can create it and then define it as a block named DVIEWBLOCK. Your block should fit within a 1x1x1-unit cube, located in the positive quadrant of the WCS, with one corner on the origin.

The command line format for DVIEW is:

Command: DVIEW

Select objects or <use DVIEWBLOCK>: (*Select objects or press ENTER.*)

Enter option

[CAmera/TArget/Distance/POints/PAn/Zoom/Twist/CLip/Hide/Off/Undo/]: (*Pick a point, select an option, or press ENTER.*)

This prompt reappears when work within an option is finished, until you press ENTER to end the command. AutoCAD uses the selected objects for the command's dynamic preview images. Any selection method may be used in building the DVIEW selection set, including Last and Previous selections.

POINT SELECTION

If you pick a point (you cannot type in coordinates or use object snaps) from the main DVIEW prompt, AutoCAD uses that point as a pivot for rotating the view direction. The screen coordinate display will change to one showing two angles, labeled Dir and Mag, which are updated as you move the screen cursor AutoCAD will show the follow-up prompt:

Enter direction and magnitude angles: (*Specify two angles.*)

The two angles can be specified by either picking a point on the screen or by typing in two angles between 0° and 360°, separated by a comma. The first angle, direction, refers to angles on your computer screen—0° is to your right, 90° is up, and so forth. The second angle, magnitude, determines how much the view rolls around a horizontal axis that passes through the selected point.

CAMERA

This option sets the viewing direction in a manner similar to the Rotate option of VPOINT, although the look and feel is completely different. With the Rotate option of VPOINT, you specify the viewing direction by first entering its angle from the X axis in the XY plane, and then you enter its angle from the XY plane.

In DVIEW, however, you can set both angles at the same time and preview the results. As you move your pointing device sideways, the viewing direction rotates in the XY plane; as you move your pointing device toward and away from yourself, the viewing direction rotates from the XY plane. As you move your pointing device diagonally, both angles of the viewing direction change and the model appears to spin in space. Actually, the camera is rotating around the model—the model is stationary (see Figure 4.91).

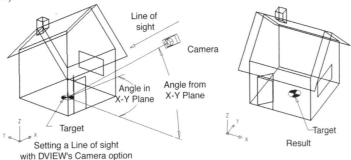

Figure 4.91

TARGET

This option sets a view direction by rotating the target about the camera, rather than the camera about the target as the CAmera option does. It works in virtually the same way, using the same prompts and status line angle read-outs, but the angles are from the camera to the target instead of from the target to the camera. If, for example, you had set the viewing direction with CAmera to 10° from the XY plane, and -40° from the X axis, the corresponding view direction angles with TArget would be -10° and 140°. Nevertheless, the view is still from the camera looking toward the target.

DISTANCE

This option does two things. First, it places the camera at a specific distance on the line of sight from the target point. Second, it turns on the Perspective View mode. The difference between normal AutoCAD views (which have parallel lines of sight), and perspective views (which have diverging lines of sight) are shown in Figure 4.92. Notice also that the UCS icon in the perspective view appears as a series of dashed lines.

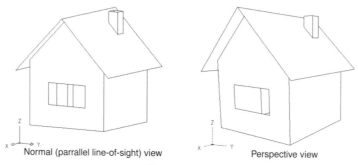

Normal (parrallel line-of-sight) view   Perspective view

Figure 4.92

When you select the Distance option from the DVIEWmenu, the following command line prompt will be shown (see Figure 4.93):

Specify new camera-target distance <current>: (*Specify a distance.*)

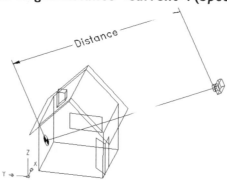

Figure 4.93

The default offered is the current distance. A slider bar, located at the top of the graphics screen, can be used to set the distance. The slider bar is labeled 0x, 1x, 4x, 9x, and 16x. These numbers represent multiplication factors of the current distance. The size of the preview image changes dynamically as you move the indicator on the slider bar with your pointing device. Moving the indicator to the right increases the distance, causing the image to become smaller; moving it to the left decreases the distance, causing the image to become larger. Also, the screen status line displays the distance as the indicator on the slider bar is moved.

Distance can also be set by typing in a number. As soon as a distance has been specified, the UCS icon will change from X and Y arrows to a perspective view box. Although the Distance option is the only way to turn on the Perspective mode, you can fine-tune the appearance of the perspective view with the Zoom and POints options. The Off option in the main DVIEWmenu turns off the Perspective mode.

POINTS

The POints option is for setting the exact location of both the target point and the camera point. When this option is selected, AutoCAD will anchor a rubberband line at the current location of the target and display the prompt:

Specify target point <current>: (*Specify a point.***)**

The X,Y,Z coordinates of the current target location (which is approximately in the middle of the objects you selected as DVIEW preview objects) are offered as the default location. You can use any method to specify a new target point, including object snaps and point filters. If the Perspective mode is on, AutoCAD will temporarily turn it off to allow pointing.

After the target point has been specified, AutoCAD prompts for a new camera point:

Specify camera point <current>: (*Specify a point.***)**

The current camera point is offered as a default value. A rubberband line, anchored at the target point you just set, is shown as an aid in locating the new camera point. On the left-side of Figure 4.94, the target point is placed in the middle of the open door's edge, and the camera point is placed on top, and in the middle of the window. The resulting view is shown on the right.

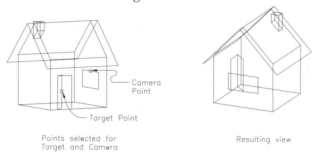

Points selected for
Target and Camera

Resulting view

Figure 4.94

You will notice that although the two selected points are in line with the view direction, the viewing location is farther away from the house than the camera point. This is because normal views, those with parallel lines of sight, are based on a view direction rather than a specific camera point. If you used the Distance option of DVIEW after setting these same target and camera points, the default distance would be the target-to-camera distance, and the perspective view would be based at the camera point. With this particular model, such a view would contain just a few lines.

PAN

Moves the preview image across the screen without changing the zoom level. The command line prompts are:

Specify displacement base point: (*Specify a point.*)

Specify second point: (*Specify a point.*)

The crosshair cursor will lock on to the base point and the preview objects will slide around on the screen as you select the second point.

ZOOM

DVIEW has two slightly different versions of Zoom, depending on whether the Perspective Viewing mode is on or off. Within parallel projection views, AutoCAD bases the zoom in the center of the viewport and displays the command line prompt:

Specify zoom scale factor <1>: (*Specify a value or press ENTER.*)

The current magnification level will be multiplied by the value entered. Because the default value is always 1, pressing ENTER will leave the view unchanged. You can type in a value or use the slider bar located at the top of the viewport (see Figure 4.95). Moving your pointing device to the right will increase magnification, while moving it to your left will decrease magnification.

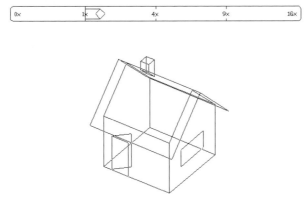

Figure 4.95

When the Perspective Viewing mode is on, AutoCAD uses the metaphor of camera lens length, rather than a magnification factor of the current zoom level. The command line prompt will be:

Specify lens length <current mm>: (*Specify a value or press ENTER.*)

Similar to a telephoto camera lens, a large lens length will magnify the preview objects and reduce the field of view, whereas a small lens length simulates a wide angle lens—reducing the size of the objects and increasing the field of view (see Figure 4.96). A lens length of 50 mm will give a view similar to what an unaided human eye would see.

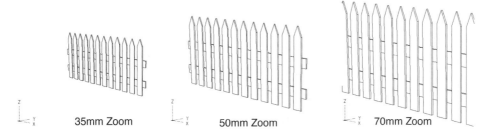

35mm Zoom   50mm Zoom   70mm Zoom

Figure 4.96

Pressing ENTER will leave the lens length unchanged. You can type in a new lens length or use the slider bar near the top of the viewport, which has the same graduations as the Nonperspective mode zoom slider bar. Moving the bar's indicator to the right increases the lens length, whereas moving it to the left decreases the lens length.

TWIST

This option revolves the view around an axis that is parallel to the line of sight and located in the center of the viewport. The follow-up prompt is:

Specify view twist angle<current>: (*Specify an angle or press ENTER.*)

The current twist angle is offered as a default option. AutoCAD anchors a rubberband line in the center of the viewport, and you can show AutoCAD a new angle with your pointing device or type in a new angle. Angle direction is counterclockwise, with 0° at the 3 o'clock position (see Figure 4.97). When the twist angle is 0°, all lines pointing in the Z axis direction will be vertical in the viewport. A 180° twist will turn the view upside down.

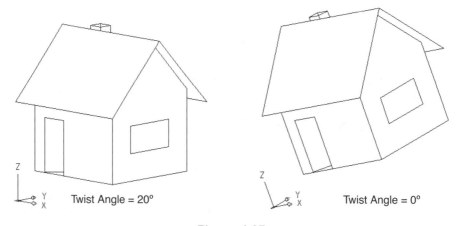

Figure 4.97

CLIP

Clipping planes are invisible walls perpendicular to the line of sight that are used to hide sections of the 3D model. There are two different clipping planes:

- A back clipping plane, which obscures everything from the clipping plane away from the camera point.

- A front clipping plane, which obscures everything from the clipping plane toward the camera point.

The target and the location of a back clipping plane is shown on the left in Figure 4.98, while the resulting view is shown on the right.

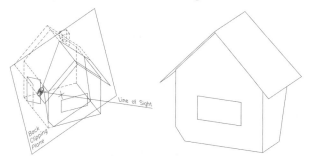

Figure 4.98

Figure 4.99 is similar to Figure 4.109, but a front, rather than a back, clipping plane is used to illustrate the differences between a front and back clipping plane.

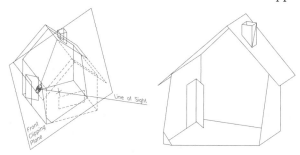

Figure 4.99

You can use either clipping plane or both at the same time. Once a clipping plane has been established, it remains in effect even if you change the line of sight and even after you exit DVIEW. Clipping planes can be used in either parallel projection views or in perspective views. In fact, a front clipping plane located at the camera point is automatically created when the Perspective mode is activated, and it remains in effect as long as the Perspective mode is on. This is not something that you normally notice, but it does affect the front clipping plane prompts.

When you select the CLip option from the main DVIEW prompt, AutoCAD responds with the follow-up prompt of:

Enter clipping plane option [Back/Front/off] <Off>: (*Select an option or press ENTER.*)

- Off

- Turns off all clipping planes.

- Back

- Manages back clipping planes. It uses the follow-up prompt of:

Specify distance from target or [ON/OFF] <current>: *(Select an option, specify a distance, or press ENTER.)*

- Distance from target sets the location of the back clipping plane relative to the target, with the current location offered as a default location. The distance can either be positive, which is toward the camera, or negative, which locates the clipping plane behind the target. AutoCAD displays a slider bar at the top of the viewport that you can use in setting the clipping plane location. As you move the slider bar, AutoCAD dynamically shows the results of the back clipping plane and displays the distance on the status line.

- ON activates the back clipping plane.

- OFF turns the back clipping plane off.

- Front

- Manages front clipping planes. It uses the follow-up prompt of:

Specify distance from target or [set to Eye/(camera)/ON/OFF] <current>: *(Select an option, specify a distance, or press ENTER.)*

- Distance from target sets the location of the front clipping plane relative to the target, with the current location offered as a default location. The distance can either be positive, which is toward the camera, or negative, which locates the clipping plane behind the target. AutoCAD displays a slider bar at the top of the viewport that you can use in setting the clipping plane location. As you move the slider bar, AutoCAD dynamically shows the results of the front clipping plane and displays the distance on the status line.

- ON activates the front clipping plane.

- OFF turns the front clipping plane off.

- Eye places the front clipping plane at the camera. This option is provided because of the relationship that the front clipping plane has with perspective views. When the Perspective View mode is active, a front clipping plane is placed at the camera. It cannot be turned off, but it can be moved away from the camera. In fact, when the Perspective View mode is on, the only options of the front clipping plane prompt are:

Specify distance from target or [set to Eye (Camera)] <current>: *(Enter an E, specify a distance, or press ENTER.)*

Hide

Hides objects and portions of objects that are behind opaque objects. This option allows you to perform the same hidden-line removal operation as the HIDE command without leaving the DVIEW command.

Off

Turns off the Perspective View mode.

Undo

Cancels the previous DVIEW operation. You can use this option repeatedly to step back through DVIEW operations.

SUMMARY OF DVIEW OPTIONS

- Options may be selected by typing in the entire option name or just the uppercase letters in the option name.

Option	Result
Pick	Sets a view direction by using the selected point as a pivot.
CAmera	Sets a view direction by rotating the camera from the XY plane, and in the XY plane from the X axis. This is similar to VPOINT.
TArget	Sets a view direction by rotating the target from the XY plane and in the XY plane from the X axis. The angles are reversed from those of the CAmera option.
Distance	Turns on the Perspective View mode and sets the camera distance from the tar get.
POints	Sets specific points of the target and camera locations.
PAn	Moves the preview object images within the viewport.
Zoom	Magnifies or reduces the size of the preview object images.
TWist	Rotates the view about an axis parallel to the view direction and located at the center of the viewport.
CLip	Manages front and back clipping planes.
Hide	Performs a hidden-line removal operation within DVIEW.
Off	Turns off the Perspective View mode.
Undo	Reverses the previous DVIEW operation.

Tips: How much of your model you select for the preview image depends on the model's complexity and size, as well as on the speed of your computer system. During dynamic previews, the model is constantly erased and redrawn as the view changes. A large preview selection set, coupled with a slow computer system, can cause the time required for redrawing the model to be significant. Ideally, you would like the model to smoothly rotate and move about in space as you change the view. In reality, however, the model's image flickers on and off as you change the view, with the percentage of off time increasing with the number of preview objects. Furthermore, because the model is always shown as a wireframe, you may have difficulty visualizing a complicated model. Therefore, you will want the smallest object selection set that will give you a representative picture of the entire model.

When using the CAmera or TArget options for setting view directions, you can prevent the images from wildly spinning about by locking in one of the view direction angles. Then the image will rotate in just one plane, and it is easier to keep oriented.

The Select Point option can be a good way to establish a viewing direction, although it takes a while to get used to it. The point you select as a pivot point should be approximately in the center of the preview objects.

DVIEW's Twist option is useful for straightening views in which the model has become tilted after using 3DORBIT or the Select Point, CAmera, or Target options to establish a view direction.

3D ORBIT RELATED COMMANDS

A set of related commands is available in AutoCAD 2002 for establishing views in 3D space based on the 3DORBIT command. We will discuss these commands as if they were options of a single command, rather than as individual ones, because they are interrelated. Once any of them has been initiated, you can switch to another through the right-click shortcut menus shown in Figure 4.100.

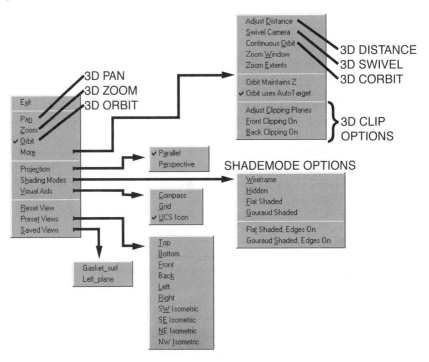

Figure 4.100

With the exception of CAMERA, all of these commands operate interactively in real-time—that is, you set the parameters of the command by moving your pointing device, and the model continually changes in accordance with your movements. If the current viewing mode is 2D wireframe, AutoCAD automatically switches to SHADEMODE's 3D wireframe mode. If SHADEMODE's hidden line mode, or one of its shaded modes in effect, that mode is retained during 3DORBIT and its related commands. Images, though, sometimes degrade to a less complex image—such as from a shaded image to a wireframe, or even from a wireframe to a bounding box— depending on your computer's video system and the complexity of the model.

3DORBIT

This command sets viewpoints by dynamically moving the camera about in 3D space as you view the model. See Chapter 2 for a complete description of 3DORBIT.

3DCORBIT

You can set your model into a continuous rotation with this command. To start the rotation, move your pointing device as you depress its pick button. The image will rotate about an axis that is perpendicular to the pointing device movement, and at a rate proportional to the speed of its movement. The continuous rotation will begin when you release the pick button, and end when you press it again or press ENTER or ESC. As your model is rotating, you can display the shortcut menus to change the shading mode and the projection type, and to turn clipping planes on and off.

3DSWIVEL

When you set view directions with this command, the camera acts as if it was being swiveled about from a fixed location. Similar to views in a camera viewfinder, objects appear to move to the left when you rotate the camera to the right, and they move up when you rotate the camera down.

3DPAN

Similar to the PAN command, 3DPAN shifts the image in the current viewport in real time as you move your pointing device without changing the zoom level. Unlike the PAN command, though, 3DPAN works within perspective views.

3DZOOM

This command enlarges the image in the current viewport in real time as you move your pointing device up and decreases its size as you move your pointing device down. Unlike the ZOOM command, 3DZOOM works within perspective views.

CAMERA

Similar to the POints option of the DVIEW command, CAMERA sets the points for the camera and the target. It uses command line prompts. First, you will be asked to specify the camera location, and then to specify the target location.

3DDISTANCE

You can move the camera toward or away from the target with this command. This is done in real time, with upward movements of your pointing device decreasing the camera-to-target distance and downward movements increasing it. You cannot set the camera-to-target distance by specifying points. Your most frequent use of this command will be to fine-tune perspective views.

3DCLIP

This command performs the same function as the CLip option of the DVIEW command—it manages both front and back clipping planes. These are invisible planes that are perpendicular to the viewport's line of sight. Front clipping planes clip away, or hide, everything from the plane to the camera, whereas back clipping planes hide everything from the plane away from the camera.

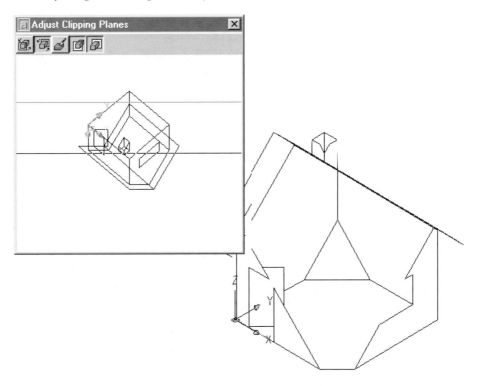

Figure 4.101

3DCLIP starts by opening a small window titled Adjust Clipping Planes over the AutoCAD window. As shown in Figure 4.101, the view direction in this small window is perpendicular to the current viewport's view direction, and the two clipping planes appear as horizontal lines.

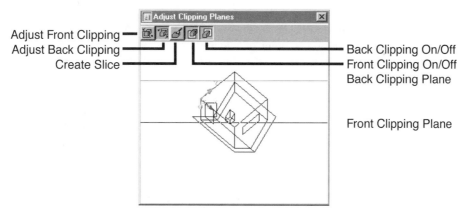

Adjust Front Clipping
Adjust Back Clipping
Create Slice

Back Clipping On/Off
Front Clipping On/Off
Back Clipping Plane

Front Clipping Plane

Figure 4.102

Options are selected by picking a button from the toolbar at the top of the Adjust Clipping Planes window (see Figure 4.102).

- Adjust Front Clipping: Select this option to drag the front clipping plane toward or away from the target. As you move the clipping plane in the Adjust Clipping Planes window with your pointing device, the effects of the clipping plane location will be displayed in the current viewport.

- Adjust Back Clipping: Use this option to move the back clipping plane toward or away from the target.

- Create Slice: This option locks the front and back clipping planes in their current position relative to each other, and then allows you to move both planes simultaneously.

- Front Clipping Plane On/Off: This option toggles the front clipping plane on and off. When clipping plane is off, it has no effect. When the button is on, it appears to be recessed.

- Back Clipping Plane On/Off: Use this option to turn the back clipping plane on and off.

OTHER 3DORBIT MENU OPTIONS

You can access the shortcut menus shown in Figure 4.111 from any of the 3DOR-BIT-related commands. Descriptions of the menu options that we have not covered are given below.

- Projection: You can set the viewing mode to be either parallel projection or perspective with this option. The viewing mode selected remains in effect until it is changed by this option or by an option of the DVIEW command.

- Shading Modes: These menu options allow you to set any of the SHADE-MODE viewing modes, except 2D wireframe.

- Visual Aids: These three menu options toggle the grid, the compass, and the UCS icon on and off. During the 3DORBIT commands, AutoCAD's grid is displayed as rows and columns of lines, rather than dots. See the discussion of the 3DORBIT command in Chapter 2 for a description of the compass.

- Reset View: This option restores the view that was current when 3DORBIT or any of its related commands was initiated.

- Preset Views: You can set one of the six standard orthographic views or one of the four isometric views that look down on the XY plane with this option. The UCS orientation is not modified when you select an orthographic view.

- Saved Views: This menu option will exist only if you have previously saved one or more views with the VIEW command. When you select this option, the names of saved views will be displayed, and you can restore any view by selecting it.

Tip: If you select objects prior to initiating 3DORBIT or any of its related commands, only those objects will be displayed in the dynamic previews of the 3DORBIT options. If no objects are selected, all objects will be in the previews.

CREATING PERSPECTIVE VIEWS AND CLIPPING PLANES

Setting up a perspective view is a straightforward process, but it involves some planning, several steps, and probably some fine-tuning. View direction is obviously an important parameter for perspective views. Less obvious, though, is that the distance between the camera and the target is important. When the camera is too close to the target, the view will be distorted; if it is not close enough, the perspective mode will not be apparent (see Figure 4.103). When clipping planes are not involved, the actual location of the target is not important—only its distance from the camera. When clipping planes are used, though, the location of the target is important because the locations of clipping planes are based on the target.

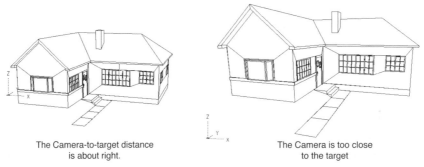

The Camera-to-target distance is about right.

The Camera is too close to the target

Figure 4.103

When clipping planes are not used, you can take the following steps to set up a perspective view by using options of DVIEW.

1. Establish a viewing direction. You can do this with DVIEW, or with VPOINT or 3DORBIT before you start DVIEW.

2. Use DVIEW's Distance option to set the camera's distance from the tar get and turn on the Perspective mode. A good distance to start with is roughly twice the width or height of your model, whichever is greater.

3. Use DVIEW's Zoom option to adjust the appearance of the view. If the image of the model is too large or severely distorted, return to the Distance option and increase the target-to-camera distance. If the image is too small or does not have a perspective view appearance, decrease the target-to-camera distance.

4. Use the Zoom, PAn, and Hide options as needed to adjust and inspect the view, and repeat the Distance option until you achieve the view you want. If you need to change the viewpoint, use DVIEW's Select Point, CAmera, or TArget options, or 3DORBIT —not VPOINT (because it will turn off the perspective viewing mode).

If you would rather not use DVIEW, you can take the following steps using the commands associated with 3DORBIT to set a perspective view without clipping planes.

1. Set a view direction with VPOINT or 3DORBIT.

2. Select Perspective from 3DORBIT's Projection shortcut menu.

3. Use 3DZOOM and 3DPAN to set the size and position of the model's image. Unlike the Zoom option of DVIEW, 3DZOOM will not distort the appearance of your model. If you have problems even locating your model, use the Zoom Extents option from the More shortcut menu.

4. If the view is severely distorted or does not look like a perspective view, use 3DDISTANCE to change the camera-to-target distance. In this command, moving the cursor up increases the perspective effect, whereas moving the cursor down decreases it. Use the 3DZOOM and 3DPAN commands again to obtain the image size and position you want.

Rooms and building interiors will often have obstructions, such as walls and ceilings, that must be removed with clipping planes at the time the perspective view is set up. The steps for setting perspective views that require clipping planes are:

1. Use CAMERA or DVIEW's POints option to set the target and camera locations. This also sets the viewing direction and the camera-to-target distance. A good distance to start with is one that is approximately twice the width or height of the viewing area. The target point is espe cially important because clipping plane locations are based on it.

2. Turn on the Perspective view mode by selecting the Distance option of DVIEW, or by selecting Perspective from the 3DORBIT Projection

shortcut menu.

3. Use 3DCLIP or the CLip option of DVIEW to set up front and back (if needed) clipping planes to remove obstructions from the view. Zoom and pan as necessary to adjust the appearance of the view.

4. If necessary, change the camera-to-target distance with the Distance option of DVIEW or 3DDISTANCE, and use DVIEW's or 3DORBIT's options to fine-tune the view.

The Perspective mode is intended for viewing, not working. You can no longer use your pointing device to select points. If you want to draw a line within the Perspective View mode, for instance, you will have to type in the line's coordinates. Any attempt to pick a point brings up the message: "Pointing in perspective view not allowed here." Furthermore, the ZOOM and PAN commands do not work in perspective views. If you need to do either, you must invoke DVIEW and use its Zoom and Pan options, or use 3DZOOM and 3DPAN. Therefore, if you intend to work on your model some more, you should set up multiple viewports—one or more for perspective viewing, and one or more for working.

You can turn off the Perspective View mode through DVIEW's Off option, by selecting Parallel from 3DORBIT's Projection shortcut menu, or by invoking AutoCAD's VPOINT or PLAN commands.

Use care when setting a back clipping plane. If you set it too far in front of the target, you can inadvertently hide everything, ending up with a blank viewport. You will also end up with a blank viewport if the front and back clipping planes are both on and are both in the same location.

TRY IT! - ROOM VIEWS

You will set up some views of the 3D room you have been working on throughout this chapter. If you have not done so already, insert the window casing you made with TABSURF and the table you made with wide extruded polylines and RULESURF into the room. Then insert the lamp and teapot you made using REVSURF and EDGESURF on top of the table. When you view the room 330° in the XY plane from the X axis and 10° from the XY plane, without hidden line removal, your room should appear similar to the one shown in Figure 4.104 (although we have added a chair and a chest of drawers to the room, along with some molding, like that around the window, to the room).

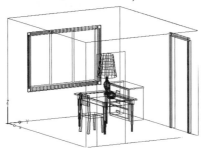

Figure 4.104

In hidden line views from this viewing angle, most of the room interior is hidden, as shown in Figure 4.105. Therefore, we will use a front clipping plane to cut through one of the room's walls.

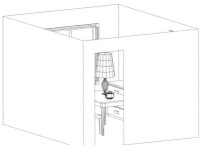

Figure 4.105

To refresh your memory and help locate points: The corner of the room between the window and the blank wall is at the WCS origin and the room is 12 feet long in the X direction, 10 feet in the Y direction, and 8 feet tall. Your first step in this exercise will be to set the camera at the point coordinates of 8',-20,5' (or 96,-240,60 if you prefer to use inches rather than architectural units), and set the target at the point coordinates of 6',10',4' (or 72,120,48). You can do this with command line input through either the CAMERA command, or the Points option of DVIEW.

Next, use either 3DCLIP or the CLip option of DVIEW to set a front clipping plane. With either method, you will see parts of the room disappear as you move the clipping toward the target. When the clipping plane is about 9' (or 108) in front of the target and hidden line removal is on, your room should look similar to the one in Figure 4.106.

Last, you will set up a perspective view of the room, as it would be seen when entering through the doorway on the east side of the room. Turn off the front clipping plane by using either 3DORBIT's More shortcut menu or the CLip option of DVIEW. Then use either the POints option of DVIEW or the CAMERA command to set the target at -5',10',3' (or -- 60,120,36), and the camera at 12',2'6,6' (or 144,30,72). We intentionally set the target outside the room to reduce distortion. Next, turn on the perspective viewing mode with either the Distance option of DVIEW, or by selecting Perspective from 3DORBIT's Projection shortcut menu. If you use the DVIEW Distance option, accept the default distance. Zoom and pan as necessary to get the view you want. Your results should resemble Figure 4.107 when hidden line removal is on.

Figure 4.106

Figure 4.107

The completed 3D room is in file 3d_ch4_16.dwg on the CD-ROM that accompanies this book. Both the clipping plane view and the perspective view have been saved in this file. You can restore them with the VIEW command.

House Exterior Views

In this exercise, you will use the DVIEW command to set up an exterior perspective view for a surface model house. Find and open the file 3d_ch4_18.dwg. This file contains the 3D surface model house shown in Figure 4.108. From the plan view, the size of the house is such that it fits in a 34–foot by 34–foot rectangle.

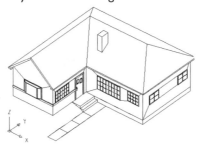

Figure 4.108

Start the DVIEW command and select the entire house for the preview objects. Then use DVIEW's CAmera option to set a viewpoint that is rotated 0° from the XY plane, and minus 50° from the X axis in the XY plane. From this viewing direction the house will look similar to the one in Figure 4.109.

Figure 4.109

As we do not need clipping planes, the location of the target is not critical. Therefore, you can bypass the Points option and use DVIEW's Distance option to set the target-to-camera distance to 70 feet (840 inches). The resulting perspective view, with hidden line removal in effect, is shown in Figure 4.110.

Figure 4.110

If you would like a more dramatic view, you can decrease the target-to-camera distance. Figure 4.111 shows a perspective view with the same viewpoint, but with a camera-to-target distance of 50 feet (600 inches). You can also use any of DVIEW's three options for setting view directions to see how the house looks from other viewpoints.

Figure 4.111

You should also experiment with the 3DORBIT command in setting perspectives. Return to the original isometric view of the model, then start 3DORBIT. Rotate the view direction so that it is similar to the one in Figure 4.109 and then select Perspective from the Projection shortcut menu. Use 3DDISTANCE to increase or decrease the perspective effect, and 3DZOOM and 3DPAN to adjust the size and position of the image.

Both of these perspective views are saved in file 3d_ch4_18.dwg and can be restored with AutoCAD's VIEW command or from the Saved Views shortcut menu of 3DORBIT. We will use this 3D surface model house again in Chapter 9 of this book, as we experiment with renderings. The roof, chimney, windows, and most other surfaces on this house are 3D faces. Most of the outside walls, however, were made with PFACE because it is easier to work with some types of rendering materials when they are attached to a single object than when they are attached to multiple objects. Each outside wall is a single polyface mesh object, and the lines extending from some window corners are face edges within a polyface mesh. Although these edges could have been made invisible, they will not show up in renderings, and they are sometimes useful during object selections.

TRY IT! - CONSTRUCTING AN AIR INLET SURFACE MODEL

In this exercise, you will construct the surface model of an air inlet with a cross-section shape that changes from an ellipse to a circle. It will demonstrate how symmetry can be used to easily make seemingly difficult surfaces. The initial set up of your drawing is not critical, other than to use inches as the unit of measure. You may also want to set the snap distance to 0.5, and turn on the snap mode, because the dimensions of the model are in increments of 0.5 inches. Start the model by drawing the wireframe objects (including the centerline) shown in Figure 4.112. The planes of the 180° arc and the two 180° ellipses are perpendicular to the WCS XY plane. The height of the small ellipse is 2.0, and the height of the large ellipse is 2.5. The two splines, which are on the WCS XY plane, can be easily drawn by specifying just their start points and endpoints and setting their end tangents in the negative Y direction for the large end, and in the positive Y direction for the small end.

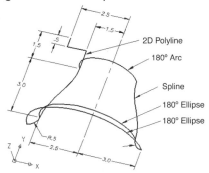

Figure 4.112

Now create three surfaces through AutoCAD's REVSURF and EDGESURF commands. Use Figure 4.113 in conjunction with the table at the bottom of the page for selecting the surface boundary objects, and for assigning an appropriate value to AutoCAD's Surftab1 and Surftab2 system variables.

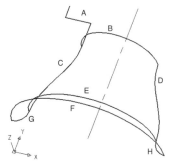

Figure 4.113

Your three surfaces should look similar to those in Figure 4.114.

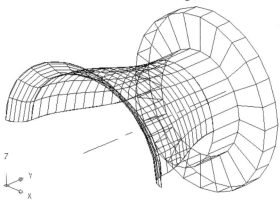

Figure 4.114

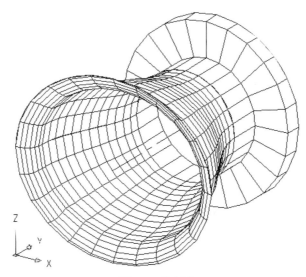

Finish the model by using AutoCAD MIRROR3D command to make mirror copies of the two EDGESURF surfaces across the WCS XY plane. Your completed surface model should look similar to the one in Figure 4.115.

This model is in file 3d_ch4_19.dwg on the CD-ROM that accompanies this book.

Figure 4.115

Command	Boundary Object(s)	Surftab1 Value	Surftab2 Value	Remarks
REVSURF	A	24	6	Revolve 360°, with the centerline as the axis.
EDGESURF	B,C,D,E	12	12	Order of boundary object selection is not important.
EDGESURF	E,F,G,H	12	6	Pick E or F as the first boundary object.

COMMAND REVIEW

3DCLIP

Similar to the **CLip** option of **DVIEW, 3DCLIP** establishes and manages front and back clipping planes.

3DDISTANCE

The camera's distance from the target is set by this command. Unlike the **Distance** option of **DVIEW**, it does not turn on the **Perspective** mode.

3DFACE

This command makes three- and four-edged planar faces with visible or invisible edges.

3DORBIT

This command moves the camera to dynamically set viewpoints in 3D space.

3DPAN

This command dynamically shifts images without changing their zoom level. Unlike the **PAN** command, shaded, hidden line, and perspective views can be panned.

3DSWIVEL

The camera swivels from a fixed position to dynamically set viewpoints in 3D space.

3DZOOM

Unlike the **ZOOM** command, **3DZOOM** can perform real-time zooms in perspective views and with hidden line and shaded views.

CAMERA

Similar to the **POints** option of **DVIEW, CAMERA** sets specific camera and target points.

EDGE

This command is an automatically loaded AutoLISP program for changing the visibility of 3D face edges.

HIDE

This command performs a hidden line removal operation. Objects behind surface objects, in relation to the current view direction, are hidden from view. There are no parameters or options for **HIDE**, and a drawing regeneration (**REGEN** command) is required to restore the Unhidden Viewing mode.

PAN

This command moves the image of objects within the current viewport without changing the view direction or magnification level.

PFACE

This command also makes planar surfaces that have three and four edges. However, the faces are tied together as a single object, and their interior edges are invisible.

PROPERTY

This command activates a dialog box used to change the properties of objects. Items such as 3D faces can have their point coordinates and edge visibility controlled by this dialog box. Other items such as polygon meshes can have their mesh pattern opened or closed. Smoothing and desmoothing of polygon meshes can also be controlled through this dialog box.

RENDER

The RENDER command creates shaded images similar to the Gouraud option of SHADEMODE. RENDER, though, can incorporate multiple lights as well as create shadows and other special effects. However, you cannot work within RENDER's images. They are strictly for presentations.

SHADE

Although the SHADEcommand has been replaced by SHADEMODEand is no longer documented, it can still be invoked from the command line. It has no prompts or options. The resulting viewing modes are equivalent to SHADEMODE's options of Hidden, Gouraud shading, Gouraud shading plus edges, or flat shading plus edges, depending on the setting of the Shadedge system variable.

SHADEMODE

SHADEMODE is a multipurpose view mode command. It gives you a choice of four different techniques for filling in surfaces with their object color, a hidden line mode, and a special wireframe view mode.

SPLINEDIT

This command edits 3D wireframe spline objects.

VPOINT

This command sets a view direction angle.

ZOOM

This command changes the magnification level of objects shown in the current viewport.

SYSTEM VARIABLE REVIEW

BACKZ

This variable stores the distance of the back clipping plane relative to the target. It is a read-only system variable.

FRONTZ

This read-only variable stores the distance of the front clipping plane relative to the target.

HIDEPRECISION

This system variable, introduced in Release 14, controls the type of arithmetic AutoCAD uses for hidden-line calculations. When it is set to its default value of 0, AutoCAD uses single-precision arithmetic in hidden-line calculations. When it is set to

1, double-precision arithmetic is used.

LENSLENGTH

This read-only variable stores the current lens length used as a zoom level for perspective views.

SHADEDGE

This system variable, which can be set to integer values of 0 through 3, control the shading mode of the **SHADE**command. The results are equivalent to options of the **SHADEMODE**command, as follows:

0 =   **Gouraud shading.**

1 =   **Gouraud shading plus edges**

2 =   **Hidden**

3 =   **Flat shading plus edges**

SPLFRAME

This variable controls the visibility of invisible 3D face edges. When it is set to 1, invisible edges of 3D faces become visible. When Splframe is set to 0, its default setting, invisible edges of 3D faces are invisible. A drawing regeneration (**REGEN** command) is required for either change to take affect.

SURFTAB1

This variable sets the number of faces along the boundary edge. Surftab1 requires an integer that is not smaller than 2 nor larger than 32,766. This variable also controls the number of faces in the M direction for such commands as **REVSURF**.

SURFTAB2

This variable sets the number of faces in the N direction for **REVSURF**.

SURFTYPE

This variable controls the type of equation used to smooth polygon mesh surfaces.

SURFU

This variable sets the number of faces in the M direction of smoothed polygon meshes. The maximum value allowed is 200 and the minimum is 2.

SURFV

This variable sets the number of faces in the N direction of smoothed polygon meshes. The maximum value allowed is 200 and the minimum is 2.

TARGET

This read-only variable stores the coordinates of the target point.

VIEWDIR

This variable stores the view direction in the form of a point offset from the point coordinates in Target. It is a read-only system variable.

VIEWMODE

This read-only variable stores current viewing modes in a bit-code. The integer stored is the sum of the following:

1   perspective view active

2   front clipping plane on

4   back clipping plane on

8   UCS follow mode on

16   Front clipping plane not at the eye (camera) point

VIEWTWIST

This read-only variable stores the twist angle of the current viewport.

WORLDVIEW

This variable controls which coordinate system—the world coordinate system (WCS), or the user coordinate system (UCS)—is used during DVIEW and VPOINT. When Worldview is set to 1, the default setting, AutoCAD changes to the WCS during the duration of DVIEW and VPOINT. When Worldview is set to 0, AutoCAD remains in the current UCS during DVIEW and VPOINT.

CHAPTER REVIEW

Directions: Answer the following questions with a short answer.

1.   List the common characteristics of AutoCAD surface objects.

2.   List some characteristics of extruded surfaces.

3.   How do you assign extrusion thickness to an object?

4 .   What are tessellations, and how do you control their quantity?

5.   What are the two methods for making 3D face edges invisible?

6.   How can you temporarily make invisible edges on 3D faces visible?

7.   What is one advantage that polyface mesh surfaces, made with the PFACE command, have over 3D faces? What is one disadvantage of the command compared with 3DFACE?

8.   Is it possible for a polyface mesh surface to extend beyond the apparent boundary defined by its vertices?

9.   What happens when you use AutoCAD's EXPLODE command on a polyface mesh surface?

10.   Excluding regions, thickness, and circles, what are the three AutoCAD object types for surfaces?

11.   Name the two system variables that control the number of faces in a polygon surface mesh.

12.   How does the TABSURF command differ from the RULESURF command?

13.   What are the requirements of the boundary objects for the EDGESURF command?

14.   How do you know which edges on a surface made with EDGESURF are affected by Surftab1 and which edges are affected by Surfab2?

15.   How is the number of mesh faces controlled within the 3DMESH command?

16.   List two reasons for using the shaded viewing modes of the SHADEMODE command.

17.   What are the three options in DVIEW that set viewing directions?

18.   Name the command and the DVIEW option that set the target and camera points?

19.   How does the Perspective Viewing mode differ form AutoCAD's normal Parallel Line of Sight Viewing mode? What are the two ways to turn on the Perspective Viewing mode?

20.   What is a major drawback to the Perspective mode? How do you turn off the Perspective Viewing mode?

21.   What are clipping planes? What are the two types of clipping planes? Name the command and the DVIEW option that manage clipping planes.

22.   Match an AutoCAD command on the left with a surface type or characteristic on the right.

　　　　 _____ a. 3DFACE   1. A planar version of 3D solids.

　　　　 _____ b. 3DMESH   2. Gives extrusion thickness to wireframe objects.

　　　　 _____ c. EDGESURF   3. Planar surface with three or four vertices.

　　　　 _____ d. ELEV   4. Planar surface with unlimited vertices.

　　　　 _____ e. PFACE   5. Surface defined on a point-by-point basis.

　　　　 _____ f. REGION   6. Surface mesh between four wireframe objects.

　　　　 _____ g. REVSURF   7. Surface mesh between two wireframe boundaries.

　　　　 _____ h. RULESURF   8. Surface mesh made by extruding a boundary curve.

　　　　 _____ i. DVIEW   9. Surface mesh made by revolving a boundary curve.

23. Which of the following object types will not accept thickness?

    a. 2D polyline   e. spline

    b. 3D polyline   f. text

    c. arc   g. xline

    d. line

24. Although you can create as many 3D faces as you want during one session of the 3DFACE command, the 3D faces will be separate objects having three or four edges.

    a. true   b. false

25. The EDGE command enables you to make invisible edges on 3D faces visible, but it cannot make invisible edges of polyface meshes visible.

    a. true   b. false

26. AutoCAD is not able to make true rounded surfaces.

    a. true   b. false

27. The AutoCAD commands that create polygon mesh surfaces delete the wire frame objects used as boundaries.

    a. true   b. false

28. Which of the following pairs of boundary objects can be used by RULESURF?

    a. A circle and an ellipse.   d. Two lines.

    b. A line and a circle.   e. Two points.

    c. A line and a point.

29. Surfaces created by the REVSURF command always start at the boundary object.

    a. true   b. false

30. Which of the following statements related to modifying polygon mesh surfaces are true?

    a. Grips can be used to move individual mesh nodes.

    b. Meshes can be smoothed to approximate B-spline surfaces.

    c. The PEDIT command can modify polygon mesh surfaces.

    d. Two individual surfaces can be joined to form one surface.

    e. You can remove unwanted mesh nodes.

31. The HIDE command will hide text if it is behind an opaque object.

    a. true   b. false

32.   How can you correct a perspective view is overly distorted and has pronounced V-shaped corners?

33.   When would you use 3DZOOM, rather than real-time ZOOM, to increase or decrease image size in the current viewport?

34.   List the differences between SHADEMODE's 3D wireframe mode and the default 2D wireframe mode.

35.   How does SHADEMODE's flat shading differ from its Gouraud shading?

Creating Solid Primitives

LEARNING OBJECTIVES

This chapter will introduce you to creating solid model primitives in addition to performing extrusion and revolution operations. When you have completed Chapter 5, you will:

- Be familiar with the properties of AutoCAD 3D solid objects.

- Know how to create 3D solids having basic (primitive) geometric shapes.

- Know how to create 3D solids by extruding and revolving profile objects.

SOLID MODELING

In Chapter 3 we started out making 3D wireframe models, in which an object is represented solely by its edges. The wireframe model of a cube, for instance, consists of eight equal-length lines. A round hole through the cube would have to be simulated by drawing a couple of circles on opposite sides of the cube, since there is absolutely nothing between the lines of the wireframe.

In Chapter 4, we began making surface models, which have infinitely thin surfaces stretched between their edges. A surface model cube will have six square surfaces. Although it may look like a real cube, it is actually an empty shell—there is nothing under the surface. A round hole through the cube would have to be modeled by making round openings on opposite surface faces, with a tube between them to represent the hole's surface. Notice also that both wireframe and surface models are generally composed of many separate AutoCAD objects.

Now we will move on to solid models, which have mass in addition to surfaces and edges. No lines or surface objects are needed in making the cube, and you can make a hole through the cube without having to patch and add surfaces. Furthermore,

AutoCAD solid models are generally a single object.

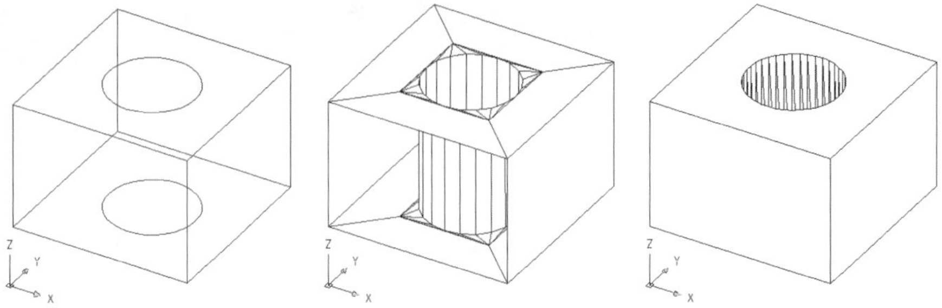

Wireframe
model cube

Surface
model cube
(with one surface removed)

3D Solid
model cube

Figure 5.1

Of course, it is still a computer simulation of a solid cube, and at times the differences between a wireframe cube, a surface model cube, and a solid model cube are not apparent on your computer display. The differences are in the ways the objects are built and in the data connected with the objects. A solid model cube is a single object, even if it has a hole through it. Furthermore, AutoCAD keeps track of mass property information (such as volume, surface area, center of gravity, and moments of inertia) on the cube.

Also, curved and rounded surfaces on solids are more accurate than those on AutoCAD surface models. AutoCAD surface models, you will recall, approximate curved and rounded areas using small, flat faces. This is not done on solid models—their surfaces are always mathematically correct (although AutoCAD does temporarily convert them to faceted surfaces during hiding and shading operations). Because of the extra data kept on solid objects, they require more computations by the computer, more computer memory, and larger file sizes.

When modeling objects that are solid, such as the cube we have been talking about, you will probably find that making them as a solid model is easier and more intuitive than making them as a surface model. Solid modeling uses basic building blocks that are modified, combined, and even used to modify each other.

However, collections of objects, especially if they consist of many flat panels, such as the surface model 3D room we made in Chapter 4, are usually better made as surface models than as solid models. Furthermore, you can make surface models of some shapes that cannot be done as a solid model. The gracefully curving teapot spout we made with EDGESURF, for instance, cannot be made as an AutoCAD solid model.

AutoCAD's solid modeler is based on a program named the ACIS Geometric Modeler, which is often referred to as the ACIS kernel. Autodesk is one of many

CAD/CAM companies that license the ACIS kernel as the engine for their solid modeling programs.

One consequence of the ACIS kernel is that you will occasionally see references to it in manuals and on-screen error messages. Another consequence of the ACIS kernel is that solid models made with Autodesk's Advanced Modeling Extension (AME) are not directly usable by AutoCAD's ACIS-based solid modeler, although they can be converted to AutoCAD 3D solids. AME was an add-on 3D solid modeler used with AutoCAD Releases 11 and 12.

Models created with AME are listed as blocks by AutoCAD. They can be converted to AutoCAD 3D solids through the AMECONVERT command. This command will prompt you to select objects, and you can use any object selection method to pick an AME solid or region. There are no additional prompts or input. Only objects created with AME versions 2 and 2.1, however, can be converted.

REGIONS

Regions are a unique, 2D, closed AutoCAD object type. They have a surface that reflects light in renderings and can hide objects that are behind it, just as 3D faces and polygon meshes do; but unlike those surface entity types, a region can contain interior holes and their edges can take on any shape—they are not limited to segmented, straight-line edges. Some AutoCAD editing operations, including fillet, chamfer, stretch, and break, do not work on regions. They can be hatched, though.

Regions are often classified as a 2D version of 3D solids, because the Boolean operations commonly used to modify solids work equally well on regions. AutoCAD is also able to report mass property information, such as area, perimeter length, centroid, and moments of inertia, on regions, just as it does on solids.

Regions are especially useful for their mass property information, which is often the basis for stress and weight calculations. They are also useful as profile objects for making extruded and revolved 3D solids, as we will see shortly. Furthermore, because they can have curved boundary edges, as well as interior holes, regions are useful as surfaces.

Existing objects are used to create a region. The objects may be either a single closed object, such as a circle, ellipse, 3D face, polygon, closed 2D polygon, or a closed spline, or a collection of open objects that form a closed area. AutoCAD refers to both types of these closed areas as loops. When a collection of open objects are used to form a loop, they must be in the same plane and connected end to end. No gaps or intersections are permitted. Lines, arcs, open 2D polylines, and open splines may be used in any combination to make a loop, but it cannot contain a 3D polyline. Other excluded entity types are polygon meshes and polyface meshes.

The command for creating regions is REGION. Its command line format is:

Command: REGION

Select objects: (*Use any object selection method.*)

M loop(s) extracted.

N Region(s) created.

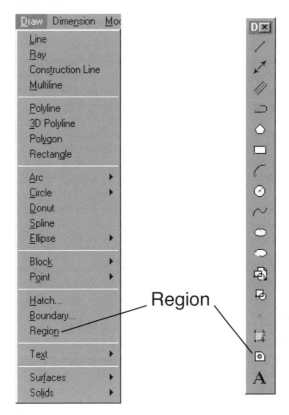

Figure 5.2

Notice that AutoCAD reports on the number, M, of loops found, and on the number, N, of regions created from those loops. If any closed but unacceptable objects were found, such as a 3D polyline, AutoCAD will display a message saying that they were rejected.

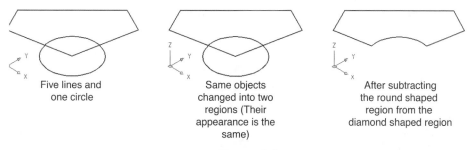

Five lines and one circle

Same objects changed into two regions (Their appearance is the same)

After subtracting the round shaped region from the diamond shaped region

Figure 5.3

The resulting region takes on the current entity properties and the current layer. Loops within loops are transformed into separate regions—they are not holes in the outer region. You must use the SUBTRACT command, which we will discuss later, to turn them into holes.

PRIMITIVE 3D SOLIDS

AutoCAD has six different commands for making 3D solids in basic geometric shapes—blocks, wedges, cylinders, cones, spheres, and tori (doughnut-shaped objects). These solid shapes are often called primitives because they are used as building blocks for more complex solid models. They are seldom useful by themselves, but these primitives can be combined and modified into a wide variety of geometric shapes.

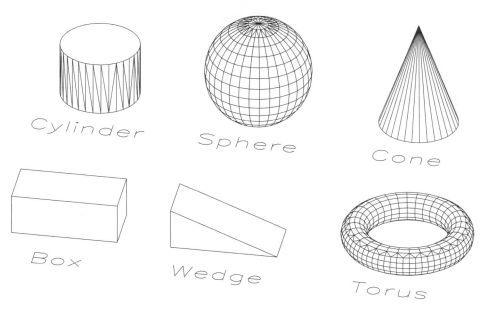

Cylinder   Sphere   Cone

Box   Wedge   Torus

Figure 5.4

Figure 5.5 shows the screen pull-down menus and the Solids toolbar buttons for starting the commands that create 3D solid objects.

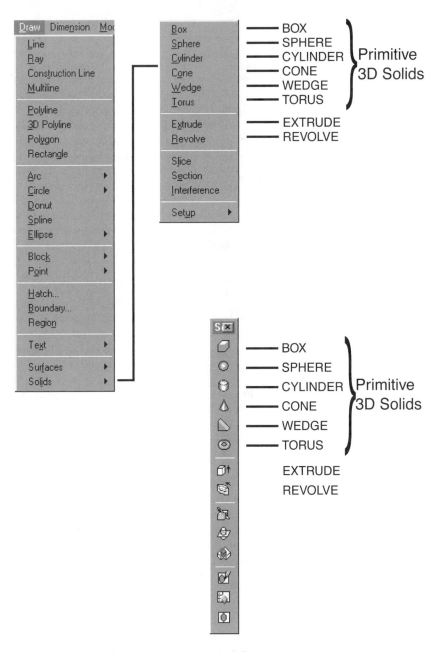

Figure 5.5

THE BOX COMMAND

This command makes brick-shaped solid objects. They will have six rectangular sides, which are either perpendicular or parallel to one another. Boxes are probably the most often used primitive, as many of the objects we model are made up of rectangles and squares. AutoCAD always positions boxes so that their sides are aligned with the X, Y, and Z axes and refers to the X direction as the box's length, the Y direction as width, and the Z direction as height.

After you initiate the command, you are given the choice to base the box on a corner or on its center. The command line prompts and input are:

Command: BOX

Specify corner of box or [Center] <0,0,0>: (*Specify a point, enter C, or press ENTER .*)

CORNER OF BOX

Pressing ENTER will place one corner of the box at the coordinate system origin, whereas specifying a point will base one corner of the box on that point. The follow-up prompt is:

Specify corner or [Cube/Length]: (*Specify a point or choose an option.*)

- Corner
- Specifying a point will establish the other corner of the box. This point cannot have the same X or Y coordinate as the initial point. If this point is not in the same plane as the initial corner, AutoCAD will use this point to set the length, width, and height of the box and will end the command. If the point is in the same plane as the initial corner, AutoCAD will prompt for a height:

Specify height: (*Specify a distance.*)

- A negative value will draw the box in the minus Z direction.
- Cube
- This option creates a box with a width and height equal to its length. AutoCAD will prompt for a length:

Specify length: (*Specify a distance.*)

- A positive value will draw the box in the positive X, Y, and Z directions from the initial corner point. A negative value will draw the box in the negative X, Y, and Z directions from the initial corner point.

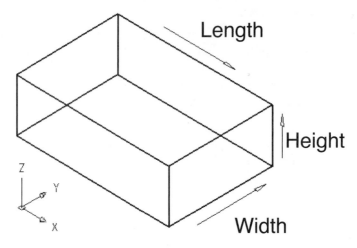

Figure 5.6

- Length
- If you select the Length option after specifying an initial point, AutoCAD will issue separate prompts for the length, width, and height of the box:

Specify length: (*Specify a distance.*)

Specify width: (*Specify a distance.*)

Specify height: (*Specify a distance.*)

- You can specify distances by pointing or by typing in values. A negative value will draw that particular dimension of the box in the negative direction of the axis.

CENTER

Selecting the Center option from the main BOX prompt will base the box on its geometric center point, rather than on a corner. AutoCAD will then prompt for the center point:

Specify center of box <0,0,0>: (*Specify a point or press ENTER.*)

Specify corner or [Cube/Length]: (*Specify a point or choose an option.*)

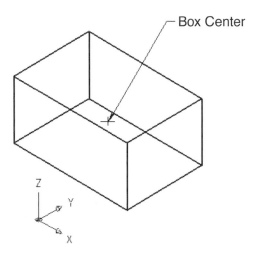
Box Center

Figure 5.7

- Corner of Box
- Specifying a point establishes one corner of the box. This point cannot have the same X or Y coordinate as the center point. If this point is not in the same plane as the center of the box, AutoCAD will use this point to set the length, width, and height of the box, and will end the command. If the point is in the same plane as the center point, it serves as the middle point for a edge of the box, and AutoCAD will prompt for a height:

Specify height: (*Specify a distance.*)

- This distance is divided by two to establish the top and bottom sides of the box in the positive and negative Z directions. Therefore, entering a negative value has no effect.
- Cube
- This option creates a box with its width and height equal to its length. AutoCAD will prompt for a length:

Specify length: (*Specify a distance.*)

- The box will be cube-shaped and centered on the center point. Negative length values have no effect.
- Length
- If you select the Length option after specifying a center point, AutoCAD will issue separate prompts for the length, width, and height of the box:

Specify length: (*Specify a distance.*)

Specify width: (*Specify a distance.*)

Specify height: (*Specify a distance.*)

- AutoCAD uses the absolute value of each distance, which can be specified by pointing or by typing in a value, to establish the dimensions of the box.

Tips: If you want a box to be skewed relative to the X, Y, and Z axes, you must rotate it after it is made, or else orient the UCS prior to making the box.

The Center option of creating a box is useful for centering the box on a particular point in space, such as the center of gravity of an existing object.

TRY IT! - CONSTRUCTING A SOLID BOX

As an example of the box primitive, we will draw the same box, shown in Figure 5.8, three different ways. Start a new drawing from scratch and follow the next series of command prompts for constructing the box primitive.

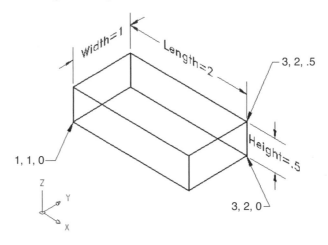

Figure 5.8

Command: BOX

Specify corner of box or [Center] <0,0,0>: 1,1,0

Specify corner or [Cube/Length]: 3,2,0

Specify height: .5

Command: BOX

Specify corner of box or [Center] <0,0,0>: 1,1,0

Specify corner or [Cube/Length]: 3,2,.5

Command: BOX

Specify corner of box or [Center] <0,0,0>: 1,1,0

Specify corner or [Cube/Length]: L

Specify length: 2

Specify width: 1

Specify height: .5

THE WEDGE COMMAND

Wedges are like boxes that have been sliced diagonally edge to edge. They have a total of five sides, three of which are rectangular. The top rectangular side slopes in the X direction, and the two sides opposite this sloping side are perpendicular to each other. The remaining two sides are right triangle–shaped.

The prompts and options used for the WEDGE command are similar to those used for boxes and, like boxes, wedges can be based either on a corner or on the center of the wedge. Their center, however, is a point that is halfway between the length, width, and height of the wedge—it is not at the center of gravity (the centroid). Thus, as shown in Figure 5.10, a wedge 4 units long, 2 units wide, and 2 units high, which is based on a center point at 0,0,0, will have its centroid at X = -0.6667, Y = 0, Z = -0.3333.

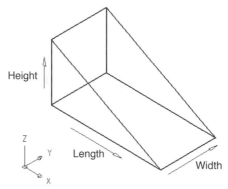

Figure 5.9

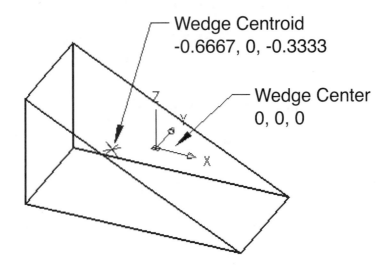

Figure 5.10

The command line prompts and input for creating a wedge primitive are:

Command: WEDGE

Specify first corner of wedge or [Center] <0,0,0>: *(Specify a point, enter C, or press ENTER.)*

CORNER OF WEDGE
Pressing ENTER will place one corner of the high part of the wedge at the coordinate system origin, and specifying a point will base one corner of the wedge on that point. The follow-up prompt is:

Specify corner or [Cube/Length]: *(Specify a point or choose an option.)*

- Corner
- Specifying a point will establish the opposite corner of the wedge. This point sets the direction of the wedge's sharp point. If this point is not in the same plane as the initial corner, AutoCAD will use this point to set the length, width, and height of the wedge and will end the command. If the point is in the same plane as the initial corner, AutoCAD will prompt for a height:

Specify height: *(Specify a distance.)*

- A negative value will draw the wedge in the minus Z direction.

- Cube
- This option creates a wedge with a width and height equal to its length. AutoCAD will prompt for a length:

Specify length: (*Specify a distance.*)

- A positive value will draw the wedge in the positive X, Y, and Z directions from the initial corner point. A negative value will draw the wedge in the negative X, Y, and Z directions from the initial corner point.
- Length
- If you select the Length option after specifying an initial point, AutoCAD will issue separate prompts for the length, width, and height of the wedge:

Specify length: (*Specify a distance.*)

Specify width: (*Specify a distance.*)

Specify height: (*Specify a distance.*)

- A negative value will draw that particular dimension of the wedge in the negative direction of the axis.

CENTER

Selecting the Center option from the main WEDGE prompt will base the wedge on its center point, rather than on a corner. AutoCAD will then prompt for the center point:

Specify center of wedge <0,0,0>: (*Specify a point or press ENTER.*)

Specify opposite corner or [Cube/Length] <corner of wedge>: (*Specify a point or choose an option.*)

- Corner of Wedge
- Specifying a point establishes one sharp corner of the box. It cannot have the same X or Y coordinate as the center point. If this point is not in the same plane as the center of the box, AutoCAD will use this point to set the length, width, and height of the wedge and will end the command. If the point is in the same plane as the center point, it serves as the middle point for a edge of a box that the wedge fits in, and AutoCAD will prompt for a height:

Specify height: (*Specify a distance.*)

- This distance is divided by 2 to establish the high point and the flat side of the wedge. A positive value will make the sloping surface of the wedge face up, whereas a negative value will make a wedge with its sloping surface facing down.

- Cube
- This option creates a wedge with its width and height equal to its length. AutoCAD will prompt for a length:

Specify length: (*Specify a distance.*)

- A positive length will make a wedge with its sloping surface facing up, while a negative length will make a wedge having its sloping surface facing down.
- Length
- If you select the Length option after specifying a center point, AutoCAD will issue separate prompts for the length, width, and height of the wedge:

Specify length: (*Specify a distance.*)

Specify width: (*Specify a distance.*)

Specify height: (*Specify a distance.*)

- Negative values affect the orientation of the sloping surface on the wedge.

TRY IT! - CONSTRUCTING A SOLID WEDGE

We will use the Center option to draw the wedge shown in Figure 5.11 three different ways. Start a new drawing from scratch and follow the next series of command prompts for constructing the wedge primitive.

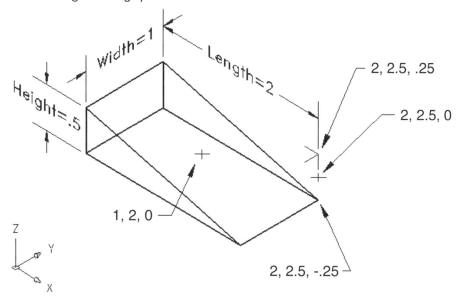

Figure 5.11

Command: WEDGE

Specify first corner of wedge or [Center] <0,0,0>: C

Specify center of wedge <0,0,0>: 1,2,0

Specify opposite corner or [Cube/Length]: 2,2.5,0

Specify height: .5

Command: WEDGE

Specify first corner of wedge or [Center] <0,0,0>: C

Specify center of wedge <0,0,0>: 1,2,0

Specify opposite corner or [Cube/Length]: 2,2.5,.25

Command: WEDGE

Specify first corner of wedge or [Center] <0,0,0>: C

Specify center of wedge <0,0,0>: 1,2,0

Specify opposite corner or [Cube/Length]: L

Specify length: 2

Specify width: 1

Specify height: .5

THE CYLINDER COMMAND

Cylinders will probably be your second most often used primitive. They can have a cross-section that is either circular or elliptical. The command line format for the command is:

Command: CYLINDER

Current wire frame density: ISOLINES=(*current*)

Specify center point for base of cylinder or [Elliptical] <0,0,0>:
  (*Specify a point, enter E, or press ENTER.*)

The message regarding wireframe density refers to the number of parallel lines that AutoCAD will use to delineate the side surface of the cylinder. The default number, which is controlled by the Isolines system variable, is 4.

CENTER POINT

Specifying a point or pressing ENTER to accept the default 0,0,0 point will make a cylinder having a circular cross-section centered on the selected point. AutoCAD will prompt you to specify a radius or a diameter for the cylinder, and then for the location of the other endpoint of the cylinder.

Specify radius for base of cylinder or [Diameter]: (*Specify a distance or enter D.*)

If you specify a distance, AutoCAD will use it for the cylinder's radius. If you type in a D, AutoCAD will prompt you for the diameter of the cylinder.

After entering the diameter or radius, AutoCAD displays the following prompt:

Specify height of cylinder or [Center of other end]: (*Specify a distance or enter C.*)

- Height

- If you specify a distance, by either pointing or entering a value, AutoCAD will draw the cylinder in the Z direction to that height. Entering a negative value will draw the cylinder in the minus Z direction.

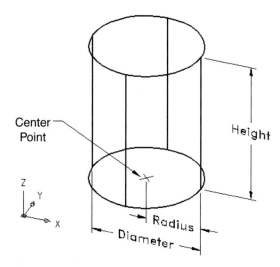

Figure 5.12

- Center of Other End

- This option will base both the length and orientation of the cylinder on a single point.

Specify center of other end of cylinder: (*Specify a point.*)

- The center line of the cylinder will extend from the initial base point to the specified point.

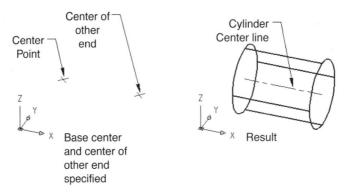

Figure 5.13

ELLIPTICAL

This option makes a cylinder with an elliptical cross-section. You can base the ellipse either on its center point or at one of its axis endpoints.

Specify axis endpoint of ellipse for base of cylinder or [Center]: *(Specify a point or enter C.)*

- Axis Endpoint

- Specifying a point will set one end of one axis of the ellipse-shaped base. AutoCAD will then prompt for the other end of the axis and for a point to set the end of the other axis of the ellipse.

Specify second axis endpoint for base of cylinder: *(Specify a point.)*

Specify length of other axis for base of cylinder: *(Specify a distance.)*

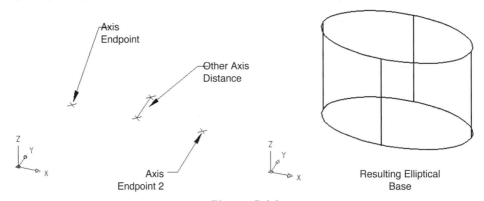

Figure 5.14

- AutoCAD will draw a rubberband line as you specify a point for the second endpoint, and then anchor a rubberband line on the midpoint of this axis as you specify the length of the other axis. The distance you specify, whether by pointing or by entering a value, is half of the length of the second ellipse axis.

- Center

- This option bases the elliptical cross-section at its center point. AutoCAD will prompt you for the center point, the endpoint of one axis, and a distance from the center point to the other axis endpoint:

Specify center point of ellipse for base of cylinder <0,0,0>: (*Specify a point or press ENTER.*)

Specify axis endpoint of ellipse for base of cylinder: (*Specify a point.*)

Specify length of other axis for base of cylinder: (*Specify a distance.*)

- AutoCAD will anchor a rubberband line on the center to help locate the axis's other endpoint, and the half-length of the second axis.

- The Axis endpoint and Center options for cylinders with elliptical cross-sections both use the same options for establishing the other end and the orientation of the cylinder. Moreover, these are the same options used for cylinders with circular cross-sections.

Specify height of cylinder or [Center of other end]: (*Specify a distance or enter C.*)

- Height

- If you specify a distance, by either pointing or entering a value, AutoCAD will draw the cylinder in the Z direction to that height. Entering a negative value will draw the cylinder in the minus Z direction.

- Center of Other End

- This option will base both the length and orientation of the cylinder on a single point. The follow-up prompt is:

Specify center of other end of cylinder: (*Specify a point.*)

- The center line of the cylinder will extend from the initial base point to the specified point.

 Tip: If the wireframe views of your cylinders don't have enough lines for you to make out their surfaces clearly, change the setting of the Isolines system variable from its default of 4 to a higher value, such as 8 or even 12 (see Figure 5.15). The higher the number, the more lines AutoCAD uses to delineate curved and rounded surfaces on solids. A regen is required before new settings of Isolines take effect.

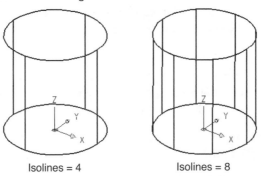

Isolines = 4   Isolines = 8

Figure 5.15

TRY IT! - CONSTRUCTING A SOLID CYLINDER

First we will make a cylinder with a round cross-section pointed in the Y direction. See Figure 5.16 for the results. Start a new drawing from scratch and follow the next series of command prompts for constructing the cylinder primitives.

Command: CYLINDER

Current wire frame density: ISOLINES=12

Specify center point for base of cylinder or [Elliptical] <0,0,0>: 1,1,0

Specify radius for base of cylinder or [Diameter]: .5

Specify height of cylinder or [Center of other end/]: C

Specify center of other end of cylinder: 1,3,0

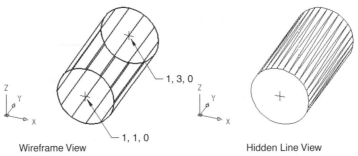

1, 3, 0

1, 1, 0

Wireframe View   Hidden Line View

Figure 5.16

Next we will make a cylinder with an elliptical cross-section pointed in the Z direction, as shown in Figure 5.17.

Command: CYLINDER

Current wire frame density: ISOLINES=12

Specify center point for base of cylinder or [Elliptical] < 0,0,0>: E

Specify axis endpoint of ellipse for base of cylinder or [Center]: 1,1,0

Specify second axis endpoint of ellipse for base of cylinder: 2.5,1,0

Specify length of other axis for base of cylinder: .5

Specify height of cylinder or [Center of other end/]: 1.5

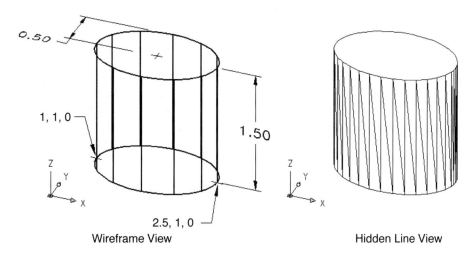

Wireframe View   Hidden Line View

Figure 5.17

THE CONE COMMAND

Cone primitives are a close relative to cylinders. They have the same round, or elliptical, cross-section, but they taper to a point rather that keeping the same cross-section size throughout their length. Therefore, the steps for making cones mirror those for making cylinders. All cones will have a sharp tip; there are no provisions for making truncated cones. The command line prompts and input for the CONE command are:

Command: CONE

Current wire frame density: ISOLINES=(*current*)

Specify center point for base of cone or [Elliptical] <0,0,0>: (*Specify a point, enter E, or press ENTER.*)

CENTER POINT

Specifying a point, or pressing ENTER to accept the default 0,0,0 point will make a cone having a circular cross-section centered on the selected point. AutoCAD will prompt you to specify a radius or a diameter for the cone and then for the location of the tip of the cone.

Specify radius for base of cone or [Diameter]: (*Specify a distance or enter D.*)

If you specify a distance AutoCAD will use it for the cone's radius. If you type in a D, AutoCAD will prompt you for the diameter of the cone.

Either option will lead to the following prompt:

Specify height of cone or [Apex]: (*Specify a distance or enter A.*)

- Height
- If you specify a distance, by either pointing or entering a value, AutoCAD will draw the cone with its tip pointed in the Z direction to that height. Entering a negative value will point the cone in the minus Z direction.

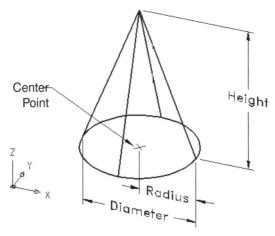

Figure 5.18

- Apex
- This option will base both the length and orientation of the cone on the location of its tip point (apex). AutoCAD will prompt for the apex point:

Specify apex point: (*Specify a point.*)

- The center line of the cone will extend from the initial base point to the specified point, as shown in Figure 5.19.

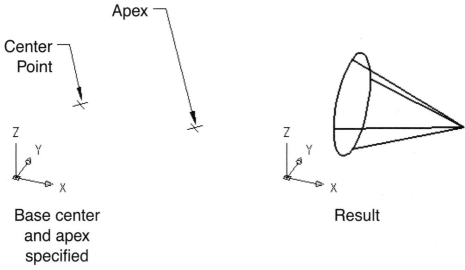

Base center and apex specified

Result

Figure 5.19

ELLIPTICAL

This option makes a cone with an elliptical cross-section. You can base the ellipse either on its center point or at one of its axis endpoints.

Specify axis endpoint of ellipse for base of cone or [Center]: (*Specify a point or enter C.*)

- Axis Endpoint
- Specifying a point will set one end of one axis of the elliptical base. Then AutoCAD will prompt for the other end of the axis and for a point to set the end of the other axis of the ellipse.

Specify second axis endpoint for base of cone: (*Specify a point.*)

Specify length of other axis for base of cone: (*Specify a distance.*)

- AutoCAD will draw a rubberband line as you specify a point for the second endpoint and then anchor a rubberband line on the midpoint of this axis as you specify the length of the other axis. The distance you specify, whether by pointing or by entering a value, is half of the length of the second ellipse axis.

- Center
- This option bases the elliptical cross-section at its center point. AutoCAD will prompt you for the center point, the endpoint of one axis, and a distance from the center point to the other axis endpoint:

Specify center point of ellipse for base of cone <0,0,0>: (*Specify a point or press ENTER.*)

Specify axis endpoint of ellipse for base of cone: (*Specify a point.*)

Specify length of other axis for base of cone: (*Specify a distance.*)

- AutoCAD will anchor a rubberband line on the center to help locate the axis's other endpoint and the half-length of the second axis.

The Axis endpoint and Center options for cones with elliptical cross-sections both use the same options for establishing the apex of the cone. Moreover, these are the same options used for cones with circular cross-sections.

Specify height of cone or [Apex]: (*Specify a distance or enter A.*)

- Height
- If you specify a distance, by either pointing or entering a value, AutoCAD will draw the cone in the Z direction to that height. Entering a negative value will draw the cone so that its tip is pointed in the minus Z direction.
- Apex
- This option will base both the length and orientation of the cone on the location of its apex. AutoCAD will prompt for the apex point:

Specify apex point: (*Specify a point.*)

- The center line of the cone will extend from the initial base point to the specified point.

TRY IT! - CONSTRUCTING A SOLID CONE

First, we will draw a cone with a base diameter of 1.5 units, 2 units tall, pointed in the Z direction. The resulting cone is shown in Figure 5.20. Start a new drawing from scratch and follow the next series of command prompts for constructing the cone primitives.

Command: CONE

Current wire frame density: ISOLINES=10

Specify center point for base of cone or [Elliptical] <0,0,0>: 1.5,1.5,0

Specify radius for base of cone or [Diameter]: D

Specify diameter for base of cone: 1.5

Specify height of cone or [Apex]: 2

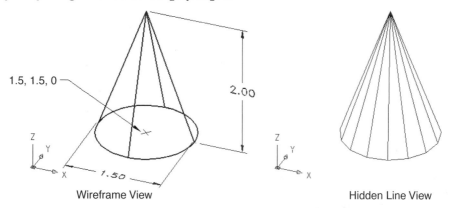

Wireframe View   Hidden Line View

Figure 5.20

Next, we will draw a cone with an elliptical cross-section, laying in the XY plane, pointed 135° from the X axis. Notice that you can draw the elliptical base on the XY plane and AutoCAD will tilt it in accordance with the cone's apex location. The results of the following command line input is shown in Figure 5.21.

Command: CONE

Current wire frame density: ISOLINES=10

Specify center point for base of cone or [Elliptical] <0,0,0>: E

Specify axis endpoint of ellipse for base of cone or [Center]: C

Specify center point of ellipse for base of cone <0,0,0>: 1.5,1.5,0

Specify axis endpoint of ellipse for base of cone: @1<0

Specify length of other axis for base of cone: .75

Specify height of cone or [Apex]: A

Specify apex point: @2.5<135

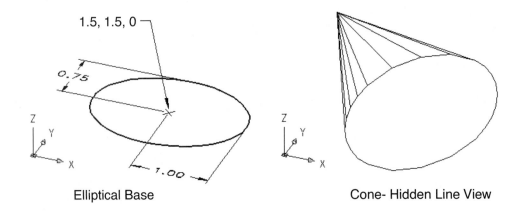

Elliptical Base   Cone- Hidden Line View

Figure 5.21

THE SPHERE COMMAND

Making spheres is the most straightforward process of all of the primitives. You specify the sphere's center point and then either the radius or diameter of the sphere. That's all there is to it. The command line format is:

Command: SPHERE

Current wire frame density: ISOLINES=(*current*)

Specify center of sphere <0,0,0>: (*Specify a point.*)

Specify radius of sphere or [Diameter]: (*Specify a distance or enter D.*)

RADIUS

Specifying a distance by either typing in a value or by pointing will set the radius of the solid sphere. AutoCAD will drag a rubberband line from the center of the sphere to help set the radius.

DIAMETER

This option bases the size of the sphere on its diameter. AutoCAD will display the prompt:

Specify diameter: (*Specify a distance.*)

The diameter may be specified either by entering a value or by pointing. AutoCAD will drag a rubberband line anchored at the center point to help set the diameter.

TRY IT! - CONSTRUCTING A SOLID SPHERE

We will make a solid sphere, having a radius of 1 unit, that will be sitting on the XY plane. Start a new drawing from scratch and follow the next series of command prompts for constructing the sphere primitive.

Command: SPHERE

Current wire frame density: ISOLINES=12

Specify center of sphere <0,0,0>: 1,1,1

Specify radius of sphere or [Diameter]: 1

Notice that the center of this sphere is one unit above the XY plane. If your sphere does not look like a sphere in wireframe views, increase the setting of the Isolines system variable. A regeneration is required before changes to Isolines take effect.

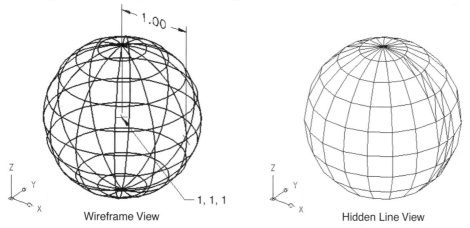

Wireframe View   Hidden Line View

Figure 5.22

THE TORUS COMMAND

Although they are not often needed, the torus is the most interesting and flexible of the six primitive solids. The basic shape of a torus is that of a doughnut, but you can eliminate the hole in the doughnut, and you can make tori shaped like footballs rather than doughnuts.

The prompts and input of the TORUS command are comparable to revolving a circle about an axis. First, you pick a center point for the torus, which establishes the rotational axis location. This central axis is always pointed in the Z direction. Hence, the circle is always revolved in a path parallel to the XY plane. After setting the axis location, you specify either the radius or the diameter of the torus. This sets the distance from the rotational axis to the center of the circle that will be revolved. Finally, you specify either the radius or diameter of the tube. This determines the size of the circle that is to be revolved. Although AutoCAD refers to the resulting object as a tube, it is completely solid, not hollow.

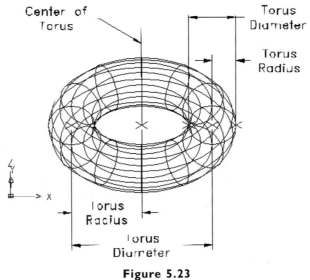

Figure 5.23

The command line format for the TORUS command is:

Command: TORUS

Current wire frame density: ISOLINES=(current)

Specify center of torus <0,0,0>: (Specify a point or press ENTER.)

Specify radius of torus or [Diameter]: (Specify a distance or enter D.)

If you specify a distance, AutoCAD will use it for the radius of the torus. If you type in a D, AutoCAD will prompt you for the diameter of the torus. Negative values are allowed. They make a football-shaped solid, which we will describe shortly. Either option will lead to the following prompt:

Specify radius of tube or [Diameter]: (Specify a distance or enter D.)

If you specify a distance, AutoCAD will use it for the radius of the tube. If you type in a D, AutoCAD will prompt you to enter the diameter of the tube. The tube radius can be larger than the torus radius, which results in a torus without a center hole. If the radius or diameter of the torus was negative, then the tube radius or diameter must be larger than the absolute value of the torus radius or diameter. For example, if you specified -2.5 as the torus radius, then the radius of the tube must be larger than 2.5.

A football-shaped solid is made when a negative number is used as a torus radius or diameter—as if an arc, rather than a circle, was revolved around the center line. The distance from the arc's quadrant to the center line—the football's radius—is equal to the sum of the torus radius and the tube radius. Thus, if in our previous example you had used a tube radius of 3.0, the radius of the football would be 0.5 (-2.5 + 3.0). The radius of the arc is equal to the tube radius, which in our example would be 3.0 units.

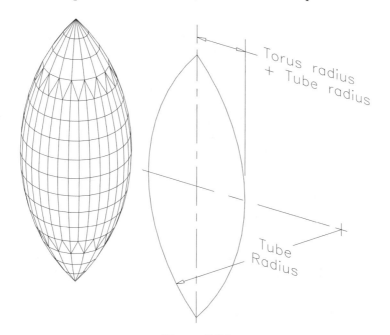

Figure 5.24

TRY IT! - CONSTRUCTING A SOLID TORUS

Start a new drawing from scratch and follow the next series of command prompts for constructing the torus primitive. First, we will make a doughnut-shaped torus having a torus radius of 2 units, and a tube radius of 0.75 units centered on the UCS origin. The resulting torus is shown in Figure 5.25 with hidden line removal on.

Command: TORUS

Current wire frame density: ISOLINES=8

Specify center of torus <0,0,0>: (ENTER.)

Specify radius of torus or [Diameter]: 2

Specify radius of tube or [Diameter]: .75

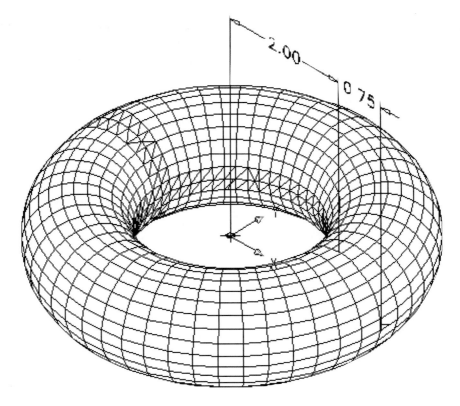

Figure 5.25

Next, we will make a torus that has no center hole. We will center it at the UCS origin, use a torus diameter of 4 units, and a tube diameter of 5 units. A hidden line view of the resulting torus is shown in Figure 5.26.

Command: TORUS

Current wire frame density: ISOLINES=8

Specify center of torus <0,0,0>: (ENTER.)

Specify radius of torus or [Diameter]: D

Specify diameter: 4

Specify radius of tube or [Diameter]: D

Specify diameter: 5

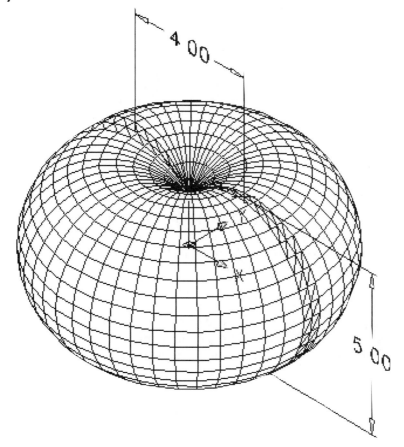

Figure 5.26

Last, we will make a football-shaped solid. The football will have a radius of 1.5 (-2.5 + 4), with an arc-shaped outline in which the arcs have a radius of 4. See Figure 5.27 for a hidden line view of the torus.

Command: TORUS

Current wire frame density: ISOLINES=8

Specify center of torus <0,0,0>: 3,2.5,0

Specify radius of torus or [Diameter]: -2.5

Specify radius of tube or [Diameter]: 4

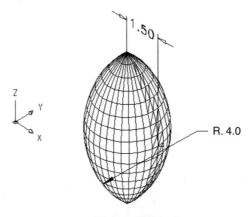

Figure 5.27

Note: While the BOX, WEDGE, CYLINDER, CONE, SPHERE, and TORUS commands can be used to easily create basic primitive shapes, more complicated shapes consisting of polyline profiles are either extruded or revolved to form the solid model. These solid modeling concepts will be discussed in the next series of pages.

PROFILE-BASED SOLIDS

In addition to the commands for making primitives, AutoCAD has two commands for making 3D solids from profile objects. Profile objects are closed planar objects. The REVOLVE command makes a 3D solid by revolving a profile object about an axis, and the EXTRUDE command makes a 3D solid by pushing the profile object in a given direction or along a path.

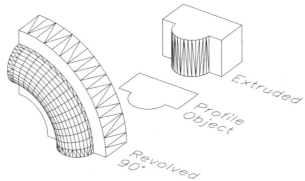

Figure 5.28

These two commands are able to make any shape that the commands for the primitive solids can make, and they can make some shapes that the primitives cannot make. They require more work though, because both of them not only need a profile object but may also need an object to serve as an axis or as a path.

PROFILE OBJECTS

The profile object must be a single, closed entity. Although you can pick several objects at one time, each object must be closed. You cannot pick a set of lines, for instance, even if they meet end to end and the last line closes with the start of the first line. The profile object must also be planar. Although it would sometimes be handy to use a closed 3D object—such as a wavy disk or helix-shaped ribbon—it is not allowed.

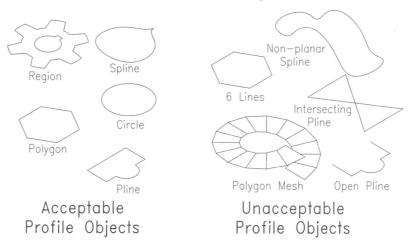

Figure 5.29

Probably the object type you will most often use as profile objects will be a 2D polyline. You can make a wide variety of shapes from 2D polylines, and they are easy to work with. As you would expect, all of the 2D polyline derivatives—doughnuts, ellipses, and polygons—can be used as profiles. Any width the polyline has is ignored, with the center of the polyline being used as the profile boundary. Consequently, doughnuts are equivalent to circles—they do not make a round profile with a hole in its center.

There is no limit to the number of vertices a polyline may have, although polylines with an extremely large number of vertices can make the resulting solid object impossibly slow to work with. For the same reason, spline-fit polylines are also generally not good profile objects. Although profile objects must be closed, 2D polylines do not have to be finished with the Close option. However, polylines with crossover, or even touching, segments, are not allowed.

Regions will probably be your second most often used object type for profiles. They can be especially useful profile objects because they can have holes in them.

Other acceptable profile objects are splines (as long as they are planar), spline ellipses, and circles. Traces also can be used, as can 2D solids (provided they do not cross over themselves). The outlines of these filled objects are used. Even 3D faces are accepted as profile objects.

Text, polyface meshes (made with PFACE), and polygon meshes, cannot be used as profile objects. Surprisingly, neither can 3D polylines, even if they are closed and confined to one plane.

The resulting solid object is in the current layer, not the layer of the profile object. The system variable Delobj determines whether or not objects used as profile objects are retained. When Delobj is set to 0, profile objects are retained; when Delobj is set to 1 (the default setting), each profile object is automatically deleted when the extruded or revolved solid is created.

THE REVOLVE COMMAND

This command transforms the path in space made when a flat profile is revolved about an axis into a solid. It is comparable to the REVSURF command, although it makes a solid object rather than a surface, and it uses a closed profile rather than a boundary curve.

REVOLVE requires three different steps. First, you select a profile object; second, you select an axis; and third, you specify the angle through which the profile is to be revolved.

Though the profile object may touch the axis, no part of it is allowed to cross it. Also, the axis will have a positive direction, which controls the rotation direction of partially revolved solids.

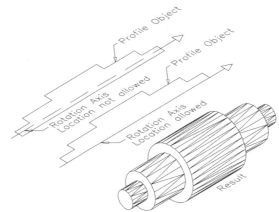

Figure 5.30

The command line format for REVOLVE is:

Command: REVOLVE

Current wire frame density: ISOLINES=(*current*)

Select objects: (*Use any object selection method.*)

Specify start point for axis of revolution or

define axis by [Object/X (*axis*)/Y (*axis*)]: (*Specify a point or select an option.*)

Although you can select more than one object, AutoCAD will revolve only the first one selected. As soon as you have selected the profile object, AutoCAD gives you four choices for defining an axis of revolution.

START POINT OF AXIS

This option for defining an axis is initiated from the main REVOLVE menu by specifying the first of two points on the axis. AutoCAD will then prompt you for a second point on the axis. The positive direction of the axis is from the first point to the second point.

Specify endpoint of axis: (*Specify a point.*)

Specify angle of revolution <360>: (*Specify an angle or press ENTER.*)

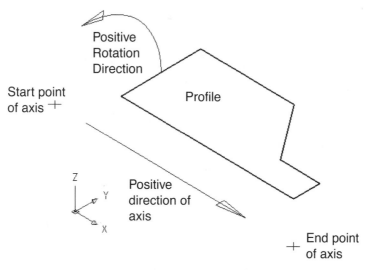

Figure 5.31

All of the options for defining an axis use the same "Angle of revolution" prompt, so we will postpone discussing it until we finish with the axis of revolution options.

OBJECT

You may use an existing line, or single-segment 2D or 3D polyline as a rotation axis. The polyline can only have one segment, and it must be a line segment. The positive direction of this line object is from the end nearest the pick point to the opposite endpoint.

Select an object: (*Select a line object.*)

Specify angle of revolution <360>: (*Specify an angle or press ENTER.*)

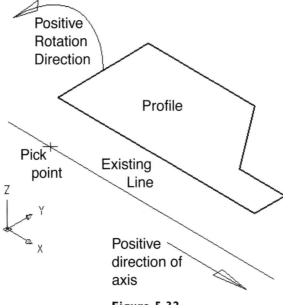

Figure 5.32

X

This option uses the X axis as a rotation axis. The positive direction of the rotation axis is the same as the coordinate system X axis.

Specify angle of revolution <360>: (*Specify an angle or press ENTER.*)

Y

Uses the Y axis as a rotation axis. The direction of the rotation axis is the same as the coordinate system Y axis.

Specify angle of revolution <360>: (*Specify an angle or press ENTER.*)

After you have specified an object to be rotated and an axis of rotation, AutoCAD asks for the angle of revolution. The revolved solid always begins at the profile and rotates through the specified angle, which can be anywhere between 0° and 360°. Rotation direction follows the right-hand rule, which means that if you look at the axis so that its positive end is pointed directly toward you, positive rotation will be counterclockwise. You can type in a negative angle to reverse the rotation direction. The default response to the "Angle of revolution" prompt is for a full 360° circle, in which case the direction of rotation is immaterial.

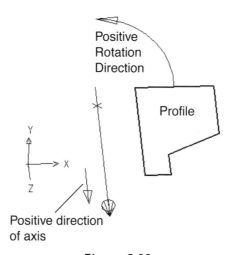

Figure 5.33

To demonstrate REVOLVE we will revolve a profile object, shown in Figure 5.34, about the Y axis and about the X axis to make two entirely different solids. This profile object is a closed 2D polyline, positioned so that one edge abuts the Y axis and the bottom edge is 1 unit from the X axis. The resulting 3D solids are shown in hidden line views in Figures 5.35 and 5.36.

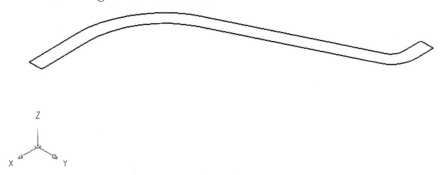

Figure 5.34

First we will revolve it -90° about the Y axis:

Command: REVOLVE

Current wire frame density: ISOLINES=4

Select objects: *(Select the closed 2D polyline.)*

Specify start point for axis or revolution or

define axis by [Object/X *(axis)***/Y** *(axis)***]: Y**

Specify angle of revolution <360>: -90

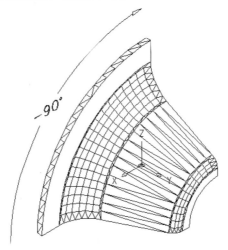

Figure 5.35

Next we will revolve it 180° about the X axis:

Command: REVOLVE

Current wire frame density: ISOLINES=4

Select objects: *(Select the closed 2D polyline.)*

Specify start point for axis or revolution or

define axis by [Object/X *(axis)***/Y** *(axis)***]: X**

Specify angle of revolution <360>: 180

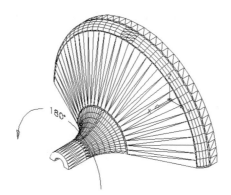

Figure 5.36

THE EXTRUDE COMMAND

The EXTRUDE command makes a solid object from the trail of a profile object moving in space. Its name comes from a manufacturing process in which material, such as aluminum, is forced through a die to form a linear shape having the same cross-section as the die. AutoCAD extrusions are more versatile than manufacturing extrusions, however, because they can either be in the profile object's Z direction or along a path defined by an existing object, and the extrusion can be tapered. The command line format for EXTRUDE is:

Command: EXTRUDE

Current wire frame density: ISOLINES=(*current*)

Select objects: (*Use any object selection method.*)

Specify height of extrusion or [Path]: (*Specify a distance or enter P.*)

All objects you select will be extruded according to the same height or path selection, with objects not meeting profile object criteria being ignored.

HEIGHT OF EXTRUSION

Specifying a distance, either by picking two points or by entering a value, extrudes the profile in its Z direction, which is not necessarily in the same direction as the current Z axis. An entity's Z direction, often called its extrusion direction, is stored in AutoCAD's database along with color, linetype, and other properties. The extrusion direction of most closed, planar objects is that of the Z axis when the object was created, and it is always perpendicular to the object. If you enter a negative number, the extrusion will be made in the object's minus Z direction. After you specify a height, AutoCAD will ask for a taper angle.

Specify angle of taper for extrusion <0>: (*Specify an angle or press ENTER.*)

The default angle of 0 forces the cross-section size of the extruded solid to remain constant throughout its length, whereas positive angles taper the extrusion inward, thus causing the cross-section size to become smaller. Negative angles, on the other hand, will taper the extrusion outward—making the cross-section size larger along the length of the extrusion. It is convenient to think in these terms of relative cross-section size, since interior holes in the profile object become larger with positive taper and smaller with negative taper.

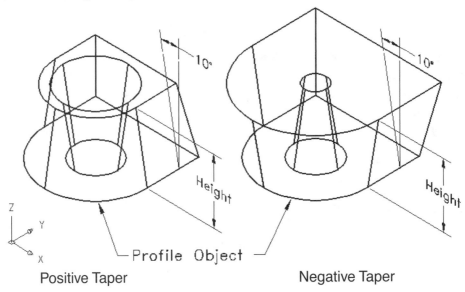

Positive Taper   Negative Taper

Figure 5.37

Taper angle is commonly called draft angle, even in some of AutoCAD's messages. It represents the angle between the extrusion direction and the resulting slanted surface of the solid. Any angle between but not including -90° and 90° is allowed, although the actual maximum angle for inward taper depends on extrusion height. It cannot be so large that the sides of the extrusion will intersect. An angle that is too large will bring up an error message saying that the extrusion will self-intersect, and no extrusion will be made.

PATH

This option uses a single-existing object to serve as a path for the extrusion. This path object determines the length, direction, and shape of the extrusion. When you select the Path option, you will not be able to taper the extrusion—its cross-section size remains constant. Allowed entity types for paths are lines, arcs, ellipses (either spline or polyline), 2D polylines, 2D polygons, 3D polylines, and splines.

Paths can be open or closed, and can even have nonplanar curves, but there are restrictions. One restriction is that arc portions of the path have a radius that is equal to or larger than the profile's width. Thus, if a profile is 1 unit wide, all arc portions on the path must have a radius of at least 1 unit. On the other hand, corners (where two straight segments having different directions are joined) are allowed, even though they could be considered as zero-radius arcs. AutoCAD simply miters the corners of the extrusion.

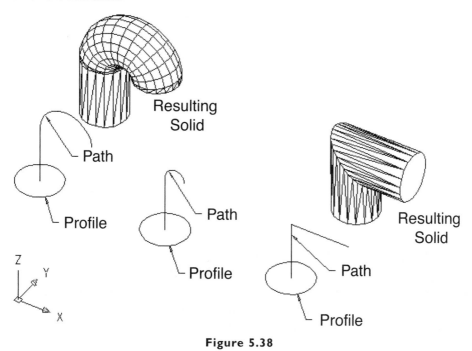

Figure 5.38

Three-dimensional curves, including helixes, can be used for paths as long as the path is made from a 3D polyline with straight segments. Spline-fit 3D polylines and non-planar spline entities cannot be used as path curves.

With one exception, extrusions will always start at the profile, even if the start of the path is not perpendicular (normal) to the profile, and will end the solid so that it is perpendicular to the end of the path, as shown in Figure 5.39. As a result, the cross-section of the solid will be foreshortened when the start of the path is not perpendicular to the profile. Although it is perfectly acceptable for the path to not be perpendicular to the profile, it cannot lie in the same plane.

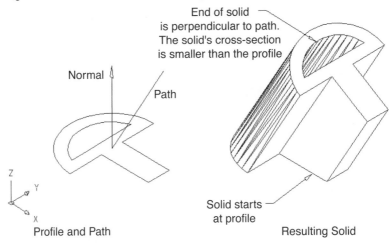

Profile and Path   Resulting Solid

Figure 5.39

The exception is that when the path is a spline (the entity type, not a spline-fit polyline), the starting end of the extruded solid will always be perpendicular to the starting end of the path. If the profile is not perpendicular to the start of the path, AutoCAD automatically rotates the profile so that it is perpendicular, as shown in Figure 5.40. The end of the extruded solid will be perpendicular to the end of the spline path, just as it is with other entity type paths.

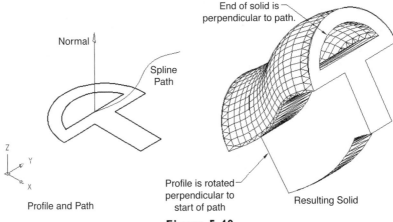

Profile and Path   Resulting Solid

Figure 5.40

An additional complication is that paths are always projected to the center of the profile. This projection of the path is different than simply moving the path. It is comparable to AutoCAD's OFFSET command, in which the offset distance is equal to the distance between the path and the center of the profile. Just as the radius of an offset arc will be larger or smaller than the original arc, arcs in the projected path can be larger or smaller than those in the original path, depending on its relative location with the profile. Furthermore, the length of the projected path may be longer or shorter than the original path. This projected path is a virtual path—no new object is created, although the extrusion is made as if one were.

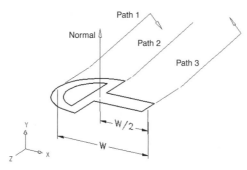

Figure 5.41

As an example, Figure 5.41 shows one profile object with three possible path objects, none of which are located on the profile. Path 2 is located halfway between the profile ends in the general direction of the path, so it will extrude the profile through its full length. Path 1 will be projected toward the center of the profile as shown by the arrow, becoming smaller, and path 3 will become longer as it is projected toward the center. The extrusions made from the profile and the three paths are shown in Figure 5.42.

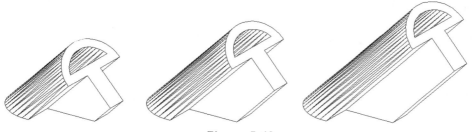

Figure 5.42

These projections and resulting change in path size also occur with closed paths. Moreover, when a closed path has sharp corners, such as a polygon-shaped path, the path is moved to a position in which the profile is located in the middle of one of the extrusion's miter joints. We will explore this effect shortly when we try making extruded solids.

Tips: Keep paths as simple as possible. It may be easier to achieve the shape you desire by editing and modifying the extruded solid later than to try to make it with an elaborate path.

Start the path in a direction that is perpendicular to the profile object.

Position each path in the center of the profile object, which is the midpoint between the two ends of the profile in the direction of the path.

We will concentrate on the Path option in these demonstrations of extruded solids. This option allows you to create unique solid forms, but it sometimes seems difficult to control.

First, we will make a coil spring by extruding a circle in a helix path. You can use cylindrical coordinates to make the path or use the AutoLISP routine HELIX.LSP listed in Chapter 3 of this book. The path helix has a constant radius of 1 and its pitch is 1. We will make two revolutions using 24 segments per revolution. This helix must be made with a 3D polyline, since AutoCAD will not accept a nonplanar spline as a path. The wire diameter of the spring will be 0.5 units, represented as a 0.5-diameter circle located at the beginning of the path, and perpendicular to the starting segment of the path.

Command: EXTRUDE

Current wire frame density: ISOLINES=4

Select objects: (*Select the circle.*)

Specify height of extrusion or [Path]: P

Select extrusion path: (*Select the 3D polyline.*)

The resulting solid is shown on the right in Figure 5.43. The spring is an approximation of a smoothly coiled spring since it is made of short, straight segments. You can increase or decrease the degree of approximation through the number of segments per revolution in the path helix, but you will need to consider your computer's speed and power as you do so.

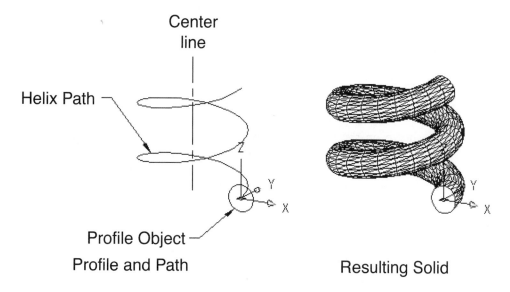

Figure 5.43

Next, we will extrude a hexagonal profile object around a hexagonal path. Both the profile and the path, shown on the left side of Figure 5.44, were made with AutoCAD's POLYGON command, using the Circumscribed about circle option. The profile is located in the middle of one segment and is normal (perpendicular) to that segment. The command line sequence of prompts and input will be:

Command: EXTRUDE

Current wire frame density: ISOLINES=4

Select objects: (*Select the profile hexagon.*)

Specify height of extrusion or [Path]: P

Select extrusion path: (*Select the path hexagon.*)

Path was moved to the center of the profile.

Profile was oriented to lie on the bisector plane.

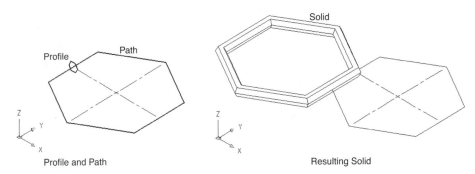

Profile and Path

Resulting Solid

Figure 5.44

The results, shown on the right in Figure 5.44, may surprise you. As AutoCAD informed you at the conclusion of the command, the path was moved so that the profile was centered in one of the vertices of the path hexagon, and the profile was rotated to be in line with the solid's miter joints. A consequence of these actions is that the cross-section through the straight segments of the solid will be a slightly compressed version of the original profile object. AutoCAD will always move a closed path so that the profile object is on a vertex point on the path and with closed paths containing sharp corners orient the profile so that it is in line with the solid's closest miter corner.

COMMAND REVIEW

BOX

This command constructs brick-shaped objects. The solid primitive created with this command will have six rectangular sides, which are either perpendicular or parallel to one another.

CYLINDER

This command is used to construct a solid shape that has a circular shape and a height dimension. A cylinder could also be constructed with an elliptical base.

CONE

This command will construct a solid primitive similar to a cylinder. The cone can have a round or elliptical cross-section, which tapers to a point.

EXTRUDE

This command constructs a solid object from the trail of a flat profile moving in space.

REVOLVE

This command constructs a solid object from a flat profile that is revolved about a path in space.

SPHERE

This command constructs a sphere when you specify a center point and either the radius or diameter of the sphere.

TORUS

This command constructs a donut-shaped solid primitive. After picking the center of the torus, you are prompted for the radius/diameter of the torus and the radius/diameter of the tube.

WEDGE

This command is used to construct a box that has been sliced diagonally from edge to edge. They have a total of five sides, three of which are rectangular.

CHAPTER REVIEW

Directions: Answer the following questions.

1. List some of the basic differences between AutoCAD solid models and surface models.

2. In what ways does a region differ from a planar AutoCAD surface made from a 3D face or even a polygon mesh?

3. Can open objects, such as lines and arcs be used as the basis for a region? If so, what are the restrictions?

4. List the six AutoCAD commands for making primitive 3D solids.

5. Why are the objects created by these commands called primitives?

6. How can you have a box primitive that is skewed relative to the principal WCS planes?

7. What are the two basic geometric shapes that the TORUS command can make?

8. What are the requirements for profile objects of extruded and revolved solids?

9. What are the REVOLVE command's four options for specifying an axis of rotation? Can the profile object extend on both sides of the axis?

10. Can you start and stop the revolved solid at any angle about the axis? How is the direction of rotation determined?

11. Under what condition is the profile object retained after it is used by the EXTRUDE or REVOLVE commands?

Directions: Circle the letter corresponding to the correct response in the following.

12. Which of the following objects cannot be used as the basis for a region?

    a. 2D polyline

    b. 3D face

    c. 3D polyline

    d. Circle

    e. Polyface mesh

    f. Spline

13. Is it possible to create a primitive cylinder that is parallel to the XY plane?

    a. yes

    b. no

14. Is it possible to create a primitive cylinder with an elliptical cross-section?

a. yes

b. no

15. Is it possible to create a primitive cylinder that is tapered?

a. yes

b. no

16. Profile-based 3D solids, made with the EXTRUDE and REVOLVE commands, can duplicate any of the geometric shapes that the commands for primitives can make.

a. true

b. false

17. Extruded solids are always linear (in a straight line).

a. true

b. false

18. Extruded solids cannot have a taper.

a. true

b. false

19. If you use the height option of the EXTRUDE command, extrusion direction is always in the current Z axis direction.

a. true

b. false

Creating and Editing Solid Models

LEARNING OBJECTIVES

This chapter will cover how to modify, edit, and display solid models. When you have completed Chapter 6, you will:

- Be able to build complex solid models by using Boolean operations to combine and modify basic 3D solid geometric forms.

- Know how to fillet and chamfer sharp corners on 3D solids, and how to slice 3D solids into two pieces.

- Know how to edit faces and edges of 3D solids, and how to hollow out 3D solids.

- Be able to control the appearance of 3D solids.

MODIFYING SOLIDS

Creating a primitive 3D solid, or even a revolved or extruded 2D solid, is usually just the first step in building a solid model. In subsequent steps the basic solids are combined and altered to achieve the shapes and forms your design calls for.

AutoCAD's tools for working on solids fall into three broad categories. First, there are Boolean operations, which use two or more existing solids to create a new solid form. Second, there are modification operations that work on a single solid object at a time. Third, there are editing operations for modifying selected faces and edges of 3D solids. We will discuss Boolean operations first. The Menu options and toolbar buttons for Boolean operations are shown in Figure 6.1.

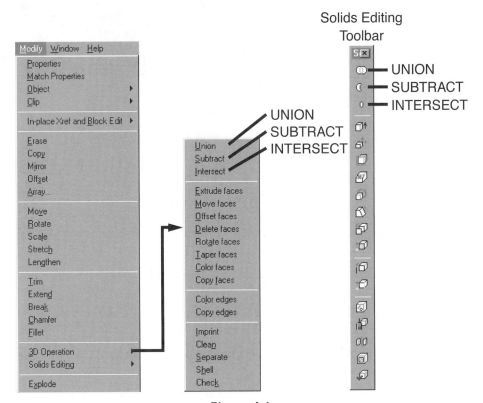

Figure 6.1

BOOLEAN OPERATIONS

Boolean operations are named after the 19th-century English mathematician George Boole, who developed theories on logic and sets that are still widely used in computers. Virtually all computer programming languages have the three Boolean logical operators of OR, AND, and XOR. Similarly, AutoCAD has three commands for performing Boolean operations on solids and regions: UNION, SUBTRACT, and INTERSECT.

All three commands are simple and straightforward, as summarized in Figure 6.2. UNION joins two or more solids, SUBTRACT removes the volume of one solid from another, and INTERSECT makes a solid from the intersecting volume of two or more solids. Although the diagram uses only pairs of solids to demonstrate the Boolean operations, they work in a similar way when more than two solids are involved. Also, although we will confine our discussion to solids, the Boolean operations work equally well, and in the same way, on regions. You cannot, however, mix regions and solids in a Boolean operation.

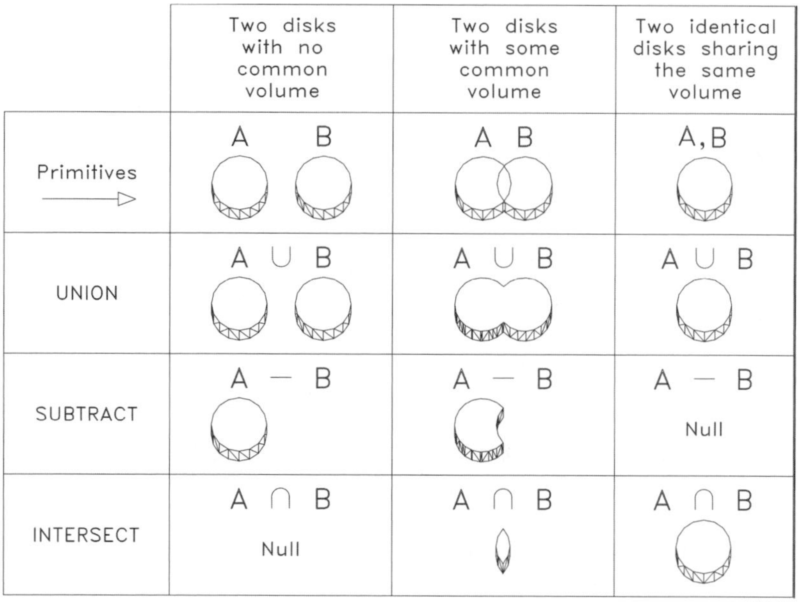

	Two disks with no common volume	Two disks with some common volume	Two identical disks sharing the same volume
Primitives ——▷	A B	A B	A , B
UNION	A ∪ B	A ∪ B	A ∪ B
SUBTRACT	A − B	A − B	A − B Null
INTERSECT	A ∩ B Null	A ∩ B	A ∩ B

Figure 6.2

Often the object obtained from a Boolean operation is referred to as a composite solid because it contains elements of at least two solids. Some solid modeling programs keep track of the original objects and can even return a composite solid back into its more basic objects. AutoCAD cannot do this, however. Once a Boolean operation has been performed on a set of solids, the solids cannot be returned to their original form (except by performing additional modification operations, or by using UNDO).

THE UNION COMMAND

UNION, which combines a set of solids into one solid, is likely to be your most frequently used Boolean operation (See Figure 6.3). When you invoke the command, AutoCAD simply prompts you to select the solid objects to be joined—there are no options or further prompts. Any of the object selection methods may be used in choosing the solids. All nonsolid objects in your selection set are ignored, but if the set does not contain at least two solids, AutoCAD will display the message: "At least 2 solids or coplanar regions must be selected."

Command: UNION

Select objects: (*Select at least two objects using any object selection method.*)

All solids within the selection set will be combined into one solid, regardless of where they are in 3D space. Regions, however, are unioned only if they are in the same plane (coplanar).

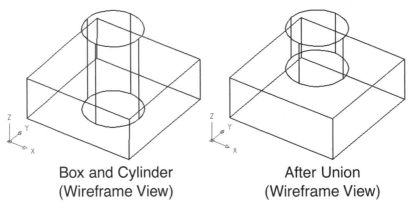

Box and Cylinder
(Wireframe View)

After Union
(Wireframe View)

Figure 6.3

If the solids overlap, the common volume is absorbed into the new solid, and AutoCAD makes any new edges and boundaries that may result from the combined solids. If the solids do not touch one another, they are still joined into one solid, although the space between them remains open. Unioned but nontouching solids are sometimes useful when you intend to use them in future Boolean operations. For instance, you may union a group of cylinders, arranged in a pattern, to be used in making holes in other solids.

The new composite solid takes its layer from the existing solid objects, not from the layer that is current, as AutoCAD does in many operations. If the original solids are in different layers, the composite solid takes the layer of the first object picked or, when a window or crossing selection is used, the layer of the newest solid.

TRY IT! - USING UNION

We will first use the UNION command on some relatively simple solids. Make two solid boxes. One should be 0.5 units long, 2 units wide, and 1 unit high. The other should be 1.5 units long, 2 units wide, and 0.5 units high. Then make a cylinder having a radius of 1, and a height of 0.5 units. Place them together as shown on the left in Figure 6.4. Notice that the 0.5-high box shares space with the other box as well as with the cylinder. Then join the three solids with the UNION command.

Command: UNION

Select objects: (*Select the three solid objects.*)

The resulting single solid is shown on the right in Figure 6–4. Both the before and after views are shown in their wireframe viewing mode.

File 3d_ch6_01.dwg on the accompanying CD-ROM contains all three solids along with the composite solid. We'll make a round hole in this solid when we explore the subtract Boolean operation.

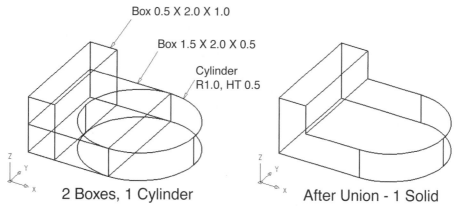

Figure 6.4

Next, we will begin to make a solid model that is a remake of a wireframe model we made in Chapter 2 as we experimented with the UCS. Make the two solid objects shown in Figure 6.5. One is an extruded solid and the other is a cylinder primitive. The easiest way to make the extruded solid is to rotate the UCS 90° about the X axis and draw the line entities using the dimensions shown. Then turn the lines into a 2D polyline, or a region, and use it as a profile for the EXTRUDE command, using a height of -2 units. Finally, move the UCS to the top of the 1-unit-long, flat area on the model (the Face option of the UCS command is a good way to move the UCS) and invoke the CYLINDER command, using a radius of 1 and a height of -0.25. The cylinder will overlap the extruded solid, but that is of no consequence.

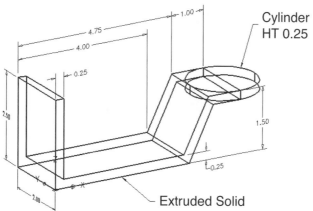

Figure 6.5

Once the two solids have been made, and are in position, use the UNION command to join them.

Command: UNION

Select objects: (*Select the two solid objects.*)

The results are shown in Figure 6.6 with the HIDE command in effect. In later applications of 3D solid commands we will add holes and chamfers, similar to those we made in the wireframe model, to this model.

Compare your solid model with the one in file 3d_ch6_02.dwg on the accompanying CD-ROM.

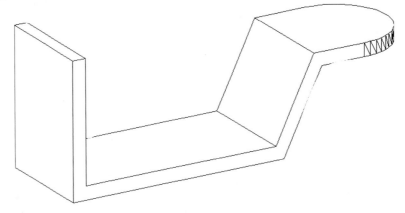

Figure 6.6

THE SUBTRACT COMMAND

The Boolean subtract operation removes the intersecting volume of one set of solids from another set of solids (See Figure 6.7). You will use this operation to trim and make holes in a solid. It is carried out with the SUBTRACT command, which first asks for a set of objects that are to have other objects subtracted from them.

If more than one object is selected, AutoCAD automatically performs a union to create a single source object. Next, you are prompted to select the objects that are to be subtracted from the source set. If you happen to include source objects in the second set, AutoCAD proceeds with the operation as if they had not also been picked for the second set.

The command line format for the SUBTRACT command is:

Command: SUBTRACT

Select solids and regions to subtract from...

Select objects: (*Select objects using any object selection method.*)

Select solids and regions to subtract...

Select objects: (*Select objects using any object selection method.*)

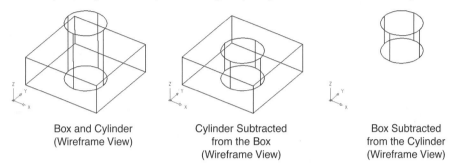

| Box and Cylinder (Wireframe View) | Cylinder Subtracted from the Box (Wireframe View) | Box Subtracted from the Cylinder (Wireframe View) |

Figure 6.7

The volume that is common to both sets is removed from the source object and the second set of objects disappears, including any volume that is outside the source object. If the two selection sets do not share any volume, then the second set disappears without subtracting anything. If the source objects are completely enclosed within the second set, both sets of objects disappear and AutoCAD displays the message: "Null solid created—deleted".

The layer of the resulting solid will be that of the source solid. If more than one source solid was selected and they are in different layers, then the resulting solid's layer will be that of the source object selected first or, if a window or crossing selection method was used, the most recently created object.

TRY IT! - USING SUBTRACT

We will make a hole in the first composite solid we made while applying the UNION command. First, use the CYLINDER command to make a cylinder with a radius of 0.5 units and a height of 1 unit, as shown on the left in Figure 6.8. Then subtract the cylinder from the composite solid.

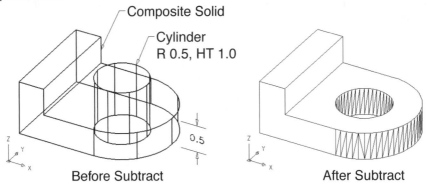

Before Subtract   After Subtract

Figure 6.1

Command: **SUBTRACT**

Select solids and regions to subtract from...

Select objects: (*Select the composite solid.*)

Select solids and regions to subtract...

Select objects: (*Select the cylinder.*)

The results, with HIDE in effect, are shown on the right in Figure 6.8. The entire cylinder has disappeared, even the part that extended above the source object.

File 3d_ch6_03.dwg on the accompanying CD-ROM contains a version of the model before SUBTRACT and a version after SUBTRACT.

Next, we will use SUBTRACT to make the holes and square notches on the solid model bracket we started with the UNION command. The holes will be based on cylinders, with the dimensions and center points shown in Figure 6.9. For convenience, we placed the large cylinder on the XY plane and made it high enough to pass through the existing composite solid. Since objects that are subtracted are not retained, it does not matter if they extend beyond the object they are to be subtracted from.

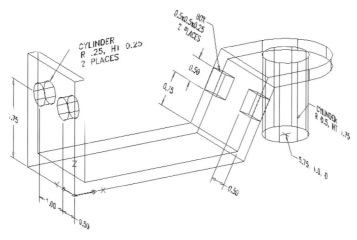

Figure 6.9

The square notches will be based on boxes that are 0.5 units by 0.5 units by 0.25 units high. You will probably find it easiest to make these in their proper position if you use the 3point or Face option the of the UCS command to position the UCS on the slanted portion of the composite solid. Once the five solid objects are made, applying SUBTRACT seems almost trivial.

Command: SUBTRACT

Select solids and regions to subtract from...

Select objects: (*Select the composite solid.*)

Select solids and regions to subtract...

Select objects: (*Select the three cylinders and two boxes.*)

The results are show in Figure 6.10 with the HIDE command on. Later, we will add the chamfers that the wireframe model had, and we will fillet the sharp bends of the bracket.

File 3d_ch6_04.dwg on the accompanying CD-ROM has these five solid objects in place on the bracket, along with a model of the bracket after the SUBTRACT operation.

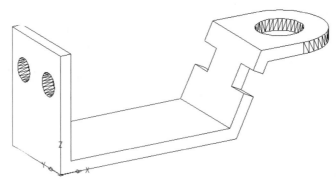

Figure 6.10

Our final exercise with the SUBTRACT command will complete the surface model of an electrical device enclosure we have been sporadically working on. We started this model as a wireframe in Chapter 3, and surfaced most of it in Chapter 4. However, the back end of the surface model, as can be seen in Figure 6.11, remains open. This area will have four cutouts for electrical connections and interfaces, which would make it very tedious to surface using 3D faces or even polymesh surfaces. Therefore, we will surface this area with a region. Even though this is not the intended purpose of region objects, they work well as planar surfaces, and are convenient to use.

Figure 6.11

Freeze the layers the surfaces are in to leave only the wireframe showing. Then position the UCS as shown in Figure 6.12.

Figure 6.12

Create and make current a new layer for wireframe objects, having a name such as WF-02, and trace the outline of the open area with four lines or a 2D polyline. Reduce the clutter on the wireframe by freezing all layers except the current one. Then, draw the cutouts using the dimensions shown in Figure 6.13.

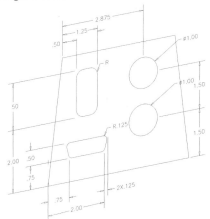

Figure 6.13

Create a new layer, having a name such as SURF-05, and make it current. If you want to retain the objects you have just drawn, set the Delobj system variable to 0. Then, invoke the REGION command and select all of the objects you have drawn. AutoCAD will report that five regions were created. Each closed object was transformed into a region, even if it was made of individual lines and arcs. Finally, subtract the four regions representing cutouts from the outer region. The command line sequence is:

Command: SUBTRACT

Select solids and regions to subtract from...

Select objects: (*Select the outer region.*)

Select solids and regions to subtract...

Select objects: (*Select the four interior regions.*)

After thawing the layers used for the other surfaces your 3D model should look like the one in Figure 6–14 when HIDE is in effect. If the circle cutouts are not round looking, increase the value of the Facetres system variable. This finishes the model.

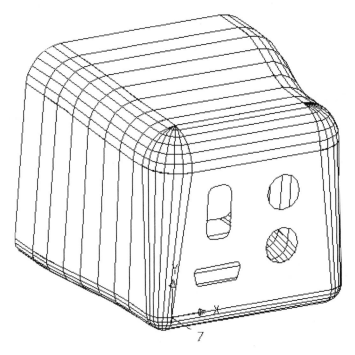

Figure 6.14

File 3d_ch6_05.dwg on the accompanying CD-ROM contains the dimensioned region object along with the completed model.

In Chapter 8 we will describe how a multiview, orthographic production drawing can be made from this surface model.

Tip: Postponing UNION and SUBTRACT operations as long as possible is a good practice to follow, because you can more easily change the components of a solid model. For instance, if you have created a cylinder that you indent to subtract from another solid to make a hole, it is easier to move or make copies the cylinder than it is to move or make copies of the hole.

THE INTERSECT COMMAND

The Boolean intersection operation, implemented with the INTERSECT command, creates a new solid object from the overlapping volume of two or more solids (See Figure 6.15). In one sense it is the opposite of the UNION command. When solids are unioned, everything is retained, with their common volume absorbed into the new composite solid. When solids are intersected, everything except their common volume is deleted.

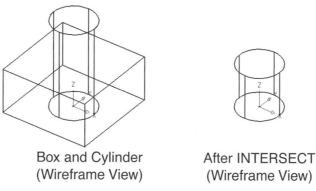

Box and Cylinder   After INTERSECT
(Wireframe View)   (Wireframe View)

Figure 6.15

Like the UNION command, INTERSECT simply prompts you to select objects, and you can use any object selection method. The command line format for the command is:

Command: INTERSECT

Select objects: (*Select at least two objects using any object selection method.*)

The resulting composite solid will take on the layer of the first object selected or the most recently created object if a window or crossing selection method was used. Only the overlapping volume is retained—everything else disappears. If there is no intersecting volume, all of the selected objects will disappear, and AutoCAD will display the message: "Null solid created—deleted."

TRY IT! - USING INTERSECT

As an exercise and as an example of what can be done with INTERSECT, we will turn the 2D drawing shown in Figure 6.16 into a 3D solid model. We will do this by making extruded solids from each of the three views, putting all three solids in one location, and using the Boolean INTERSECT command.

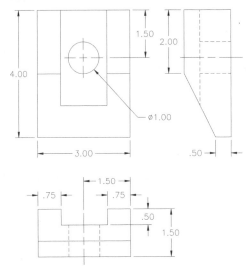

Figure 6.16

The first step is to draw the outlines of the objects as they are shown in the top, front, and right orthographic views. You need to draw only the outlines of the views, plus the circle in the top view, which represents a round hole through the object. Then, turn all of the view outlines into regions or 2D polylines and extrude the top view and the circle to a height of 1.5 units, the front view to a height of 4 units, and the side view to a height of 3 units. Actually, you can extrude these objects higher than these values because the intersecting volume is all that matters. After you SUBTRACT the extruded circle from the extruded top view, your three solids should look similar those shown in Figure 6.17 when HIDE is turned on.

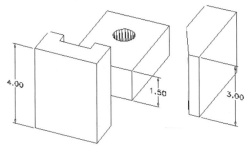

Figure 6.17

Next, use ROTATE3D to rotate the solids representing the front and side views 90°, as shown in Figure 6.18.

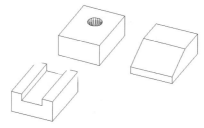

Figure 6.18

Finally, move the three solids so that they completely enclose one another as shown on the left in Figure 6.19, and invoke the INTERSECT command:

Command: INTERSECT

Select objects: (Select the three solid objects.)
The resulting completed solid model is shown on the right in Figure 6.19, with HIDE on.

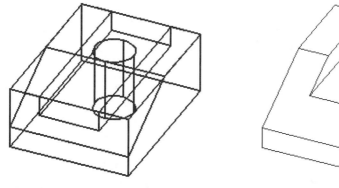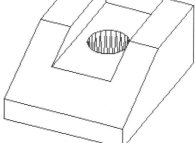

Before INTERSECT

Result

Figure 6.19

 File 3d_ch6_06.dwg on the accompanying CD-ROM has the completed model as well as all of the stages for its creation.

TRY IT! - MORE APPLICATIONS OF THE INTERSECT COMMAND

Start this next exercise in using INTERSECT by drawing the top and front views of the part shown in Figure 6.20. You should base your drawing on the dimensions given in the figure, but do not include the dimensions or the hidden lines and center lines in it. Then, use the REGION command to turn your objects into regions. There will be a total of five regions. Use AutoCAD's SUBTRACT command to turn the three circles in the top view into holes in the top view outline. You can use SHADEMODE's flat shading to verify that the results of this operation were correct.

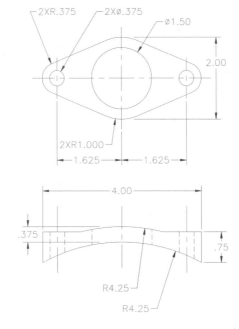

Figure 6.20

You now have a region representing the top view of the part and another representing the part's front view. Use ROTATE3D to rotate the front view region into the ZX plane. You should also move the two regions close together as shown in Figure 6.21.

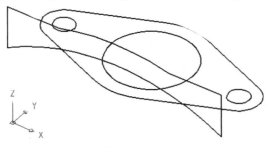

Figure 6.21

Extrude the region representing the top view of the part at least 0.875 inches in the Z direction. A longer extrusion length will not cause any problems (we used 1 inch). Extrude the region for the front view at least 2 inches. The length you use will depend on how close the region is to the extruded top view, and the extrusion direction will depend on how your UCS is oriented. If your UCS is oriented as the one shown in Figure 6.22, the extrusion direction will be in the minus Z direction.

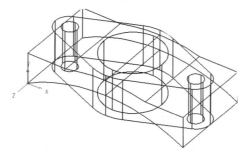

Figure 6.22

Finish the part by invoking INTERSECT and picking the two extruded 3D solids. Your finished model should look similar to the one shown in Figure 6.23 in the 2D wireframe viewing mode.

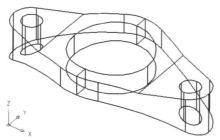

Figure 6.23

File 3d_ch6_07.dwg on the CD-ROM that comes with this book contains the completed 3D solid model for this exercise, as well as the 2D drawing, and the regions it is based on.

SINGLE-OBJECT MODIFICATION OPERATIONS

While the Boolean operations always require at least two solids, AutoCAD has other tools for modifying one solid object at a time. As you would expect, many of the usual editing commands do not work on 3D solids. You cannot use the BREAK, TRIM, EXTEND, LENGTHEN, or STRETCH commands on 3D solids. You can, however, copy, move, rotate, and erase 3D solids. If you explode a solid, the planar surfaces on it turn into regions, and the curved and rounded surfaces become bodies. (AutoCAD body objects, which are somewhat like a nonplanar version of regions, are created solely through the explosion of a 3D solid.)

The FILLET and CHAMFER commands both recognize solid objects and use a special set of prompts and options for making rounded (fillets) and beveled (chamfers) edges on solids (They can be selected from the menus in Figure 6.24). Unlike the wireframe version of these commands, they only work on a single object—you cannot fillet or chamfer two adjoining solids.

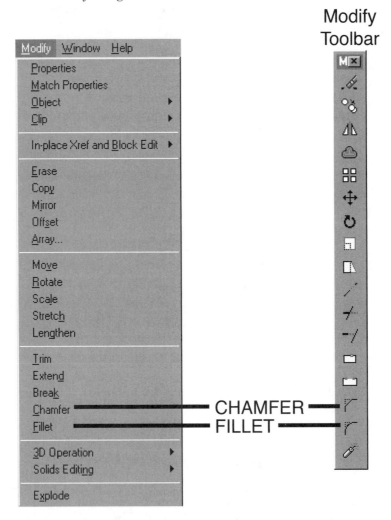

Figure 6.24

THE FILLET COMMAND

A fillet is a rounded edge between two faces of a solid. The cross-section of the fillet is that of an arc, with each end of the arc tangent to the adjoining face. When a 3D solid is selected as the first object during the FILLET command, AutoCAD displays a special follow-up prompt for solids instead of prompting for a second object. This prompt allows you to fillet any number of edges during the command and even change the radius within the command, so that some fillets will have a different radius than other fillets. Also, if you fillet the edges of three adjoining faces, AutoCAD will round the corner into a spherical shape in Figure 6.25.

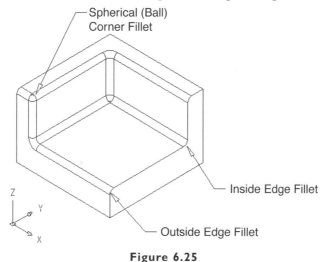

Figure 6.25

The command line format for the FILLET command on a solid is:

Command: FILLET

Current settings: Mode = trim. Radius = current

Select first object or [Polyline/Radius/Trim]: *(Select a 3D solid.)*

Enter fillet radius <current>: *(Enter a distance greater than zero.)*

Select edge or [Chain/Radius]: *(Select an edge, enter C or R, or press ENTER.)*

The message about Trim mode that appears when the command starts is of no consequence for 3D solids. The edge that you used to select the solid will be highlighted and will be filleted if you press ENTER. Notice that AutoCAD redisplays the current radius after you select a solid and allows you to change the radius. Unlike the fillet radius for wireframe objects, the fillet radius for solid objects must be larger than zero.

SELECT EDGE

Selecting an edge by picking a point on it will highlight the edge and redisplay the prompt. After ENTER is pressed, AutoCAD will report how many edges were selected and fillet them. Once an edge has been selected, there is no way to deselect it.

CHAIN

This option allows you to select several edges with one pick, provided the edges are tangent to one another. The follow-up prompt displayed is:

Select edge chain or [Edge/Radius]: (*Select an edge or enter E or R.*)

- Select Edge Chain (See Figure 6.26)
- Select an edge to be filleted. All edges tangent to the selected edge and to each other will be highlighted for filleting, and the chain follow-up prompt will be redisplayed. If there are no edges tangent to the selected edge, only the selected edge will be highlighted.

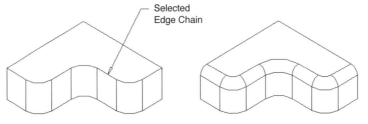

Selected
Edge Chain

Figure 6.26

- Edge
- This option returns to the single-edge mode prompt.
- Radius
- Allows you to change the fillet radius. All subsequent edges selected will use this radius. The follow-up prompt is:

Enter fillet radius <current>: (*Specify a distance or press ENTER.*)

RADIUS

This option of the single-edge mode prompt allows you to change the fillet radius. All subsequent edges selected will use this radius. The follow-up prompt is:

Enter fillet radius <current>: (*Specify a distance or press ENTER.*)

In the next example, we will fillet three adjoining edges of a wedge to illustrate how AutoCAD handles filleted corners. We will use a radius of 0.25 units for the top edges, and a radius of 0.5 for the front vertical edge.

Command: FILLET

Current settings: Mode = TRIM. Radius = 0.1250

Select first object or [Polyline/Radius/Trim]: (*Pick point 1.*)

Enter fillet radius <0.1250>: .25

Select an edge or [Chain/Radius]: (*Pick point 2.*)

Select an edge or [Chain/Radius]: R

Enter fillet radius <0.2500>: .5

Select an edge or [Chain/Radius]: (*Pick point 3.*)

Select an edge or [Chain/Radius]: (*ENTER.*)

3 edge (s) selected for fillet.

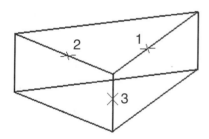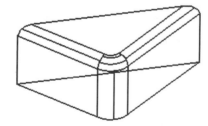

Figure 6.27

The filleted solid is shown on the right in Figure 6.27, with HIDE on.

The next application illustrates how the Chain option works. The front edge of the solid shown on the left in Figure 6.28 consists of three 180° arcs. Their ends are tangent with each other, and the ends of the two outside arcs are tangent with the straight side edges. This solid object was made by extruding a 2D polyline. Even though the curves on a solid like this are smooth, AutoCAD considers each arc in the edge to be a separate edge. Therefore, the top of this solid has six separate edges. We will fillet the front and side edges, but not the back edge.

Command: FILLET

Current settings: Mode = TRIM. Radius = 0.5000

Select first object or [Polyline/Radius/Trim]: *(Pick point 1.)*

Enter fillet radius <0.5000>: .1875

Select an edge or [Chain/Radius]: C

Select an edge chain or [Edge/Radius/]: *(Pick point 2.)*

Select an edge chain or [Edge/Radius]: *(ENTER.)*

5 edge (s) selected for fillet.

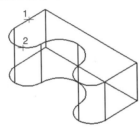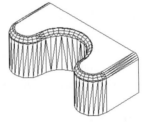

Figure 6.28

The filleted solid, with HIDE on, is shown on the right in Figure 6.28.

Our last application will demonstrate the power that the FILLET command has when used on 3D solids. The object on the left in Figure 6.29 is made of a box and a wedge that have been unioned into a single 3D solid object. We'll fillet the edges identified with numbers to create a complex blend between four edges.

Command: FILLET

Current settings: Mode = TRIM. Radius = 0.0000

Select first object or [Polyline/Radius/Trim]: *(Pick point 1.)*

Enter fillet radius: .25

Select an edge or [Chain/Radius]: *(Pick point 2.)*

Select an edge or [Chain/Radius]: *(Pick point 3.)*

Select an edge or [Chain/Radius]: (*Pick point 4.*)

Select an edge or [Chain/Radius/]: (*ENTER.*)

4 edge (s) selected for fillet.

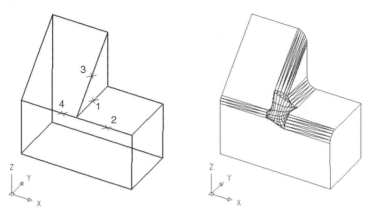

Figure 6.29

The results are shown on the right side of Figure 6.29. The blend is an extremely complex area that would be very difficult to make with 2D drafting techniques. This solid model, on the other hand, is easy to make and it can be used for numerical controlled tool path programming or sectioned to make 2D profiles for further analysis.

TRY IT! - USING FILLET

We will fillet the sharp bend corners of the slanted portion of the bracket we have been working on. We will give the inside edges a fillet radius of 0.125 and the outside edges a fillet radius of 0.375. The command line sequence of prompts and responses, in conjunction with the points shown on the left side of Figure 6.30, will be:

Command: FILLET

Current settings: Mode = TRIM. Radius = 0.0000

Select first object or [Polyline/Radius/Trim]: (*Pick point 1.*)

Enter fillet radius: .125

Select an edge or [Chain/Radius]: (*Pick point 2.*)

Select an edge or [Chain/Radius]: R

Enter fillet radius <0.125>: .375

Select an edge or [Chain/Radius]: (*Pick point 3.*)

Select an edge or [Chain/Radius]: (*Pick point 4.*)

Select an edge or [Chain/Radius]: (*ENTER.*)

4 edge (s) selected for fillet.

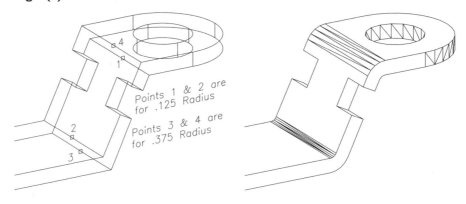

Points 1 & 2 are for .125 Radius

Points 3 & 4 are for .375 Radius

Figure 6.30

Notice that when the current fillet radius is 0, AutoCAD does not show the current setting during its prompt for a radius after a solid is selected. The resulting fillets are shown, with HIDE on, on the right side of Figure 6.30.

Compare your model with the one in file 3d_ch6_08.dwg on the accompanying CD-ROM.

THE CHAMFER COMMAND

The CHAMFER command makes beveled edges on 3D solids. While the FILLET command, when applied to 3D solids, is based solely on edges, CHAMFER is based on both edges and faces. This is because each edge of the bevel can be offset a different distance on each face. Consequently, the command asks you to identify a base surface. Chamfers are then made between the base surface and its adjoining faces (See Figure 6.31).

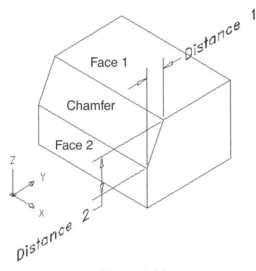

Figure 6.31

The command line format for CHAMFER is:

Command: CHAMFER

(TRIM mode) **Current chamfer Dist1 = current, Dist2 = current**

Select first Line or [Polyline/Distance/Angle/Trim/Method]: *(Select a 3D solid.)*

Base surface selection:

Enter surface selection option [Next/OK (current)] <OK>: *(Enter N or O or press ENTER.)*

The message about TRIM mode that appears when the command starts is of no consequence for 3D solids. As you will usually select the solid to be chamfered by picking an edge between two surfaces, you must specify which of the two surfaces is to be the base surface. AutoCAD will highlight the edges of one of the two surfaces, as shown in Figure 6.32. If this is the surface you want as the base surface, select the OK option. But, if you want the surface on the other side of the edge to be the base surface, select the Next option and AutoCAD will highlight its edges. You can use the Next option to alternate between these two adjacent surfaces until you select the OK option or press ENTER.

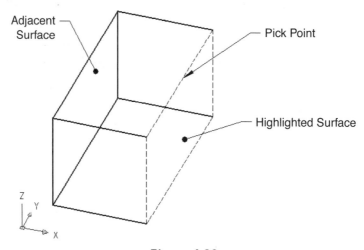

Figure 6.32

If the surface you want to be the base surface has mesh lines or isolines on it you can start the CHAMFER command by picking a mesh line or isoline. AutoCAD will accept that surface as the base surface and will skip the "Next/OK" prompt.

Once the base surface has been established, AutoCAD will prompt for the chamfer distances on the base surface and on the adjacent surfaces.

Enter base surface chamfer distance <current>: *(Specify a distance or press ENTER.)*

Enter other surface chamfer distance <current>: *(Specify a distance or press ENTER.)*

Pressing ENTER will accept the current chamfer distance. Both chamfer distances must be greater than zero. AutoCAD will then prompt for the edges to be chamfered.

Select edge or [Loop]: *(Enter L, select an edge, or press ENTER.)*

This prompt is repeated until ENTER is pressed, signaling an end to the command.

SELECT EDGE

This option selects individual edges. The edge selected must be on the base surface. AutoCAD will highlight the selected edges and chamfer them when ENTER is pressed.

LOOP

This option allows you to chamfer all edges of the base surface with one pick. The following prompt will be displayed:

Select an edge loop or [Edge]: (*Enter E, select an edge, or press* ENTER.)

- Select Edge Loop

- Select an edge on the base surface. All edges of the base surface will then be highlighted, as shown on the left in Figure 6.33, and the follow-up prompt will be redisplayed. Press ENTER to end the command and chamfer the highlighted surfaces, as shown on the right in Figure 6.33.

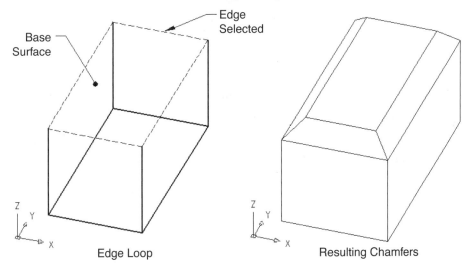

Figure 6.33

- Edge

- This option brings back the previous prompt.

TRY IT! - USING CHAMFER

We will finish the solid model we have been working on by chamfering the top corners of the left side of the bracket. These are the edges labeled with points 1 and 2 in Figure 6.34.

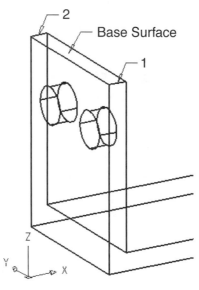

Figure 6.34

Command: CHAMFER

(TRIM mode) **Current chamfer Dist1 = 0.0000 Dist2 = 0.0000**

Select first Line or [Polyline/Distance/Angle/Trim/Method]: *(Pick point 1.)*

Base surface selection:

Enter surface selection method [Next/OK (current)] <OK>: *(The front surface will be highlighted, so press N .)*

Enter surface selection method [Next/OK (current)] <OK>: *(Now the top surface is highlighted, so press ENTER.)*

Specify base surface chamfer distance: .5

Specify other surface chamfer distance <0.5000>: *(ENTER.)*

Select an edge or [Loop]: *(Pick point 1 again.)*

Select an edge or [Loop]: *(Pick point 2.)*

Select an edge or [Loop]: *(ENTER.)*

The finished 3D solid model is shown in Figure 6.35 with hidden lines removed.

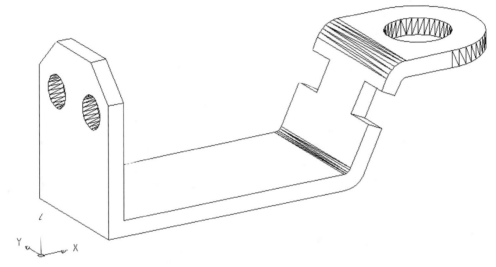

Figure 6.35

The completed model bracket is in file 3d_ch6_09.dwg on the accompanying CD-ROM. In Chapter 6 we will make a dimensioned, multiview, production drawing from this 3D solid.

THE SLICE COMMAND

This command cuts one or more solid objects into two pieces. You can keep both pieces or just the pieces on one side of the cut; and, when you elect to discard the pieces on one side of the cut, you can say which side you want saved. SLICE uses an infinitely large cutting plane. Therefore, you cannot slice through just part of an object, or have staggered sections through an object. See Figure 6.36 for a number of ways to choose this command.

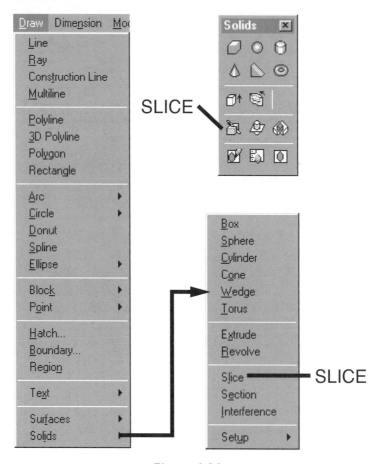

Figure 6.36

AutoCAD will not divide a sliced solid into more than one object on either side of the slicing plane, even if the sliced solid is split into several pieces. For example, if the U-shaped solid shown in Figure 6.37 is sliced so that there are two separate pieces left; the resulting two pieces comprise a single solid object, even though they are not joined.

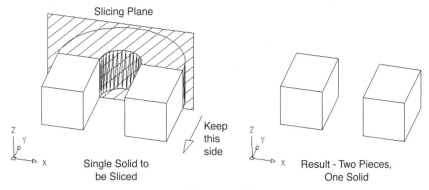

Figure 6.37

The SLICE command first asks you to select the objects to be sliced and then offers seven different options for defining the slicing plane. At the conclusion of each of those options AutoCAD allows you to keep both parts of the sliced objects or to point to the side of the slicing plane that you wish to retain. The command line format for the command is:

Command: SLICE

Select objects: (*Use any object selection method.*)

Specify first point on slicing plane by

[Object/Zaxis/View/XY/YZ/ZX/3points] <3points>: (*Specify an option, a point, or press ENTER.*)

3POINTS
With this option you define the slicing plane by specifying three points in space. Specifying a point from the main SLICE prompt initiates this option and AutoCAD will prompt you for the second and third points. If you press ENTER AutoCAD will prompt for all three points:

Specify first point on plane: (*Specify a point.*)

Specify second point on plane: (*Specify a point.*)

Specify third point on plane: (*Specify a point.*)

AutoCAD will anchor a rubberband line at the first point to help you locate the second and third points of the slicing plane.

OBJECT

This option uses an existing object to define the slicing plane. The object must be a 2D polyline, arc, circle, or planar spline. Lines, 3D polylines, and nonplanar splines are not accepted. The follow-up prompt is:

Select a circle, ellipse, arc, 2D-spline, or 2D-polyline: (*Select an object using any selection method.*)

ZAXIS

The Zaxis option for defining a slicing plane is similar to the Zaxis option of the UCS command for orienting the XY plane. Both rely on two points. The first point establishes the location of the plane and the second point controls the tilt of the plane. The plane is oriented so that it is perpendicular (normal) to a line from the first point to the second point. The follow-up prompts for the Zaxis options are:

Specify a point on the section plane: (*Specify a point.*)

Specify a point on the Z-axis (*normal*) of the plane: (*Specify a point.*)

VIEW

This option first asks for a point to establish the location of the slicing plane, and then it swivels the slicing plane so that it is perpendicular to the viewing direction.

Specify a point on the current view plane <0,0,0>: (*Specify a point.*)

XY

The XY option positions the slicing plane parallel to the current XY plane. A follow-up prompt asks for the elevation of the slicing plane.

Specify a point on the XY-plane <0,0,0>: (*Specify a point.*)

The prompt could be slightly misleading. It is asking for a point on the XY slicing plane, not on the coordinate system XY plane.

YZ

This option places the slicing plane parallel to the Y axis, and perpendicular to the current XY plane. AutoCAD will prompt you for a point to establish the plane's distance from the Y axis.

Specify a point on **YZ-plane <0,0,0>**: (*Specify a point.*)

ZX

The ZX option uses a slicing plane that is parallel to the X axis and perpendicular to the current XY plane. The follow-up prompt is for a point to set the slicing plane's distance from the X axis.

Specify a point on **ZX-plane <0,0,0>**: (*Specify a point.*)

After you have defined the slicing plane, AutoCAD will display a follow-up prompt for retaining or discarding the sliced solid's parts.

Specify a point on desired side of the plane or [keep Both sides]: (*Enter a B or specify a point.*)

- Point on desired side of the plane
- This point, which cannot be in the same plane as the slicing plane, shows AutoCAD which part of the sliced solid you want retained. The section on the other side of the slicing plane will be discarded.
- Both sides
- This option retains both sides of the sliced solid as two separate solid objects.

TRY IT! - SLICING A SOLID

In this exercise, you will use each of the seven options for defining a slicing plane to cut through the 3D solid in file 3d_ch6_10.dwg. After you open that file, make seven copies of the 3D solid to use in trying out each SLICE option. First, use the default 3points option in Figure 6.38.

Command: SLICE

Select objects: (*Select the solid.*)

Specify first point on slicing plane by [Object/Zaxis/View/XY/YZ/ZX/3points] <3points>: (*Pick point 1.*)

Specify second point on plane: (*Pick point 2.*)

Specify third point on plane: (*Pick point 3.*)

Specify a point on desired side of the plane or [keep Both sides]: (*Pick on the far side of the slicing plane.*)

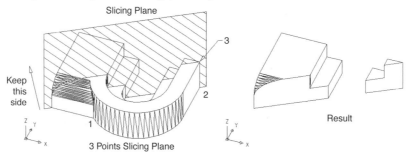

Figure 6.38

Object endpoint snaps were used to pick all three points, and the order in which the points are selected is not important. Even though the resulting solid is in two pieces, it is still a single object.

Next, use the Object option based on a circle that is off of the solid and perpendicular to the WCS to define the slicing plane (See Figure 6.39).

Command: SLICE

Select objects: (*Select the solid.*)

Specify first point on slicing plane by
    [Object/Zaxis/View/XY/YZ/ZX/3points] <3points>: O

Select a circle, ellipse, arc, 2D-spline, or 2D-polyline: (*Pick the circle.*)

Specify a point on desired side of the plane or [keep Both sides]: (*Pick on the far side of the slicing plane.*)

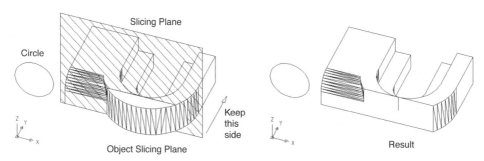

Figure 6.39

Notice that the resulting solid has an infinitely thin section at the end of the U-shaped cutout. AutoCAD has no problem with this, but manufacturing processes would.

Now, use the Zaxis option to define the slicing plane (See Figure 6.40).

Command: SLICE

Select objects: (*Select the solid.*)

Specify first point on slicing plane by [Object/Zaxis/View/XY/YZ/ZX/3points] <3points>: Z

Specify a point on the section plane: (*Pick point 1.*)

Specify a point on the Z-axis (*normal*) of the plane: (*Pick point 2.*)

Specify a point on desired side of the plane or [keep Both sides]: (*Pick on the far side of the slicing plane.*)

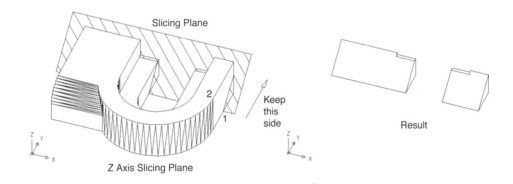

Figure 6.40

The first point was picked using an object midpoint snap and an object endpoint snap was used to pick the second point.

Next, use the View option to define the slicing plane. The viewing direction in Figure 6.41 is 290° in the XY plane from the X axis, and 40° from the XY plane.

Command: SLICE

Select objects: (*Select the solid.*)

Specify first point on slicing plane by [Object/Zaxis/View/XY/YZ/ZX/3points] <3points>: V

Specify a point on the current view plane <0,0,0>: *(Pick point 1.)*

Specify a point on desired side of the plane or [keep Both sides]: *(Pick on the far side of the slicing plane.)*

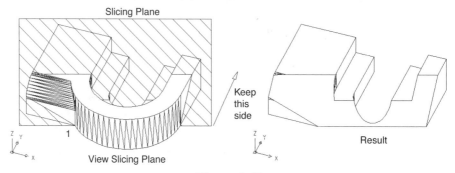

Figure 6.41

Use the XY option to cut through the 3D solid horizontally 0.375 units above the XY plane (See Figure 6.42).

Command: SLICE

Select objects: *(Select the solid.)*

Specify first point on slicing plane by [Object/Zaxis/View/XY/YZ/ZX/3points] <3points>: XY

Specify a point on the XY-plane <0,0,0>: 0,0,.375

Specify a point on desired side of the plane or [keep Both sides]: 0,0,0

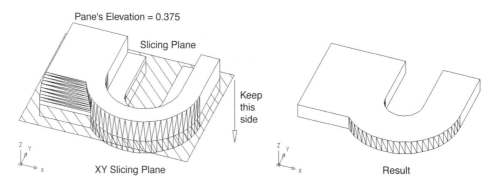

Figure 6.42

We typed in coordinates to designate which side of the slicing was to be retained, even though we could have pointed since the slicing plane was above the XY plane. However, if you use the XY plane as the slicing plane you will either have to use typed-in coordinates or an object snap.

Next, use the YZ option to cut through the solid vertically with a plane that is parallel to the Y axis (See Figure 6.43).

Command: SLICE

Select objects: (*Select the solid.*)

**Specify first point on slicing plane by
[Object/Zaxis/View/XY/YZ/ZX/3points] <3points>:YZ**

Point a point on the YZ-plane <0,0,0>: (*Pick point 1.*)

**Specify a point on desired side of the plane or [keep Both sides]:
(*Pick point on minus X side of slicing plane.*)**

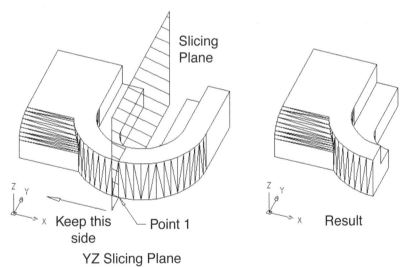

Figure 6.43

Finally we will use the ZX option to cut through the solid vertically with a slicing plane that is perpendicular to the XY plane and parallel to the X axis (See Figure 6.44).

Command: SLICE

Select objects: (*Select the solid.*)

**Specify first point on slicing plane by
[Object/Zaxis/View/XY/YZ/ZX/3points] <3points>: ZX**

Specify a point on the ZX-plane <0,0,0>: (*Pick point 1.*)

Specify a point on desired side of the plane or [keep Both sides]: (*Pick point on positive Y side of slicing plane.*)

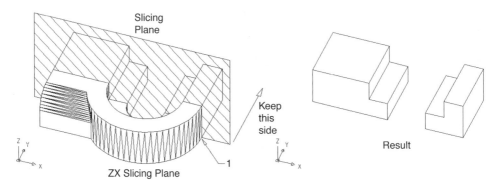

Slicing
Plane

Keep
this
side

ZX Slicing Plane

Result

Figure 6.44

EDITING 3D SOLIDS

AutoCAD 2002 includes tools that allow you to modify specific faces and edges of a 3D solid, and to hollow-out a 3D solid. Although all of these tools are options within a single command—SOLIDEDIT—you are likely to invoke the options directly from the menu and toolbar selections shown in Figure 6.45 as if they were separate commands, rather than as options of command line prompts.

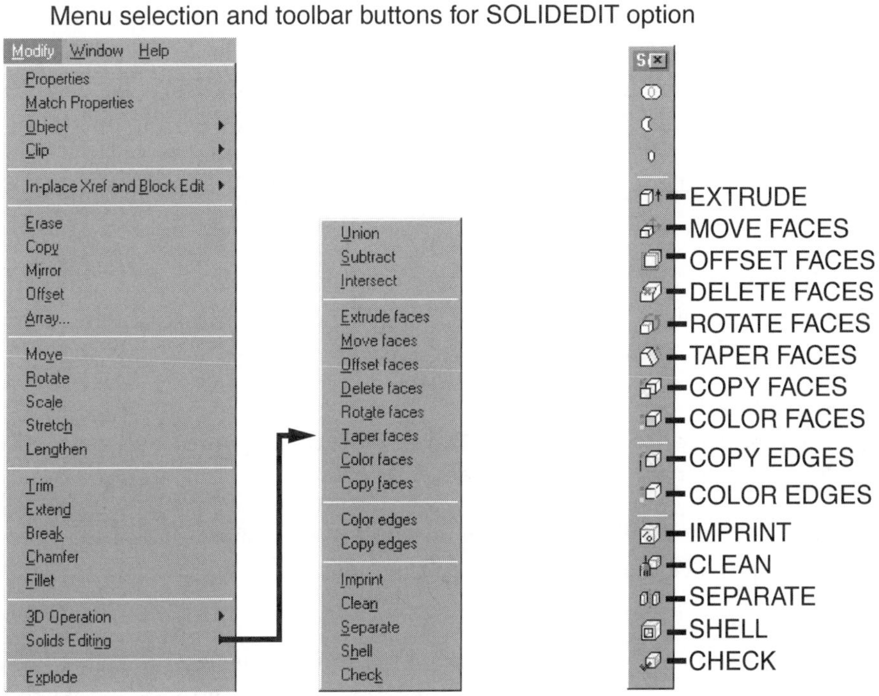

Menu selection and toolbar buttons for SOLIDEDIT option

Figure 6.45

THE SOLIDEDIT COMMAND

The options of SOLIDEDIT are divided into three categories: face, edge, and body. Using the terms we defined during the discussion of Spline Curve Basics in Chapter 3, a face is a surface area on a 3D solid that has either C0 or C1 continuity with adjacent faces. That is, it is either tangent to adjacent faces or it shares sharp edges with them. The options for editing edges operate on selected edges of faces on 3D solids, and the body editing options operate on the entire selected 3D solid. The name of this third category of editing options can be misleading because it has nothing to do with AutoCAD body object types. (Body objects are nonplanar surface objects that are created from nonplanar surfaces on a 3D solid when it is exploded.)

When SOLIDEDIT is invoked from the command line, the following message and prompt will be displayed:

Command: SOLIDEDIT

Solids editing automatic checking: SOLIDCHECK=1

Enter a solids editing option [Face/Edge/Body/Undo/eXit] <eXit>:
   (Select an option or press ENTER.)

The message referring to automatic checking shows you the current setting of the Solidcheck system variable. When Solidcheck is set to 1, the 3D solid that is being edited is automatically checked for internal errors after each editing operation. The prompt to select an editing option is redisplayed after each editing operation is completed, until you select the eXit option or press ENTER to end the command. The Undo option reverses the most recent editing operation.

FACE EDITING OPTIONS

When you select Face from the SOLIDEDIT command line prompt, a prompt listing options for editing faces will be displayed.

Enter a face editing option
   [Extrude/Move/Rotate/Offset/Taper/Delete/Copy/coLor/Undo/eXit]
   <eXit>: *(Select an option or press ENTER.)*

Press ENTER or select the eXit option to return to the main SOLIDEDIT prompt. The Undo option cancels the last face editing operation.

Face Selection

All of the face editing options start by asking you to select the faces that are to be edited, and they display the same command line prompts for you to use in selecting the faces. The first prompt is:

Select faces or [Undo/Remove]: *(Select a face.)*

The two options have no effect if a face has not been selected, so you will always select a face in response to this first prompt. The first face selected determines which 3D solid is to be edited, and AutoCAD ignores face selections on other 3D solids. Once a face has been selected AutoCAD will display the prompt:

Select faces or [Undo/Remove/ALL]: *(Select a face, enter an option, or press ENTER.)*

This prompt is repeated after each face has been selected, until you press ENTER to end the face selection process. Faces are selected individually by picking a point on an edge, isoline, or surface. You cannot use window or crossing methods to select a face. Picking points on surfaces works even when faces are invisible, as they are in a wireframe viewing mode. When a face is selected its edges will be highlighted.

When faces are stacked on top of other faces, the first pick point on a surface will select the foreground face, and subsequent picks on the same point will sequentially select the other faces in the stack, as shown in Figure 6.46.

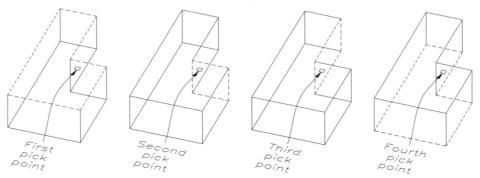

First Pick Point   Second Pick Point   Third Pick Point   Fourth Pick Point

Figure 6.46

The ALL option selects all of the faces of a 3D solid, and the Undo option cancels the last face selection. The Remove option is for deselecting faces. It displays the prompt:

Remove faces or [Undo/Add/ALL]: *(Select a face, enter an option, or press ENTER.)*

Specifying faces to be deselected works the same as when selecting faces. Undo cancels the last selection, ALL deselects all faces on the 3D solid, Add returns to the Select faces prompt, and pressing ENTER ends the face selection process.

Extrude

This option uses the edges of the selected face as a profile for an extrusion. The face must be planar. The prompts, options, and rules for extruding a face are the same as those of the EXTRUDE command. Positive extrusion distance values extrude the face away from the existing 3D solid, and negative distance values extrude the face into the 3D solid (See Figure 6.47).

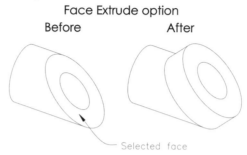

Figure 6.47

Move

Use this option to move the selected face or faces. Prompts similar to those of the MOVE command will ask you to specify base and destination points for the move. If necessary, faces adjacent to the moved face will stretch or contract to accommodate the move, as shown in Figure 6.48. Faces cannot be moved to locations that require them to become adjacent to another face. For example, although you can move the entire round hole and keyway shown in Figure 6.48 anywhere on the face it located, you cannot move it to the other horizontal face on the 3D solid.

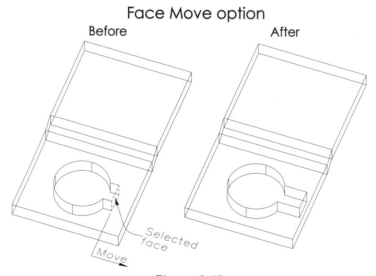

Figure 6.48

Figure 6.49 demonstrates how moving faces can change the thickness of a solid, even when several faces on different planes are involved.

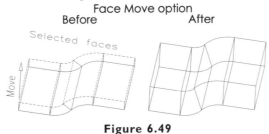

Figure 6.49

Rotate

This option rotates the selected face or faces. A command line prompt will offer you options in defining the rotation axis.

Specify an axis point or [Axis by object/View/Xaxis/Yaxis/Zaxis] <2points>: (*Enter an option, specify a point, or press ENTER.*)

- 2points
- When you specify a point, the point will define one end of the rotation axis, and you will be prompted to select a point to serve as the other end of the axis. If you select this option by pressing ENTER, command line prompts will be issued for you to specify the first and second axis endpoints.

- Axis by object
- This option uses an existing wireframe object to define the rotation axis. You will be prompted to select a curve to be used as the rotation axis. Despite the use of the word curve in the prompt, you can also select a line or a poly-line that has straight segments. The rotation axis will extend between the endpoints of these objects. If you select a circle, arc, or ellipse, the rotation axis will pass through the center of the object and will be perpendicular to the plane of the object.

- View
- This option defines a rotation axis that is parallel to the line-of-sight of the current viewport. You will be prompted to select a point that the axis is to pass through.

- Xaxis, Yaxis, Zaxis
- These options establish a rotation axis that is parallel with the X, Y, or Z axis of the current UCS. You will be prompted to select a point that the axis will pass through.

After you have defined the axis of revolution, you will be prompted to specify the rotation angle. Similar to the options for specifying an angle in the ROTATE command, you can entering an absolute angle value or specify a reference angle by picking three points.

The selected faces on the 3D solid in Figure 6.50 show how the Zaxis option for selecting a rotation axis works.

Offset

Offset moves a face perpendicularly a specified distance from its current position. For planar faces the results are similar to the Move and Extrude options, but curved faces will become larger or smaller, depending on the offset direction. After selecting the face or faces to be offset, a command line prompt will ask you to specify the offset distance. Positive distance values move faces away from the 3D solid, and negative values move faces into the 3D solid.

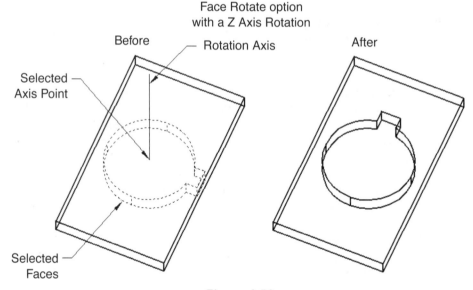

Figure 6.50

Figure 6.51 shows how the diameter of a round hole with a keyway slot is decreased by offsetting the face of the round hole and the end of the keyway, while leaving the sides of the keyway unchanged.

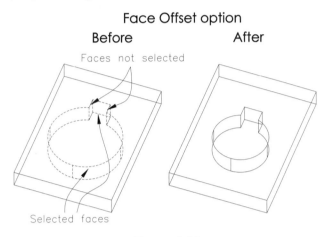

Figure 6.51

Taper

This option tilts, or inclines, selected faces. After selecting the faces to be tapered, you will be prompted from the command line to specify two points.

Specify the base point: (*Specify a point.*)

Specify another point along the axis of tapering: (*Specify a point.*)

The first point will be the swivel point for tapering the face(s), and the taper angle will be measured from a line drawn between the base point and the second point. Despite the wording of the second prompt, the second point does not define a rotation axis. Lastly, you will be prompted to specify the taper angle. An example of the face taper option is shown in Figure 6.52.

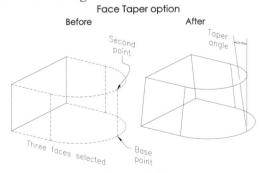

Figure 6.52

Delete

Use this option to remove faces you no longer want your 3D model to have. Once the faces have been selected, they will be deleted. No additional prompts are issued. In Figure 6.53 a round hole and a fillet have been selected and deleted from a 3D solid.

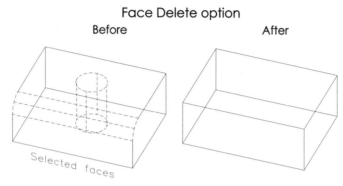

Face Delete option

Before   After

Selected faces

Figure 6.53

Copy

This option makes copies of selected faces. It issues prompts similar to those of the COPY command to place the copies by using either base and destination points, or a displacement vector. Only faces are copied—not 3D solid objects. Therefore, you cannot copy a round hole or a slot. You can only copy their side surfaces. The object type of a copied planar face will be a region, and the object type of a copied nonplanar face will be a body.

coLor

With this option you can assign an individual color to the face of a 3D solid. AutoCAD's Select Color dialog box will be displayed for you to use in choosing a color.

EDGE EDITING OPTIONS

When you select Edge from the main SOLIDEDIT prompt, AutoCAD will display the prompt:

Enter an edge editing option [Copy/coLor/Undo/eXit] <eXit>: (*Enter an option or press ENTER.*)

This prompt is redisplayed after each editing operation, until you press ENTER or select eXit to return to the main SOLIDEDIT prompt. Undo reverses the last edge-editing operation.

Copy

This option makes a wireframe copy of selected edges. You will be prompted to select the edges that are to be copied, and then, from prompts similar to those of the

MOVE command, specify the location of the copies by supplying base and destination points or a direction vector. Copied edges will be lines, circles, arcs, ellipses, or splines, depending on the shape of the 3D solid edge.

coLor
This option changes the color of selected edges. After you have selected the edges, AutoCAD will display the Select Color dialog box for you to use in specifying a color.

BODY EDITING OPTIONS
When you select Body from the main SOLIDEDIT prompt, AutoCAD will display a prompt that has options that affect the entire 3D solid.

Enter a body editing option [Imprint/seParate solids/Shell/cLean/Check/Undo/eXit] <eXit>: (*Enter an option or press ENTER.*)

When you complete a body-editing operation, this prompt will be redisplayed. Press ENTER or choose the eXit option to return to the main SOLIDEDIT prompt. The Undo option cancels the most recent body-editing operation.

Imprint
The Imprint option creates an edge on the surface of a 3D solid. Typically, you will use it to create a face that you will then modify through a face-editing operation, such as extrude or taper. The Imprint option issues four command line prompts.

Select a 3D solid: (Select the 3D solid that the imprint is to be on.)

Select an object to imprint: (*Select an object to imprint on the 3D solid.*)

Delete the source object <N>: (*Enter a N or a Y, or press ENTER.*)

Select an object to imprint: (*Select an object to imprint on the 3D solid, or press ENTER.*)

Acceptable objects to be imprinted are arcs, circles, lines, 2D and 3D polylines, ellipses, splines, regions, bodies, and 3D solids. Imprinting will occur at the intersection of the imprinting object and the selected 3D solid. Although you can select only one object to imprint at a time, the "Select an object to imprint" prompt is repeated until you press ENTER to end the command. The "Delete the source object" prompt refers to the object to imprint; not the 3D solid that receives the imprint.

Figure 6.54 shows an example of imprinting and how it can be used. The 3D solid to be imprinted and the object to imprint are shown on the left in this figure. The resulting imprints are shown in the middle of the figure. The object to imprint has been deleted. On the right side of this figure, the newly created arc-shaped face has been tapered with the Face Taper option to demonstrate that a separate face was created.

Body Imprint option

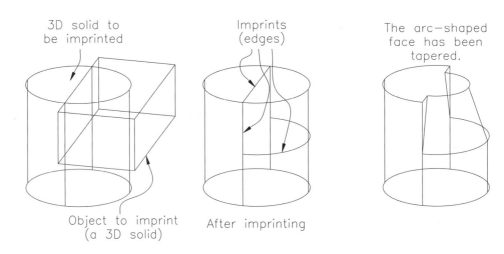

3D solid to be imprinted

Imprints (edges)

The arc-shaped face has been tapered.

Object to imprint (a 3D solid)

After imprinting

Figure 6.54

SeParate Solids
This option applies only to 3D solids that have components with empty space

between them. They have been created by unioning 3D solids that do not touch. The option will issue a prompt to select the 3D solid that is to be separated. Although the appearance of the separated 3D solids will not change, each separated component will be an individual 3D solid. This option cannot decompose a 3D solid to restore its primitives.

Shell
You can hollow out a 3D solid with this option. It works by offsetting the faces of the selected 3D solid, and deleting the solid's volume that is not between the original and offset faces. You can also create an opening in the shelled 3D solid by excluding faces from the operation. This option can be implemented only once for a particular 3D solid. The option's command line prompts are:

A positive offset distance offsets the faces into the 3D solid, and a negative distance offsets the faces away from the 3D solid. Initially, all faces on the selected 3D solid will be selected to be shelled, so you must remove the faces that you want to be open.

Selecting faces to be removed works the same as in the Face editing options. An example of a shelled 3D solid is shown in Figure 6.55.

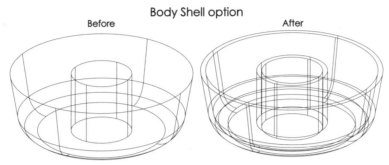

Figure 6.55

Clean

This option removes redundant and duplicate edges and vertices and unused imprints on the surface of a selected 3D solid. You will be prompted to select one 3D solid to be cleaned.

Check

This option checks for internal errors in a selected 3D solid, and issues a report that the solid is a valid ACIS solid if no errors are found. You will be prompted to select one 3D solid to be checked. When the system variable Solidcheck is set to 1, validation checks are automatically performed after each SOLIDEDIT option is completed.

 Tip: Selecting faces will present the most difficulties when you use the Face editing options of SOLIDEDIT. Selecting the faces you want to edit will be easier if you use multiple viewports and set the view directions in these viewports so that you have at least one unobstructed view of each face you intend to edit.

CONTROLLING THE APPEARANCE OF SOLID MODELS

We have been using the HIDE command on 3D solids without explanation or comment, and you have undoubtedly noticed that whenever this command is invoked, solid objects assume a different form.

Like most solid modeling programs, AutoCAD has two distinct display forms for solid models—a wireframe form and a polymesh form. Most work is done within the wireframe form, which shows just the edges of the model, along with some lines between the edges to indicate the surface of rounded and curved areas.

In their polymesh form, on the other hand, solid objects have a surface that can hide objects and reflect light. Some solid modeling programs, including the Advanced Modeling Extension (AME) that AutoCAD's built-in solid modeler replaced, require special commands to change from one form to another. But AutoCAD does this automatically. Whenever the HIDE and RENDER commands are initiated, AutoCAD automatically transforms solids into their polymesh form. The first screen regeneration returns them to their wireframe form.

You have some control over the appearance of solids, both in their wireframe and polymesh form, through four system variables—Isolines, Dispsilh, Facetres, and Facetratio. These system variables do not affect the hidden line viewing mode of the SHADEMODE command. The hidden line viewing mode of SHADEMODE always shows 3D solids in a smooth, non-polymesh form. It does not, however, do a good job in displaying profile edges of curved and rounded 3D solids, so you may prefer to use the HIDE command rather than SHADEMODE's hidden line view mode when you work with 3D solids. These system variables also control the appearance of 3D solids when they are plotted.

THE ISOLINES SYSTEM VARIABLE

AutoCAD always shows edges and changes of curvature on solids in their wireframe form as lines or as curves (depending on the object's shape). Additional lines and curves are used to define the surface of rounded portions of the solid between edges, and the quantity of these lines and curves is controlled by the value in Isolines.

Isolines, which takes an integer number ranging from 0 to 2,047, sets the lines per 360° drawn along the length of a circular solid surface. For example, when Isolines is set to its default value of 4, a solid cylinder will be displayed with four parallel lines, spaced 90° apart, running along its side surface, as shown on the left side of Figure 6.56. When Isolines is set to 16, there will be 16 lines on the cylinder, as shown in the center of Figure 6.56. If Isolines were set to 0, there would be no lines on the side surface of the cylinder—only the round ends would be shown, as in the right-hand cylinder.

Cylinder 3D Solid, Wireframe Form

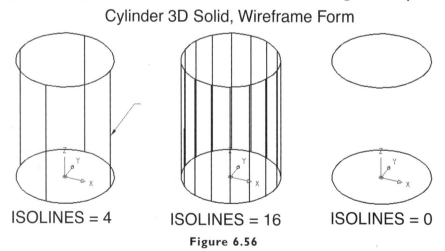

ISOLINES = 4   ISOLINES = 16   ISOLINES = 0

Figure 6.56

Only curved surfaces are affected by Isolines. Isolines are never drawn across planar surfaces, and the system variable only pertains to wireframe views—having no effect on the number of mesh lines and faces on solids when the HIDE command is invoked.

When curved surfaces on solid objects are difficult to visualize, increasing the value of Isolines, thus increasing the number of wireframe lines on the solid, can help. The current setting of Isolines, which requires a screen regeneration to take effect, applies to all solids. It is not possible for some solids to have a different Isolines setting than others.

THE DISPSILH SYSTEM VARIABLE

Dispsilh, which stands for display silhouette, also affects the display of curved and rounded surfaces on solids in their wireframe form. It is a toggle-type system variable (meaning it has only two settings, similar to an on-off switch). When Dispsilh is set to 1, AutoCAD will display lines showing the solid's profile (its silhouette) in addition to the Isolines. When Dispsilh is set to 0, its default setting, profile lines are not shown. Figures 6.57 and 6.58 show the same solid object as it appears when Dispsilh is set to 0, and to 1.

Figure 6.57

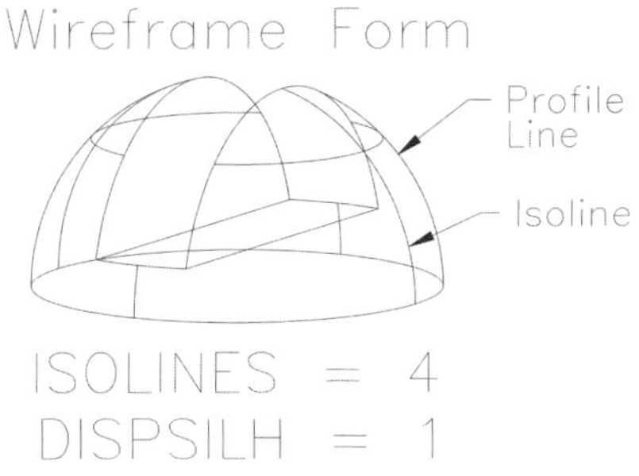

Figure 6.58

Dispsilh also affects solids when the HIDE command is invoked. When it is set to 0, AutoCAD shows curved surfaces on solids as triangular and rectangular faces with the HIDE command. Conversely, when Dispsilh is set to 1, curved surfaces are not shown as faces, although they still hide objects that are behind them. Only the silhouette of curved surfaces along with edges will be shown, resulting in a clean, uncluttered image. The same solid object is shown in Figures 6.59 and 6.60 as it appears during the HIDE command when Dispsilh is set to 0 and to 1.

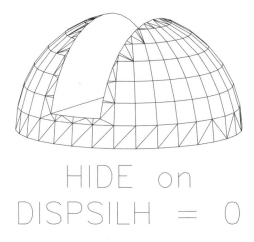

HIDE on
DISPSILH = 0

Figure 6.59

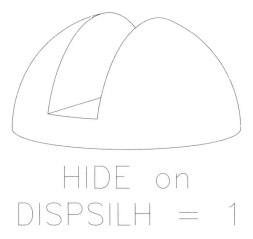

HIDE on
DISPSILH = 1

Figure 6.60

THE FACETRES SYSTEM VARIABLE

A third system variable affecting the appearance of 3D solids is Facetres, which controls the size of faces when solids are in their polygon mesh form. Facetres takes a value ranging from 0.01 to 10, with a default of 0.50. Larger values make smaller faces, and thus smoother rounded surfaces, but require more calculations by the computer, which can significantly increase the amount of time needed to display the surfaces. Too small a value, on the other hand, can cause noticeably faceted surfaces, even to the extent that round holes and cylinders appear polygon-shaped. Figure 6.61 illustrates the affect of Facetres on a solid cylinder. While values of Facetres are given in this figure, the actual effect of this system variable depends on the relative size and degree of curvature of the solid's curves.

Cylinder 3D Solid, Polymesh Form

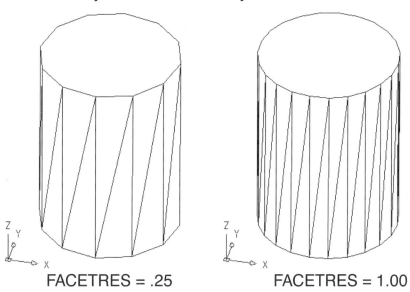

FACETRES = .25   FACETRES = 1.00

Figure 6.61

Even though no faces are shown when HIDE is in effect while Dispsilh is set to 1, a low Facetres value can make rounded edges appear as segmented lines rather than smooth curves. Of course, these faces and segmented lines are just visual devices and do not change the actual shape of the 3D model.

THE FACETRATIO SYSTEM VARIABLE

This system variable controls the appearance of cylindrical and conical 3D solids during SHADE, and during HIDE when Dispsilh is set to 0. When Facetratio is set to 0, 3D solid cylinders and cones are divided into faces around their circumference, but not along their length, during HIDE and SHADE. When Facetratio is set to 1, 3D solid cylinders and cones are also divided into faces lengthwise (See Figure 6.62). The number of lengthwise face divisions depends on the cylinder's or cone's length-to-diameter ratio, but there will always be at least two rows of faces. You are not likely to ever need to change the value of Facetratio from its default value of 0.

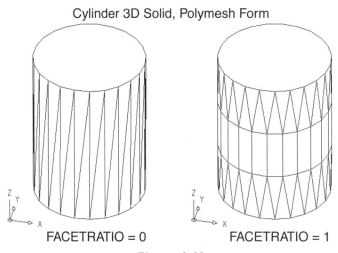

Cylinder 3D Solid, Polymesh Form

FACETRATIO = 0   FACETRATIO = 1

Figure 6.62

COMMAND REVIEW

UNION

This command combines a set of solids into a single solid.

SUBTRACT

This Boolean operation removes the intersecting volume of one set of solids from another set of solids.

INTERSECT

This command creates a new solid from the overlapping volume of two or more solids.

SLICE

This command cuts one or more solid objects into two pieces.

SOLIDEDIT

This command is used to edit faces, edges, and bodies of 3D solid objects.

SYSTEM VARIABLE REVIEW

SOLIDCHECK

When this system variable is set to a value of 1, validation and error checks are automatically performed on a 3D solid whenever a **SOLIDEDIT** editing operation is completed. If Solidcheck is set to 0, automatic validation and error checks are not performed.

ISOLINES

This system variable controls the display lines that show edges and changes of curvature on solid models in their wireframe form.

DISPSILH

This system variable is an on/off toggle that displays curved and rounded surfaces on a solid model in their wireframe form.

FACETRES

This system variable controls the size of faces when solid models are in their polygon mesh form.

FACETRATIO

This system variable controls the appearance of cylindrical and conical 3D solids during the use of the **SHADE** command and during the **HIDE** command when Dispsilh is set to 0.

CHAPTER PROBLEMS

Use the tools you have acquired from this chapter to build a solid model in each of the following exercises. You will mostly be on your own as you build these models. We will show you each model and give its dimensions, but we will just offer some suggestions for building it, rather than describing the construction steps in detail. In fact, there will be several different ways to build each model, and the methods you use will depend largely on your style of working. You should also feel free to modify and add features to these models.

Problem 6-1

First, construct the 3D model shown in an isometric view in Figure 6.63. The cylindrical geometries in this model can be built by revolving their profiles, combining cylinders, or extruding profile objects. The hexagon-shaped head part can be extruded from an AutoCAD polygon. Notice that the back side of this hexagon is slightly rounded, as shown in the detail.

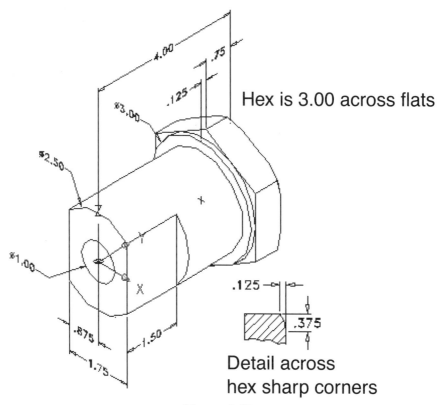

Hex is 3.00 across flats

Detail across
hex sharp corners

Figure 6.63

 On the CD-ROM that comes with this book, the model is in file 3833d_ch6_11.dwg.

Problem 6-2

Figure 6.64 shows an isometric-type view of the model for the next chapter problem. Basically, it consists of two cylinders, each of which have a tapered section and an interior hole, connected by a curved rod and two web-like stiffeners.

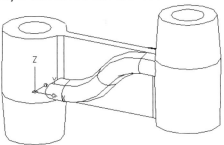

Figure 6.64

The dimensions you need to construct this model are given in the 2D drawing shown in Figure 6.65. (This drawing was made from the solid model using techniques that will be explained in Chapter 8.) The cylindrical objects in the model can be made by combining cylinders or by revolving their profiles. Once you make one, you can copy it and rotate the copy 180° to make the opposite end. The easiest way to make the connecting rod is to extrude it along the curved centerline path shown in the figure below. The stiffeners are easily made from a box-shaped solid.

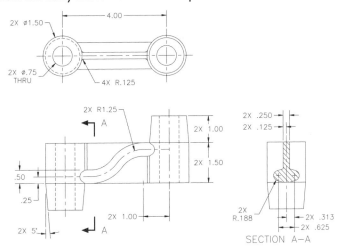

Figure 6.65

 You can compare your completed model with the one in file 3d_ch6_12.dwg on the CD-ROM that accompanies this book.

Problem 6-3

In the next solid modeling chapter problem you will create the folded sheet metal part shown in the Figure 6.66. This model is fairly straightforward to build, although you will have to move the UCS to various positions and orientations. Drawing profiles of the part sections and then extruding them works well. No dimensions are given (in an attempt to eliminate some clutter in the figure) for locating the round holes. Center each of them on the flange they are in. Notice that one hole extends down through the base of the part. The completed model is in file 3d_ch6_13.dwg on the CD-ROM that comes with this book.

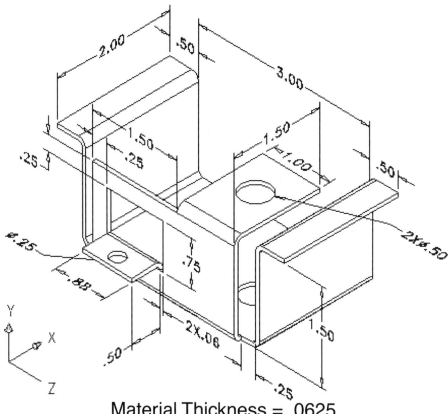

Material Thickness = .0625
Inside Bend Radius = .0625

Figure 6.66

Problem 6-4

Figure 6.67 shows an isometric view of the model for the next exercise. The model can be made by revolving a profile of its basic shape about its center axis and then adding the offset protrusion that has a hole through it. That protrusion can be made by revolving its cross-section profile or by using an extrusion that has a -5° draft angle. Revolving the profile of the part's interior section and subtracting it from the main part during a late step in the model's construction will ensure that everything is cleaned up.

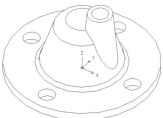

Figure 6.67

The dimensions you need to construct this model are given in Figure 6.68, which shows a cross-section of the part. Notice that the thickness of the revolved section is 0.25 units throughout. The completed model is in file 3d_ch6_14.dwg on this book's CD-ROM. After you finish your model, you might want to experiment with it by filleting some of the sharp edges and corners. You can even fillet the intersection of the cone-shaped protrusion with the main part.

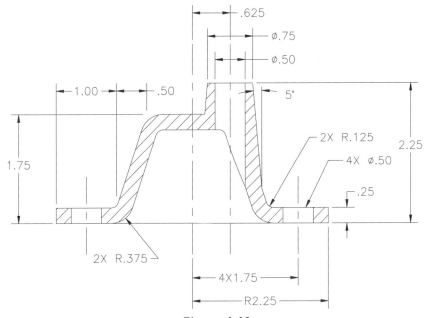

Figure 6.68

Problem 6-5

The model for the last chapter problem is shown in an isometric-type view in Figure 6.69. Though the part contains numerous components, none of them are complicated and can all be made from cylinders and extrusions. You should save the holes for a late step in the model's construction to ensure that no other part of the model extends into their space.

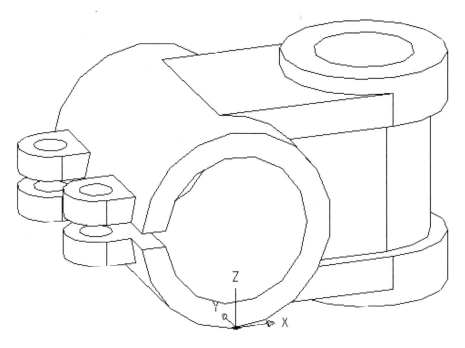

Figure 6.69

The two orthographic views in Figure 6.70 show the dimensions you need in constructing the model.

The completed solid model is in file 3d_ch6_15.dwg on the accompanying CD-ROM. Although we did not include any chamfers or fillets on our model; you may want to add some to yours.

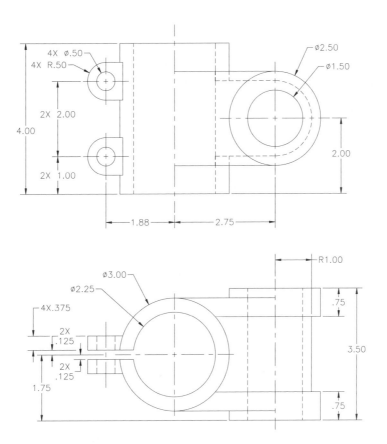

Figure 6.70

CHAPTER REVIEW

Directions: Answer the following questions.

1. List the three Boolean operations and name the two AutoCAD object types that are affected by the Boolean operations.

2. What happens when an object is subtracted from an object it does not touch?

3. What happens when an object is subtracted from an object it completely encloses?

4. How does the INTERSECT command differ from the INTERFERE command?

5. What does the Chain option of the FILLET command for 3D solids do?

6. What are the differences between the SLICE command and the SECTION command?

7. Which of the following relative positions of two 3D solids are allowed by the UNION command?

    a. The two solids are separated in space.

    b. The two solids exactly touch.

    c. The two solids overlap.

8. When you select the edge of a 3D solid as an object for the FILLET and CHAMFER commands, AutoCAD recognizes the object type and presents command line menus appropriate for filleting or chamfering 3D solids.

    a. true

    b. false

9. The Body Separate option of the SOLIDEDIT command restores the primitive components of a 3D solid.

    a. true

    b. false

10. The Face Copy option of the SOLIDEDIT command can make copies of holes and slots that are in 3D solids.

    a. true

    b. false

Directions: Match the following items as indicated.

11.   Match a system variable from the list on the left with a function or result from the list on the right.

_____ a. Dispsilh   1.   Controls the number of lines drawn across rounded and curved surfaces in the wireframe mode.

_____ b. Facetratio   2.   Controls the size of facets on curved and rounded surfaces on solids during HIDE.

_____ c. Facetres   3.   Divides facets on cone and cylinder shaped solids along their length during HIDE.

_____ d. Isolines   4.   Turns the faceted appearance of curved and rounded surfaces

Analyzing Solid Models

LEARNING OBJECTIVES

This chapter will introduce you to the methods used for analyzing solid models. An introduction to Mechanical Desktop and Autodesk Inventor is also provided in this chapter. When you have completed Chapter 7, you will:

- Understand the geometric and mass properties of 3D solids, and know how to check for interference between mating 3D solids.

- Be able to identify the advantages of using Mechanical Desktop to produce, surface, part, drawing and assembly models.

- Identify the advantages of using Inventor to produce part, drawing, assembly, and presentation models.

ANALYSIS OF SOLIDS

One of the advantages of solid models over surface models and wireframes is that you have access to data related to the model's interior. AutoCAD can give you mass property information about the model, such as its volume, center of gravity, and moments of inertia. The AREA command recognizes solid objects and returns their surface area. AutoCAD can also make sections through the model that can be analyzed separately, and it can check solids that are to fit together to ensure there is no interference. The menus and toolbars for invoking the commands for analyzing 3D solids are shown in Figure 7.1.

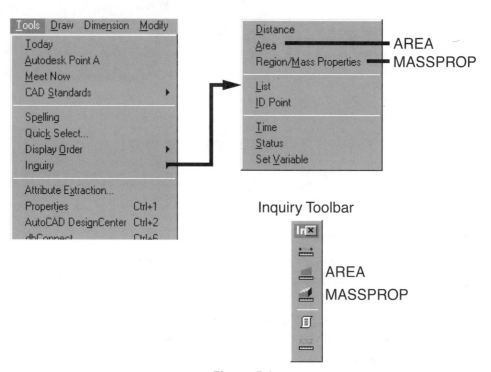

Figure 7.1

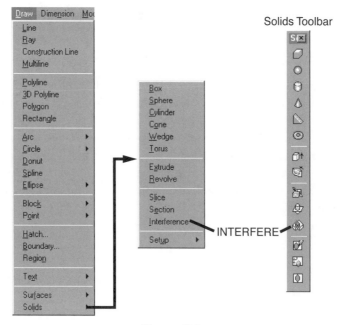

Figure 7.2

THE MASSPROP COMMAND

This command reports the mass properties of regions and solids. As you would expect, the reports for regions are slightly different from those for solids because regions are 2D objects while solids are 3D objects. The data for regions is in terms of area, while the data for solids is in terms of volume and mass. AutoCAD does not use a specific measurement system when reporting mass property data. It is simply in terms of units, and it is up to the user to decide whether they represent inches and pounds, or centimeters and grams. AutoCAD uses the current UCS for its mass property calculations. The command line format for MASSPROP is:

Command: MASSPROP

Select objects: (*Use any object selection method.*)

You can select as many objects as you choose. AutoCAD will ignore objects that are not regions or solids. If more than one solid is selected, AutoCAD will issue a report as if they were a single, combined solid. AutoCAD will also combine regions if more than one is selected, and they are coplanar. If they are not coplanar, AutoCAD will only report on those coplanar with the first object selected. If a mixture of regions and solids is selected, AutoCAD will issue separate reports on both types of objects.

After listing the mass property data, AutoCAD will ask if you want it written to a file. The file output will be in ASCII text format, with a default file name the same as that of the current drawing file and an extension of .MPR.

SOLIDS

AutoCAD reports on the following mass properties of 3D solids.

- ° Mass
  - Mass is one of the three fundamental physical dimensions (length and time are the other two). Even though there is a subtle difference between mass and weight, you can consider them to be equivalent in most cases, and weighing an object is a convenient way to determine mass. Because AutoCAD cannot weigh objects, it must determine mass by multiplying the object's density (its weight per unit volume) by its volume. However, AutoCAD always uses a density of one, and consequently will always report an object's mass to be the same as its volume.
  - Volume
  - This is the amount of space under the surface of the solid.

- Bounding Box

- These are the X, Y, and Z coordinates of the smallest box that the object could fit in. The sides of this bounding box are always parallel to the X, Y, and Z axes (see Figure 7.3).

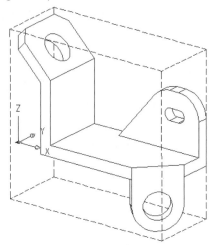

Figure 7.3

- Centroid

- This is the object's center of gravity—the point at which the object's entire mass appears to be concentrated.

- Moments of Inertia

- This is a measurement of the amount of force required to rotate the object about an axis. Consequently, there is a separate moment of inertia about each of the three principal axes of the UCS. The units of moments of inertia are mass times length squared.

- Products of Inertia

- Although similar to moments of inertia, products of inertia are relative to the principal planes rather than to the principal axes. It is expressed in terms of mass times distance to two perpendicular principal planes.

- Radius of Gyration

- This is an alternate method of expressing moments of inertia, which is sometimes applied in the design of rotating parts. It is in units of length only.

- Principal Moments and X, Y, Z Directions about Centroid

- If the coordinate system origin is moved to the solid's centroid, it can be twisted into an orientation resulting in a maximum moment of inertia about one principal axis and a minimum moment of inertia about a second principal axis. The moment of inertia about the third principal axis will have a value somewhere between the maximum and minimum. These three moments of inertia are referred to as the principal moments of inertia. In AutoCAD's mass property report they are designated by the letters I, J, and K, with the orientation of each corresponding axis shown as a unit vector.

- The principal axes are important in the design of rotating equipment, since parts rotating about a principal axis will draw minimum power during speed changes. A rotating device, such as a shaft or propeller, is said to be dynamically balanced when its center of mass lies on the axis of rotation and it is rotating about a principal axis.

REGIONS

AutoCAD reports on all of the following properties of regions when the region is on the current XY plane. If the selected region is not on the XY plane, only the area, perimeter, bounding box, and centroid are reported.

- Area

- This is the surface area of the region.

- Perimeter

- This is the length around the region. The perimeter of interior holes in the region are added to its outside perimeter.

- Bounding Box

- This is the coordinates of the smallest box that the region will fit into. The sides of the box will always be parallel to the principal axes. If the region is on the XY plane the box will become a 2D rectangle.

- Centroid

- The centroid is a point representing the geometric center of the region. Some regions, such as those shaped like the letter L, will have their centroid outside the boundary of the region (see Figure 7.4).

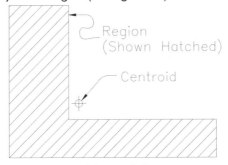

Figure 7.4

- Moments of Inertia

- This is a property often used in stress and deformation calculations. Every region will have two moments of inertia—one relative to the X axis and one relative to the Y axis. Their units are length raised to the fourth power. Since no mass is involved when the object is planar, some prefer to call this property the second moment of area, rather than moment of inertia.

- Products of Inertia

- This property is similar to moments of inertia, but it is relative to both the X and Y axes. Consequently, a region has only one product of inertia.

- Radii of Gyration

- A region's radius of gyration is the distance from a reference axis at which the entire area could be considered as being concentrated without changing its moment of inertia. Every region will have a radius of gyration about the X axis and another about the Y axis, both expressed in units of length. This property is sometimes used in calculations of rotating objects.

- Principal Moments and XY Direction about Centroid

- When the origin of the coordinate system is anchored on the centroid of a plane area, the X and Y axes can be rotated around the Z axis to an angle such that the area's maximum moment of inertia is about one axis, while its minimum moment of inertia is about the other axis. These two moments of inertia are called the principal moments of inertia, and the axes about which they are taken are called the principal axes of inertia. Product of inertia for the plane always has a value of zero relative to the principal axes of inertia.

- AutoCAD's mass property report uses the letters I and J as labels for the principal moments of inertia and indicates the orientation of the principal axes with two sets of X,Y coordinate pairs based upon the current UCS origin.

Tips: Most of us are unlikely to need any of the mass property information on solids other than volume, centroid, and bounding box. Mass moments of inertia data is required mostly in complicated calculations dealing with motions of objects and the forces causing those motions.

Moments of inertia on regions, on the other hand, are often useful in calculating stress and deflection in static objects. Those calculations are beyond the scope of this book, but they are not especially difficult, and the necessary equations are found and explained in many engineering and machinist handbooks.

Although AutoCAD always uses a density of 1 in computing the mass (weight) of solids, you can estimate the weight of your object by multiplying its volume by the density of the material you intend to use. The approximate density of some common metals are:

Material	Kilograms/Cubic Meter	Pounds/Cubic Inch
Aluminum	2,710	0.10
Brass, soft yellow	8,470	0.30
Copper	8,940	0.32
Steel, carbon	7,820	0.28
Steel, stainless	8,030	0.29
Iron, gray-cast	7,210	0.26

TRY IT! -

We will find the mass properties of a 3D wedge. First construct a solid model of the object shown in Figure 7.5. Then activate the MASSPROP command, pick the object, and observe the results.

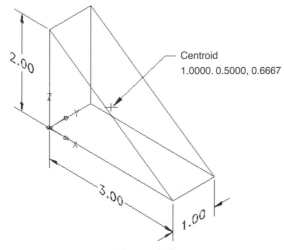

Figure 7.5

Command: MASSPROP

Select objects: (*Select the wedge.*)

—————————— SOLIDS ——————————

Mass: 3.0000

Volume: 3.0000

Bounding box: X: 0.0000 — 3.0000

Y: 0.0000 — 1.0000

Z: 0.0000 — 2.0000

Centroid: X: 1.0000

Y: 0.5000

Z: 0.6667

Moments of inertia: X: 3.0000

Y: 6.5000

Z: 5.5000

Products of inertia: XY: 1.5000

YZ: 1.0000

ZX: 1.5000

Radii of gyration: X: 1.0000

Y: 1.4720

Z: 1.3540

Principal moments and X-Y-Z directions about centroid:

I: 0.6825 along [0.9056 0.0000 -0.4242]

J: 2.1667 along [0.0000 1.0000 0.0000]

K: 1.9842 along [0.4242 0.0000 0.9056]

Write to a file ? <N>: N

THE AREA COMMAND

AutoCAD's AREA command is able to recognize solids and compute their surface area. The command line format to use AREA on 3D solids is:

Command: AREA

Specify first corner point or [Object/Add/Subtract]: O

Select objects: (*Select a 3D solid.*)

You must select the solid by picking a point on it. You can pick only one object. AutoCAD will report the solid's surface area. It will also return the solid's perimeter length, which has no meaning for a 3D solid, as being equal to zero. The AREA command also recognizes regions returning their area and perimeter length.

APPLICATION

We will find the surface area of the wedge that we found the mass properties of:

Command: AREA

Specify first corner point or [Object/Add/Subtract]: O

Select objects: (*Select the 3D wedge.*)

Area = 14.6056, Length = 0.0000

THE INTERFERE COMMAND

The INTERFERE command is sometimes referred to as a semi-Boolean operation because it is able to create a new object from the interaction of two or more existing objects. It is intended, however, to be an inspection tool for checking interference, or overlapping volume, between adjacent and mating solid objects, rather than a modification tool. Consequently, the creation of a new solid object from the overlapping volume is optional—AutoCAD will ask if you want an object made from the interference volume. Also unlike the Boolean operations, INTERFERE asks for two selection sets, although the second one is optional. The command line sequence of prompts is:

Command: INTERFERE

Select first set of solids: (*Use any object selection method.*)

Select second set of solids: (*Select a solid or press ENTER.*)

INTERFERE only works on 3D solids; it does not accept regions. If you have only one set of solids, AutoCAD will check the solids in the set with each other. If you have two sets, each solid in the first set is compared to every solid in the second set, but the objects within each selection set are not compared to each other. When you just want to check two solids with each other, it doesn't matter if they are both in one set of solids, or if they belong to two different sets, although AutoCAD will report the results in a slightly different manner.

AutoCAD will report the number of solid objects that were compared, the number of interfering solids, and the number of interfering pairs. Also, all of the interfering solids will be highlighted. AutoCAD will then ask if you want solid objects made from the overlapping volume:

Create interference solids ? [Yes/No] <N>: (*Enter Y or N, or ENTER.*)

If you enter a Y, a new solid will be created in the current layer, and the command will end. If you enter N or press ENTER, AutoCAD will highlight the interfering solids. When there are more than two interfering solids, AutoCAD will ask if you want the individual pairs highlighted:

Highlight pairs of interfering solids ? [Yes/No] <N>: (*Enter Y or N, or press ENTER.*)

If you enter N or press ENTER, the command will end. If you enter Y, AutoCAD will highlight one pair of overlapping solids. All of their edges will be highlighted—not just the interference edges. If there is more than one pair, AutoCAD will display the prompt:

Enter an option [Next pair/eXit] <Next pair>: (*Enter X or N, or press ENTER.*)

AutoCAD will cycle through the interfering pairs of solids, highlighting each in turn as you enter N or press ENTER, until you enter X to end the command.

Tips: Pinning down interference locations between even as few as three or four solids can be confusing. Often you will find it easier to select one key solid and compare it to the other solids one by one, rather than all at the same time.

Having AutoCAD create a solid from the common volume is sometimes helpful. It allows you to not only see where the interference is but also measure how much it is. These solids should be in a separate layer, so that they will stand out and can be discarded when you are finished with them.

TRY IT! CHECKING FITS AND POSITIONS

We have an existing bracket made from a hollowed-out wedge, which has two round holes in it. A mating part will fit in the middle of this bracket. The mating part also has two round holes, through which two round pins, slightly smaller than the holes, will be inserted for holding the two parts together. These parts are shown, in an exploded view, in Figure 7–6. We will use INTERFERE to check on fits and positions before we go further with the design of these parts.

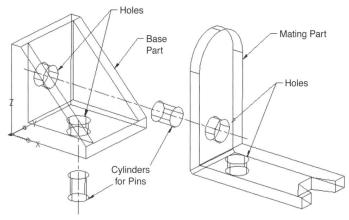

Figure 7.6

After moving the parts into their positions, we will compare just the base part with the mating part as shown in Figure 7.7.

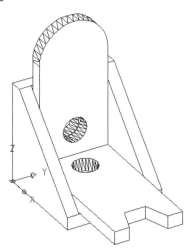

Figure 7.7

Command: INTERFERE

Select first set of solids: (*Select both the base part and the mating part.*)

Select second set of solids: (*ENTER.*)

Comparing I solid with I solid.

Solids do not interfere.

Since those two parts check out, we will move on to the holes. First, we will check the horizontal holes and pin:

Command: **INTERFERE**

Select first set of solids: (*Select the horizontal cylinder.*)

Select second set of solids: (*Select the base part and the mating part.*)

Comparing I solid with I solid.

Solids do not interfere.

So far, so good. Now, all that remains is to check the vertical holes and pin.

Command: **INTERFERE**

Select first set of solids: (*Select the vertical cylinder.*)

Select second set of solids: (*Select the base part and the mating part.*)

Comparing I solid against 2 solids.

Interfering solids (*first set*): I

   (*second set*): I

Interfering pairs: I

Create interference solids ? {Yes/No] <N>:Y

Figure 7.8 shows a close-up of the holes with the interfering volume moved up for improved visibility. To correct this interference, you could try moving the cylinder, but probably the best thing would be to move the hole in the mating part. You could use the Face Move option of SOLIDEDIT to do this.

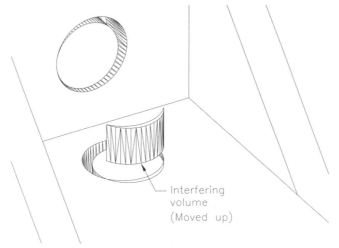

Interfering
volume
(Moved up)

Figure 7.8

This model, including the interference volume, is in file 3d_ch7_01.dwg on the CD-ROM that accompanies this book.

THE SECTION COMMAND

The SECTION command creates a planar cross-section through one or more solids. The resulting cross-section is region object type. The region takes the current layer and is left in the solid. If more than one solid is selected, a separate region is created for each solid, even if the resulting regions are not connected. The command line format for SECTION is:

Command: SECTION

Select objects: (*Use any object selection method.*)

Specify first point on section plane by
  [Object/Zaxis/View/XY/YZ/ZX/3points] <3points>: (*Specify an option, a point, or press ENTER.*)

These options for positioning the cross-section are identical to those in the SLICE command, which we covered in the section on modifying solid objects. Please see the discussion of that command for descriptions and examples of the options for SECTION.

Tip: Use a separate layer for cross-sections so that they can be easily separated and moved out from the base solids.

TRY IT! - USING SECTION

We will make a cross-section through the bracket, its mating part, and the cylindrical pins we used in exploring the INTERFERE command. We will locate the cross section parallel to the X axis, perpendicular to the XY plane, and through the center of the parts in the Y direction.

Command: SECTION

Select objects: (*Select all four objects.*)

**Specify first point on section plane by
    [Object/Zaxis/View/XY/YZ/zX/3points] <3points>: ZX**

Specify a point on the ZX plane <0,0,0>: 0,1

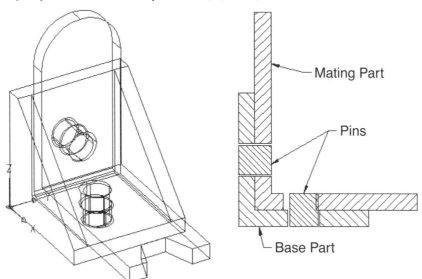

Figure 7.9

The solids and the resulting cross-sections, moved out and away from the solids, are shown in Figure 7.9. They have also been crosshatched. The cross-section for the base part, and the cross-section for the mating part are each a single region, even though they are in three separate pieces.

Compare your section with the one in file 3d_ch7_01.dwg on the accompanying CD-ROM.

3D APPLICATIONS USING MECHANICAL DESKTOP

Mechanical Desktop is a feature-based, parametric solid modeling program designed for the individual who already has a background in AutoCAD. The term feature-based means that each component or feature of a solid model must be constructed one at a time. The term parametric deals with the ability to control the size of features through dimensions. These dimensions can be absolute values or could include formulas. A few of the highlights of Mechanical Desktop will be discussed and how they compare with creating 3D solid models in AutoCAD.

MECHANICAL DESKTOP AND SOLID MODELING

Solid modeling in Mechanical Desktop begins with creating a sketch. The sketch is next converted into a profile. This profile holds intelligent information such as parallelism, concentricity, and other such constraints. Once the sketch is solved through a combination of constraints and parametric dimensions, a 3D feature is created from the profile. Typical features include extrusions, revolved solids, and sweeps. Next, other parametric features, such as holes, fillets, and chamfers, are used to form detailed features of the 3D solid model. Figure 7.10 represents a base and center plate that were both extruded. A chamfer feature was added along one edge. Then holes were placed in the vertical plate while slots were added and arrayed along the bottom plate.

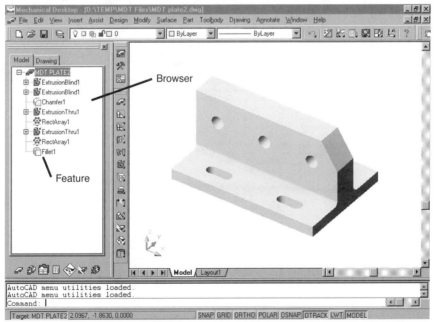

Figure 7.10

One of the more powerful functions of solid modeling in Mechanical Desktop is the ability to edit features any time during the design process. In Figure 7.10, the three holes along the vertical center plate need to be increased in diameter. The editing abilities of the Mechanical Desktop allow for changes to be made throughout the design cycle. When editing features, the original parametric dimensions are exposed in Figure 7.11. This allows for changes to be made. After the changes were made, the part is updated to reflect the new dimension value (see Figure 7.12). (Note: All three holes update since the original hole was arrayed in a rectangular pattern. Changing the parent hole affects the others.

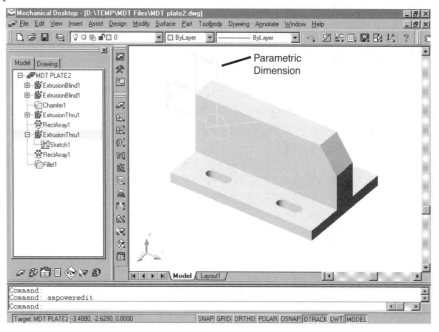

Figure 7.11

Figure 7.12

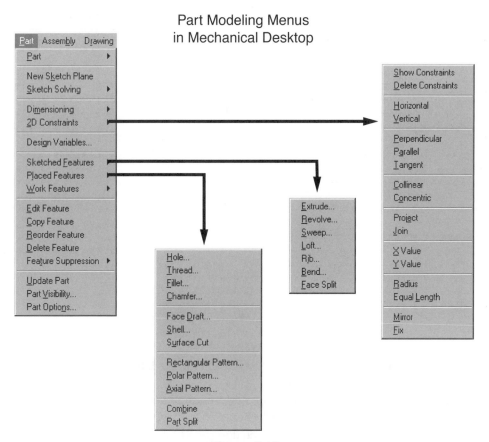

Figure 7.13

Other aspects of solid modeling in the Mechanical Desktop include the ability to place holes, threaded features, revolved and swept features, create shells or thin walls inside or outside an object, just to name a few. These operations just described are illustrated in the menu structures in Figure 7.13. Other capabilities of the part modeling process include the ability to assign parameters or mathematical formulas to certain features of a model. The parameters are then merged with an Excel spreadsheet. The parameters are copied numerous times and values changed. Clicking on a given set of parameters that have been changed edits the solid model to these changes automatically. This describes the Table Driven parts feature of Mechanical Desktop.

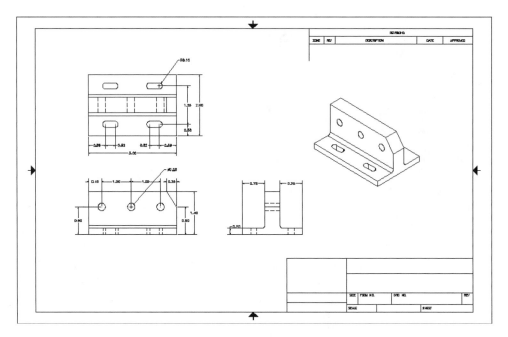

Figure 7.14

MECHANICAL DESKTOP AND ASSOCIATIVE 2D DRAWING GENERATION

Another feature of Mechanical Desktop is the ability to link the 3D solid model with a 2D orthographic view drawing. In Mechanical Desktop, unlike AutoCAD, changes to the 3D model are automatically reflected in the 2D drawing. Also, dimensions in the 2D drawing automatically update their values to dimensionally reflect the latest changes. In the same way, changes may be made to certain 2D features thus affecting and updating the3D model automatically. Figure 7.14 illustrates an orthographic drawing complete with isometric view completely generated from a 3D model in Mechanical Desktop.

Drawing and Annotation Menus
in Mechanical Desktop

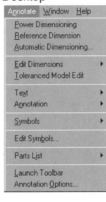

Figure 7.15

Two menu items dealing with drawing creation are illustrated in Figure 7.15. The Drawing pulldown menu area is used for generating the orthographic and isometric views. In addition to these, section, detail, broken, and auxiliary views are also supported. The Annotation pulldown menu area is where dimensions are added to the 2D view. This menu area also contains areas for generating centerlines and leaders in addition to such items as surface finish, geometric dimensioning and tolerancing, and welding symbols. A parts list can also be generated and balloons applied while in the drawing mode.

MECHANICAL DESKTOP AND ASSEMBLY MODELING

The power of Mechanical Desktop continues with the ability to merge external parts into one drawing for the purpose of forming an assembly drawing similar to the illustration in Figure 7.16. Assembly constraints such as mating, flush, angle, and insertions assist you in properly building the assembly for the purpose of performing interference calculations.

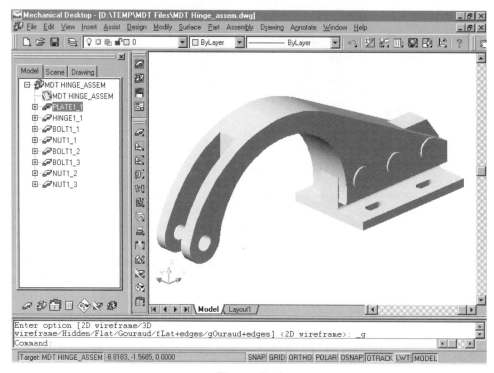

Figure 7.16

Once all parts are assembled, an explosion factor can be applied for exploded views of the assembly similar to Figure 7.17; this is commonly referred to as a Scene. Also included in this process is the ability to add trails to help you interpret how a part will fit into another.

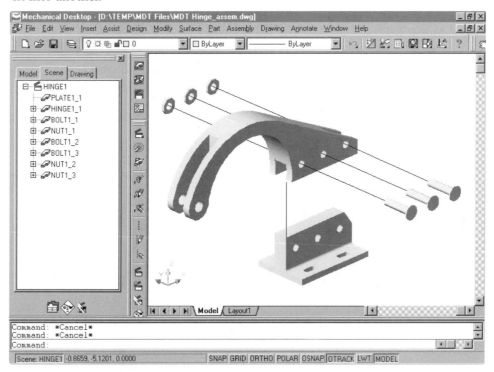

Figure 7.17

414

The Scene is then used to generate a 2D drawing consisting of a front orthographic view and the exploded assembly in isometric. Notice that the trails are present in all views with the addition of part balloons to identify each part by number. A bill of materials can be generated and located near the edge of the title strip in Figure 7.18.

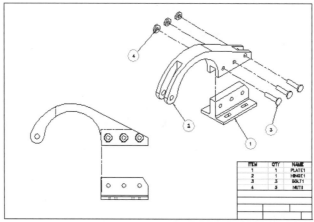

Figure 7.18

Menu items displayed in Figure 7.19 illustrate the location of commands used to create the assembly, choose the proper constraints, and generate a scene in the case of an exploded assembly model.

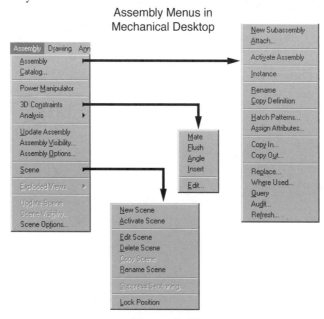

Figure 7.19

The 2 cylinder horizontal engine in Figure 7.20 illustrates a more complex assembly. The assembly modeler in Mechanical Desktop allows you to identify an item in the browser and change the visibility to view internal parts such as the piston in Figure 7.21.

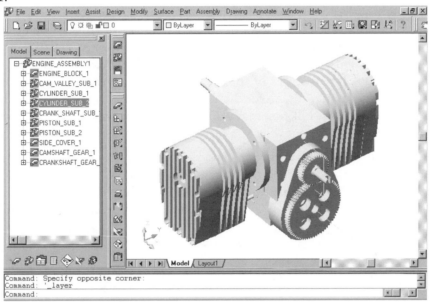

Figure 7.20

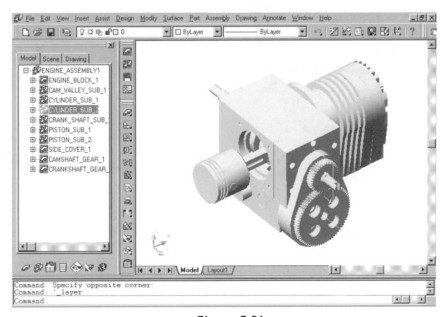

Figure 7.21

MECHANICAL DESKTOP AND SURFACE MODELING

Surface modeling inside the Mechanical Desktop allows for yet another powerful application especially when dealing with objects that have many free-form surfaces or are too difficult to be modeled in a solid modeler. You have already been exposed to typical free-form surfaces in Chapter 4 such as of boat hulls, automobile panels, jet sky bodies, and such consumer products as power tool housings and plastic dish detergent bottles. These free-form surfaces are technically referred to as NURBS surfaces (non-uniform rational B-splines), which allow you to design curves that meet a specific application. Typical types of surfaces include ruled, revolved, and blended. Expanded editing commands include the ability to trim surfaces and to find the intersection of two surfaces. Filleting is available to produce rounds over sharp corners, which is especially important when dealing with plastic injection molding processes.

Illustrated in Figure 7.22 is a wireframe model of a wheel hub with surfaces applied. Actually one of the voids in the wheel hub was surfaced in Mechanical Desktop and then arrayed a total of five times to complete the object in the figure. Like solid and assembly models, surface models can also be shaded for better clarity as in Figure 7.22.

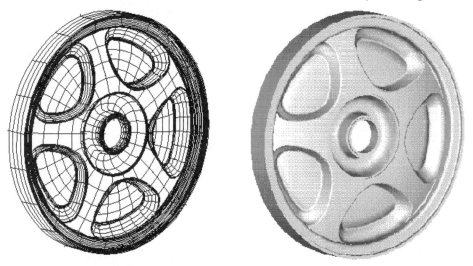

Figure 7.22

Figure 7.23 shows a consumer plastic bottle that was first constructed as a series of splines in wireframe mode. Swept surfaces were used over the splines to create the bottle profile. Ruled surfaces were used to create the bottom and top of the bottle. The surfaced version of the bottle is displayed in Figure 7.24.

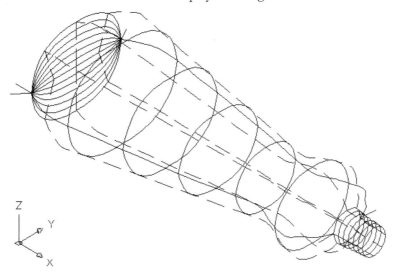

Figure 7.23

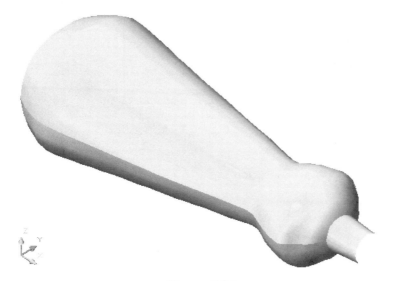

Figure 7.24

The menu items in Figure 7.25 show the robust a segment of the used for surface modeling in Mechanical Desktop. Of particular interest are the important surface editing commands such as breaking, joining, adjusting, and performing intersecting and projecting trim operations on surfaces.

Surface Modeling Menus
in Mechanical Desktop

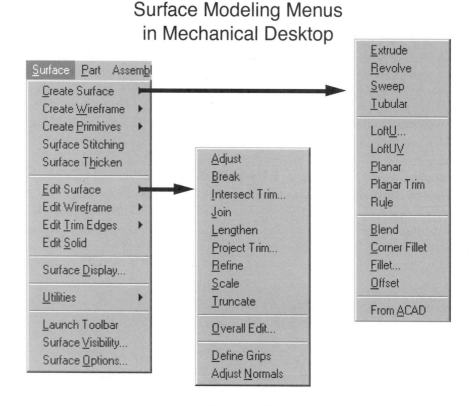

Figure 7.25

3D APPLICATIONS OF INVENTOR

Inventor is yet another feature-based, parametric solid modeling package created by Autodesk, Inc. Inventor however has no AutoCAD interface and was designed as a stand-alone product. This means an individual does not need to be familiar with AutoCAD yet still succeed in building 3D solid models. Therefore, one of the biggest advantages of using Inventor is its ease of use.

When Inventor is first launched, the screen illustrated in Figure 7.26 displays. In addition to the typical New and Open buttons, notice the Project icon. This is a very important feature in Inventor and is a method of keeping your designs and assemblies organized.

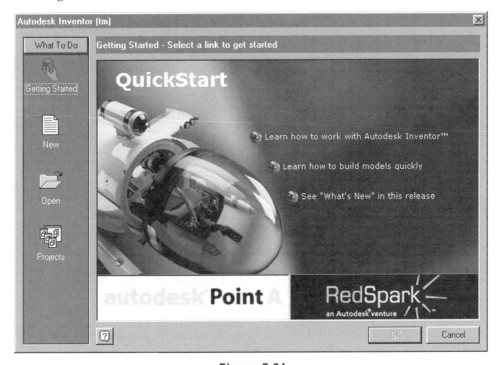

Figure 7.26

Clicking the New button in the previous figure will display the New File dialog box in Figure 7.27. Unlike Mechanical Desktop where part, assembly, exploded views, and drawings all share the same file type, namely .DWG, Inventor uses various file types. The following chart identifies and explains the purpose of each:

.ipt	3D solid model parts
.iam	Assembly models
.idw	2D drawing information
.ipn	Presentation models

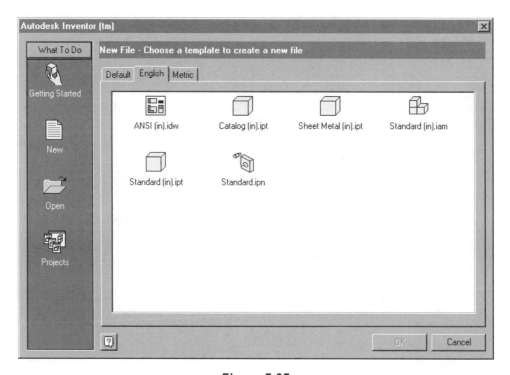

Figure 7.27

Even though you are dealing with various file types, all files are completely associative with each other. This means if you change a part that is part of an assembly, the part will automatically change in the assembly.

INVENTOR AND PART MODELING

A typical Inventor screen is illustrated in Figure 7.28. Various panel bar menus are available to easily begin creating 3D models. A browser is also available to keep track of all features as they are created. Inventor also uses a series of context menus that are activated when you click on the right mouse button. This is a very efficient way of accessing commands. Inventor uses the same concepts and workflow as with Mechanical Desktop. You first start with a sketch, apply constraints to better position the sketch and dimension the sketch.

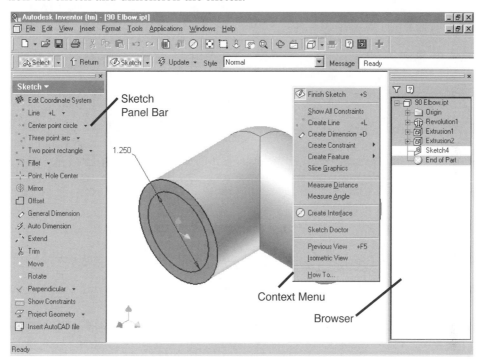

Figure 7.28

Whenever a sketch is finished, the Features menu displays. This allows you to create the desired feature from the sketch. In Figure 7.29, a circle has been extruded and cut to form the hole in the 90 degree pipe elbow.

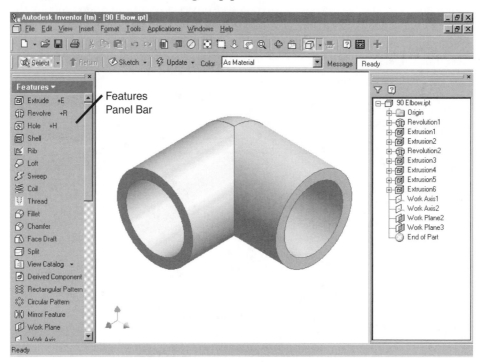

Figure 7.29

INVENTOR AND ASSOCIATIVE 2D DRAWING GENERATION

If a 2D drawing needs to be generated from the 3D model, this is easily accomplished in Inventor (see Figure 7.30). Various views including orthographic, isometric, broken, auxiliary and section views can be generated in Inventor. Dimensions too are easily added. If changes occur in the 3Dmodel, these changes will be reflected in the 2D drawing as well.

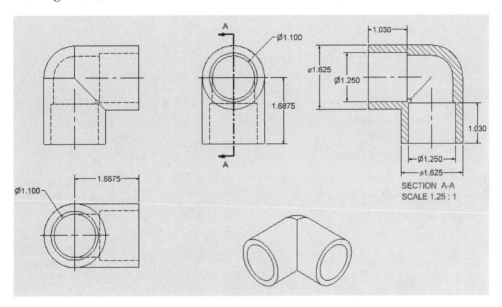

Figure 7.30

Panel bar menus that deal with 2D drawing creation from a 3D model are illustrated in Figure 7.31. The Drawing Management panel is used to create different views and to create additional sheets inside of the same drawing file. This sheet method is similar to using Layouts in AutoCAD. The Drawing Annotation panel bar menu is used for placing dimensions, centermarks, hole and thread notes, and leaders. It is through this panel bar that you add various symbols to your drawing such as welding, surface texture, and geometric dimensioning and tolerancing. You also place parts lists and balloons through this panel bar menu.

Inventor Drawing Management and Annotation Menus

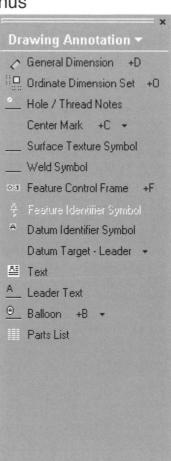

Figure 7.31

INVENTOR AND ASSEMBLY MODELING

Figure 7.32 is a typical example of an Inventor screen complete with an assembly model. The panel bar changes to Assembly; use this to place individuals components into an assembly or to create a new component right inside of the assembly. The ability to keep parts together is a function of its constraints. Typical constraints include mates and flushes, angle, tangent, and inserts. On the right side of the Inventor screen in the typical assembly model are all individual parts that make up the assembly. You can turn parts on or off for better clarity.

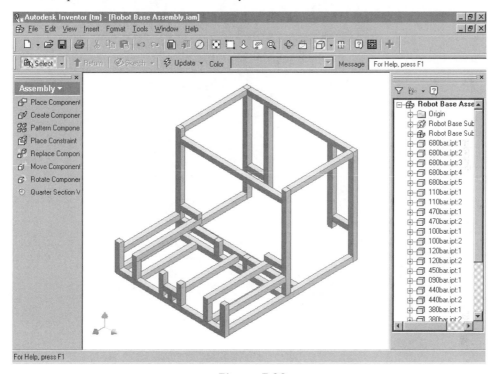

Figure 7.32

With the assembly model constructed, a 2D drawing is generated of the assembly illustrated in Figure 7.33. Notice the standard border and title block information available for use in Inventor. Through the Drawing Annotation panel bar, you can automatically generate balloons to identify each part and have the balloons relate to a parts list.

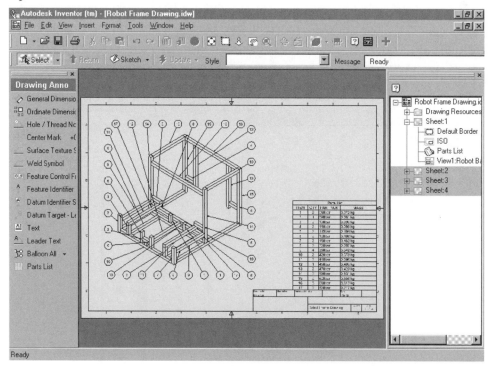

Figure 7.33

Another very powerful feature assembly modeling in Inventor is the ability to dynamically drag the assembly as a test for functionality. Figure 7.34 illustrates a heavy machinery scoop controlled by a series of four air cylinders. In the assembly browser, one of the air cylinders is assigned a property called adaptive. This means that all air cylinders will be free to move and open the mouth of the scoop. The results are illustrated in Figure 7.35. Grabbing and dragging the face of one of the scoops will allow the whole mechanism to dynamically open or close. Making one of the air cylinders adaptive affects all air cylinders in the assembly model of the scoop.

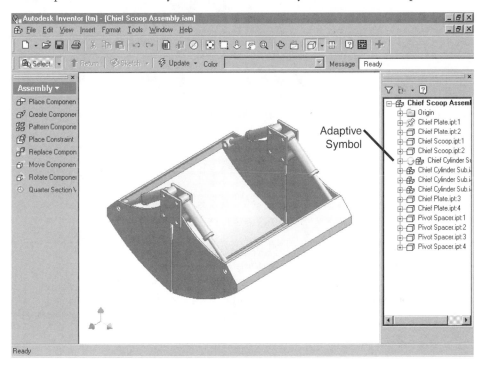

Figure 7.34

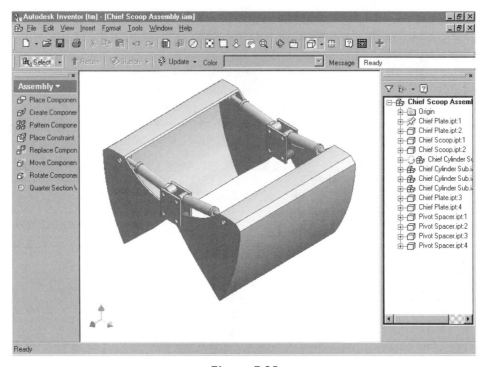

Figure 7.35

Figure 7.36 illustrates a robot assembly. In order to make internal workings of the assembly visible, Inventor allows segments such as the lexan plastic cover to be assigned a clear color. Now internal sub assemblies such as the robot frame, gripper, and motor drives are visible in addition to viewing the clear plastic covering.

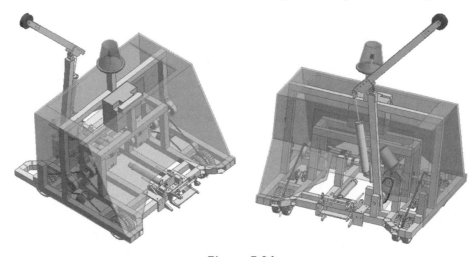

Figure 7.36

INVENTOR AND PRESENTATION MODELS

Another method of viewing the internal features of an assembly model is to produce an exploded view. Figure 7.37 illustrates an exploded view, which is called a presentation model. The Presentation panel bar allows you to create the view and control the explosion factor. After the presentation model is created, you can modify the position of individual parts by tweaking them until you reach the desired results. You can even animate the presentation model for the purpose of dynamically illustrating how component parts will fit together. This animation can be saved as an avi file for later viewing. A 2D drawing file can even be made of the presentation model complete with balloons and parts list.

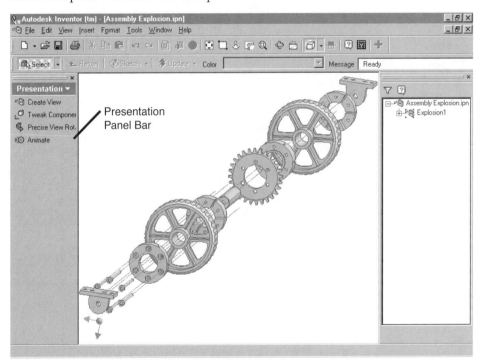

Figure 7.37

INVENTOR AND SHEET METAL MODELS

An extra set of tools is available in Inventor to create sheet metal parts. You first make changes to a series of settings such as material type, thickness, and bend radius. Next you create a number of individual faces of the sheet metal part and connect these faces by adding bends between them. Corner seams are then created and shapes are cut with enhanced tools located on the Sheet Metal panel bar. Figure 7.38 illustrates a basic sheet metal model consisting of a few of the operations previously mentioned.

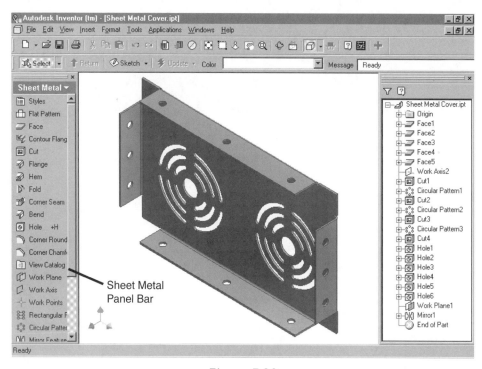

Figure 7.38

Once the sheet metal model is made, it is very easy to generate a flat pattern illustrated in Figure 7.39.   Also, the flat pattern is associated with the 3D sheet metal model; modifications made to the 3D model will automatically be reflected in the flat pattern.

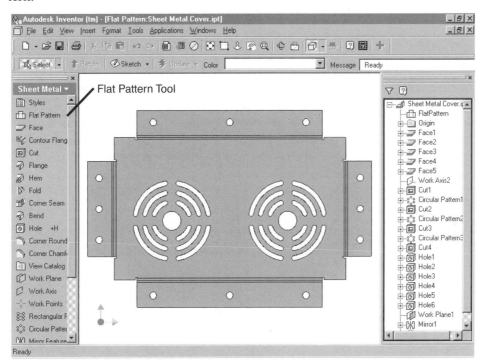

Figure 7.39

CONCLUSIONS

Ranking the software packages places AutoCAD as an introductory 3D solution. This is due to its ability to apply surfaces to wireframe information and to create solid models. 2D orthographic views can then be created from the 3D solid models although the 2D views are not associated with the 3D model. In other words, if changes are made to the 3D solid model, the 2D views do not update to these changes. Assemblies are handled with such commands as ALIGN and MOVE. Constructing 3D models using AutoCAD is designed for the individual who is already skilled in the 2D uses of the package.

Mechanical Desktop is positioned as a mid-range solid and surface modeler. Because of its ability to associate the solid model and its orthographic views, changes can be made in the 3D solid model and updated automatically in the 2D orthographic view drawing. Mechanical Desktop also has superior capabilities when dealing with shelling, filleting, chamfering, sweeping, and lofting operations. Once a group of solid models has been created, they can be assembled to generate exploded view drawings complete with part identification balloons and a bill of material. Changes to individual parts will be reflected in the assembly model. Mechanical Desktop is best suited for individuals that are already well skilled in AutoCAD.

Inventor is the stand-alone mid-range solid modeler that has an impressive sheet metal module. As with Mechanical Desktop, the Inventor solid model is associated with the 2D orthographic drawing view generated from it. Inventor has the same complex shelling, filleting, chamfering, sweeping, and lofting operations as Mechanical Desktop. Inventor has been optimized to work with large, complex assembly models. Presentation models can be generated from assemblies along with 2D drawings that contain balloons and bill of material information. Inventor's adaptive engine allows parts to adapt to others when being assembled. Inventor is designed for ease of use and where large assembly models are involved.

COMMAND REVIEW

AREA

This command is used to calculate the area and perimeter of objects.

INTERFERE

This command also creates a new solid object from the intersection of two or more existing solids, but it does not erase the original solids. Also, it operates on pairs of solids, even when more than two solids are involved.

MASSPROP

This command is used to calculate the mass properties of regions and solids. Typical information found in the mass property calculation includes mass, volume, bounding box, moments of inertia, products of inertia, radii of gyration, and principal moments and X-Y-Z directions about a centroid.

SECTION

This command uses the same options that **SLICE** does to make a section through a solid object. The object type of the section is a region.

CHAPTER REVIEW

Directions: Answer the following questions.

1. What command would you use to find the surface area of a 3D solid?

2. What is a centroid?

3. What is the use of the INTERFERE command?

4. Identify the property that is not calculated when using the MASSPROP command on a 3D solid model.

   a. Centroid

   b. Mass

   c. Volume

   d. Weight

Paper Space and 2D Output

LEARNING OBJECTIVES

Our focus in Chapter 8 will shift from creating 3D models to techniques for making 2D drawings of 3D models. When you have completed this chapter, you will:

- Understand the purpose of AutoCAD's paper space and be familiar with its properties.

- Understand the differences between tiled viewports and floating viewports, as well as know how to create and manage paper space layouts and floating viewports.

- Know how to set up multiview drawings of wireframe and surface models in paper space.

- Be able to control the display of objects and layers within floating viewports.

- Know how to add annotation and dimensions to drawings of 3D models.

- Be able to create 2D and 3D wireframes from 3D solids, and be able to create 2D drawings of a 3D model as it is seen from any viewpoint.

- Be able to use AutoCAD's specialized commands for creating multiview 2D drawings from 3D solid models.

PAPER SPACE VERSUS MODEL SPACE

Everything we have discussed so far in this book has related to model space. Model space, as we have seen, is a fully 3D environment in which you can construct a model having height, length, and width. Furthermore, you can set viewpoints from any point in space to look at this model, and as shown in Figure 8.1, you can divide the screen into multiple viewports to simultaneously view the model from several different viewpoints.

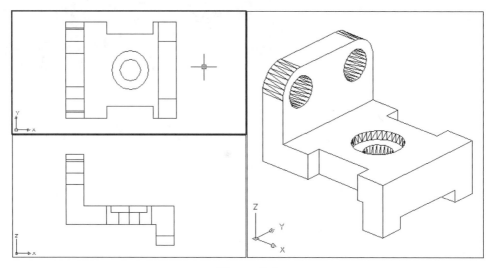

Figure 8.1

Model space, however, is not suitable for making 2D drawings from 3D models. First, regardless of how many viewports are on the computer screen, only the current viewport can be plotted. Consequently, printing multiple orthographic views of the model—showing its top, front, and side—is virtually impossible within model space. Also, adding notes and dimensions and controlling which objects are to be shown is awkward. Finally, plots having an accurate scale from any viewpoint other than the plan view of the WCS are difficult to set up.

These problems are taken care of in paper space. Paper space is an entirely different universe from model space. It is a 2D universe located in front of model space, as if it were a piece of paper. You can write notes, draw borders, and add title boxes in paper space. Moreover, you can see into model space through paper space viewports.

The purpose of paper space is to annotate and make 2D prints of objects that were created in model space. Model space is for modeling; paper space is for output. Paper space allows you to make standard multiview drawings that are directly linked to 3D models. Any changes made to the 3D model will automatically show up in the 2D drawing.

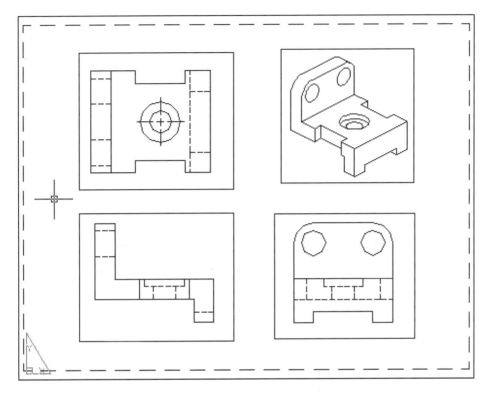

Figure 8.2

To accomplish its objectives, paper space uses special viewports, such as those shown in Figure 8.2, that are called floating viewports to distinguish them from the tiled viewports of model space. Floating viewports:

- Can be copied, moved, stretched, scaled, and erased.
- Can have gaps between viewports; conversely, viewports can overlap one another.
- Reside in a layer. If a viewport's layer is frozen or turned off, the viewport's border disappears while the contents of the viewport remain.
- Can control hidden-line removal during plotting viewport by viewport.
- Can precisely set the scale of the 3D model for each viewport.
- Can control which layers are to be visible in each viewport.

The Tilemode system variable controls whether paper space is available. When Tilemode is set to 1, AutoCAD operates in model space. When it is set to 0, AutoCAD can operate in either paper space or floating model space. The name of the variable is derived from the tiled viewports used in model space.

In an AutoCAD drawing, you will seldom set the value of Tilemode directly. Instead, you will simply click one of the tabs located on the bottom edge of AutoCAD's graphics window. The tab labeled Model sets Tilemode to 0, and the other tabs set Tilemode to 0. By default, there are two tabs for paper space—one labeled Layout1 and the other labeled Layout2. These tabs open specific setups of paper space, and we will describe shortly how you can both create and delete setups and rename their tabs.

The UCS icon assumes a different form in paper space. As shown in Figure 8–2, it is shaped like a 30–60° drafting triangle, rather than like two wide arrows. A small x on the short side of the triangle indicates the direction of the X axis. Aside from its different form, the icon acts the same as the model space icon.

Although paper space is distinctly different than model space, layers and text styles are the same in paper space as they are in model space. Although you can draw 3D objects in paper space, you cannot see them from 3D viewpoints—DVIEW, VPOINT, and 3DORBIT are not allowed in paper space. It is possible, however, to go through a floating viewport into model space and use these commands. You'll sometimes have reason for doing this, and we will describe how this is done later in this chapter.

Even though floating viewports display model space objects, AutoCAD does not completely recognize them from paper space. You cannot select model space objects to be moved, erased, stretched, and so forth. AutoCAD does, however, recognize most object snap points on model space objects from paper space. This is useful for aligning views and referencing points on models from paper space.

PAPER SPACE LAYOUTS

The steps you will use in creating a multiple-view 2D drawing in paper space from a 3D model will be:

1. Choose the paper size that you will use when plotting the drawing. Your choice will depend on the paper sizes that your plotter can handle, the number of views that the drawing will have, the scale that views will have, and your drafting standards. You can later change the paper size if necessary, even if the drawing has floating viewports.

2. Add a border and a title block.. Paper space is designed to work on a one-to-one scale for these items, just as if you were working by hand on a sheet of drafting paper. For instance, if the drawing is to be printed on 34-by-22-inch paper, you would make your border fit that paper size; if the notes are to be printed in one-eighth-inch-high letters, you would set the text height to one-eighth of an inch. Virtually all plotters require space to grip the edges of the paper, so you will have a printable space and margin within the paper that you must stay within. (This step can be delayed until after step 6, if you prefer.)

3. Create one floating viewport for each view of the model. The total number of viewports will depend on the shape of the model and on your drafting standards.

4. Set an appropriate scale and viewpoint within each viewport. Most viewports will have the same scale, but it is entirely possible for details to be shown in viewports using a larger (or even smaller) scale.

5. If the views are to be orthographic—such as the front, top, and side—align the model between viewports.

6. Add dimensions and other necessary annotation to the model and to the drawing.

7. Plot the finished drawing of the model.

When you are working with solid models, AutoCAD has two commands—SOLVIEW and SOLDRAW—that can make steps 2, 4, and 5 almost automatic. You will have to perform these steps manually, however, when you are working with wireframe and surface models.

AutoCAD 2002 allows you to have more than one paper space setup of the steps we have just described within a single drawing file. Each setup is referred to as a layout. For example, you can have one layout for a fully dimensioned, multiview, orthographic drawing of your model on 34-by-22-inch paper, another layout for an isometric drawing on 17-by-11-inch paper, another layout for an undimensioned multiview, orthographic drawing on 11-by-8.5-inch paper, and so on. A tab for each layout is located on the bottom edge of the graphics window, as shown in Figure 8.3, and you access a particular layout by simply clicking its tab.

By default AutoCAD has two layouts, which are named Layout1 and Layout2, that have identical settings. You can modify the setup of these layouts, as well as create additional layouts, delete layouts, and change their names. Layouts are created and managed by the LAYOUT command. Once a layout has been created, you will use the PAGESETUP command to set the paper size for the layout. AutoCAD 2002 also has a layout wizard that not only creates a new layout but also helps you select a plotter, set the paper size, and create floating viewports for the layout. You can initiate the layout wizard by invoking LAYOUTWIZARD.

440

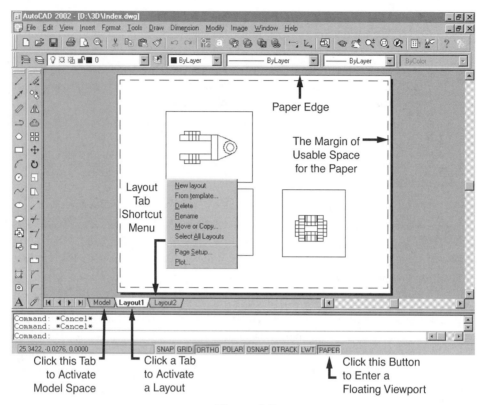

Paper Edge

The Margin of
Usable Space
for the Paper

Layout
Tab
Shortcut
Menu

New layout
From template...
Delete
Rename
Move or Copy...
Select All Layouts
Page Setup...
Plot...

```
Command: *Cancel*
Command: *Cancel*
Command:
```

25.3422, -0.0276, 0.0000   SNAP GRID ORTHO POLAR OSNAP OTRACK LWT PAPER

Click this Tab
to Activate
Model Space

Click a Tab
to Activate
a Layout

Click this Button
to Enter a
Floating Viewport

Figure 8.3

THE LAYOUT COMMAND

The command line format of LAYOUT is:

Command: LAYOUT

**Enter layout option
[Copy/Delete/New/Template/Rename/SAveas/Set/?] <Set>: (Enter an
option or press ENTER.)**

Copy

This option makes a new layout that has the parameters of an existing layout. You will
be prompted from the command line to specify the name of the layout that is to be
copied, with the current layout as the default, and to specify a name for the copy.

Delete

Use this option to delete an existing layout. A command line prompt will ask you to
enter the name of the layout you want to delete. You cannot delete the Model layout.

New

A new layout is created by this option. You will be prompted from the command line for the new layout's name.

Template

This option creates a new layout based on a file that establishes a specific paper size. A Select File dialog box will be displayed for you to use in choosing a template. The files listed in this dialog box can be either template files (DWT) or drawing files (DWG). The template files for ANSI and ISO standards will insert a border and title block in the layout and create one floating viewport with edges that fit within the title box. After selecting a file, you will be prompted to enter a name for the new layout.

Rename

Use this option to rename a layout. Command line prompts will ask to enter the layout name that you want to change, with the current layout as the default, and a new name for that layout.

SAveas

This option saves the settings of a layout to a template or drawing file. You will be prompted from the command line for the name of the layout to save, and from the Save Drawing File dialog box to specify a name and file type for the saved layout.

Set

This, the default option, accesses a layout. A command line prompt will ask you to enter the name of the layout that you want to become the current layout.

 Tips: Rather than use LAYOUT, you will probably use the shortcut menu displayed when you right-click the tab for the current layout, as shown in Figure 8.3. This menu even contains an option changing the order of the layout tabs. Also, you can make any layout become the current layout by clicking its tab.

The initial parameters of a layout are controlled by the cluster of buttons labeled General Plot Options in the Display tab of the Options dialog box (see Figure 8.4). Checkboxes in this cluster turn on and off the display of the Model and Layout tabs, the display of the layout's paper and the simulated shadow of the paper, and the display of the limits of printing margins on the paper. For 3D models, you will probably turn off the automatic creation of a floating viewport in a new layout because the single floating viewport in the current layer that AutoCAD makes will seldom be what you want. An option for automatically invoking the PAGESETUP command in new layouts is also in this cluster of checkboxes.

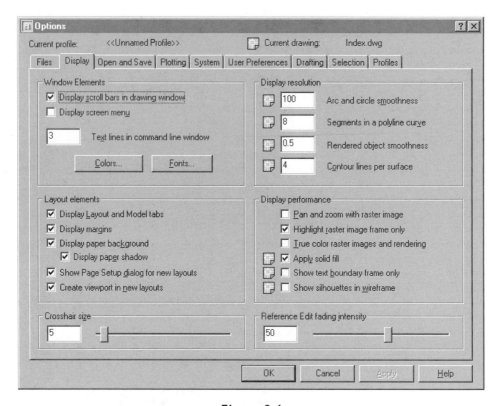

Figure 8.4

THE PAGESETUP COMMAND

Once you have created a new layout, you will use the PAGESETUP command to establish the paper size and plotting parameters of the layout. By default this command is automatically invoked when a new layout is created. You can also invoke the command from the Layout tab shortcut menu.

All of the PAGESETUP options are selected within the Page Setup dialog box. This dialog box is virtually the same as the one used by the PLOT command. Therefore, we will discuss only those options that pertain to layouts.

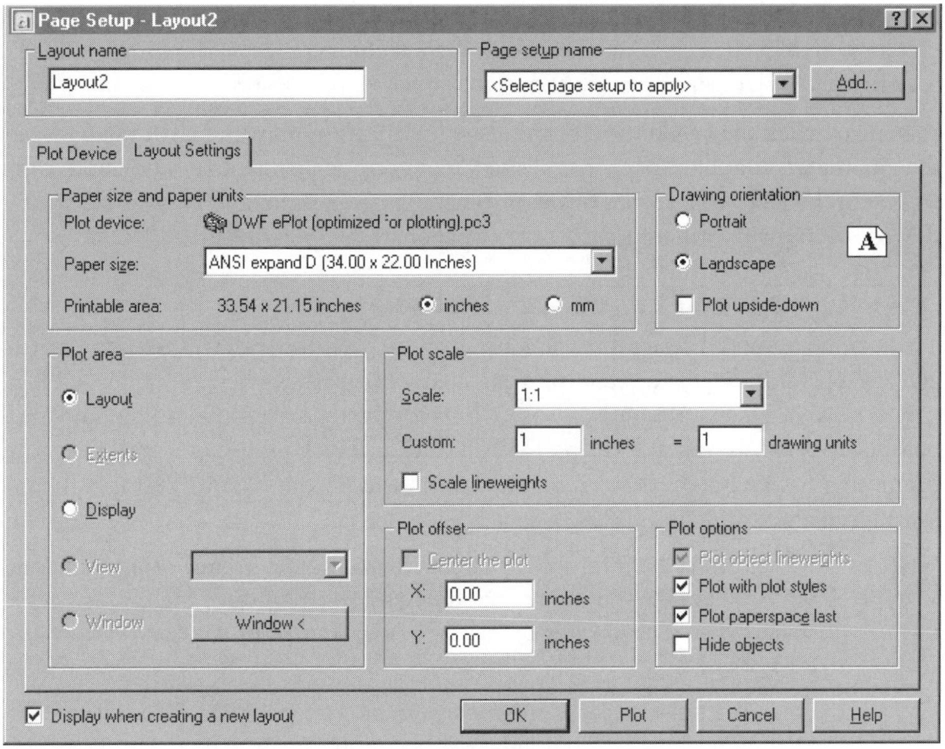

Figure 8.5

The Page Setup dialog box has two tabs—one is labeled Plot Device and the other, which is shown in Figure 8.5, is labeled Layout Settings. The current paper size is shown in this dialog box, and you can select another size through the Paper Size list box. If this list does not have a paper size suitable for creating a drawing of your model, you must open the Plot Device tab and select another printer. You will always select Layout as the area to plot rather than Extents or Display.

The cluster of list and edit boxes labeled Plot Scale are for setting the overall scale of the drawing; not the scale within the floating viewports. The only time you will change the scale of the plot from one-to-one is when you want to plot the drawing on a paper size that is different than the specified paper size. For instance, if you want to use on A-size (11-by-8.5-inch) paper to plot a layout that is based on C-size (22-by-17-inch) paper, you would set the scale to 1:2.

You can save a page layout, to be restored later in the same layout or to be used in another layout. To save the current layout, click the Add button. The User Defined Page Setup dialog box will be displayed for you to enter a name for the setup. When you click the OK button of the dialog box, the setup will be saved, and it will appear in the Page Setup Name list box. To restore a page setup, simply click its name in the list box. Saved page setups are useful when you need to plot a drawing using parameters that are different than those of the layout, such as in the example of plotting a C-size drawing on A-size paper mentioned in the previous paragraph.

When you click the OK button of the Page Setup dialog box, the layout will assume the parameters that have been established, including the paper size. The UCS origin will be in the lower left corner of the printable paper margin, and you should start in that location for plotting and most other paper space operations.

THE LAYOUTWIZARD COMMAND

This command uses a series of dialog boxes to lead you through the steps in creating and establishing the parameters of a layout. You can start the wizard by selecting Insert/Layout/Layout Wizard or Tools/Wizard/Create Layout from the AutoCAD pull-down menu, as well as entering LAYOUTWIZARD on the command line. After prompting for a name to be used for the layout, the Layout Wizard displays the following sequence of dialog boxes.

- Printer. Select a printer for the layout from a list of currently configured printers. If the printer you want is not displayed, you must leave the Layout Wizard and use the Control Panel to add the printer to the list of Windows configured printers.

- Paper Size. Select a paper size for the layout from a list box. Only paper sizes suitable for the selected printer will be available for selection.

- Orientation. Choose between landscape and portrait for the orientation of the drawing on the paper.

- Title Block. Select a border and title block for the layout. The list those that are available include title blocks that conform to American (ANSI), international (ISO), Japanese (JIS), or German (DIN) drafting standard. They can be inserted as either a block or as an XREF. You should take care that the title block you choose will fit the paper size you have chosen. You can skip this step if you do not want to insert a title block.

- Define Viewports. In this dialog box you can create one or more floating viewports. You can also set the scale of the model within the viewports. Select the radio button labeled None to skip this step. The Standard 3D Engineering Views radio button creates four viewports showing top, front, side, and isometric views of the 3D model.

- Pick Location. The dialog box for this step contains just a single button labeled Select Location. If you click that button, the dialog box will be temporarily dismissed and you will be prompted to pick the area that the floating viewport, or viewports, specified in the previous step are to fit within. Otherwise, AutoCAD will fit them within the printable margins of the paper.

- Finish. Select the button labeled Finish to complete the layout creation, or select Cancel to not create a new layout.

 Tip: The Layout Wizard is a good tool for creating layouts, especially if you are not experienced in working with them, because this single command accomplishes tasks that would otherwise require several commands.

WORKING WITH FLOATING VIEWPORTS

CREATING FLOATING VIEWPORTS

Even though, as we have seen, AutoCAD can create floating viewports for you, you will probably prefer to create them yourself, so that you can control their parameters more closely. In AutoCAD 2002, the MVIEW or VPORTS commands are used to create floating viewports. Furthermore, VPORTS has two different formats. If you invoke the command from the AutoCAD View/Viewports pull-down menu, from a toolbar button, or by entering VPORTS on the command line, a dialog box will be displayed for you to use in creating floating viewports. On the other hand, if you start VPORTS from the command line and precede its name with a hyphen, options identical to those of MVIEW will be offered in command line prompts. You can also select specific options from the View/Viewports pull-down menu and the Viewports toolbar, as shown in Figure 8.6. Moreover, a command for changing the shape of floating viewports—VPCLIP—can be invoked from the Viewports toolbar.

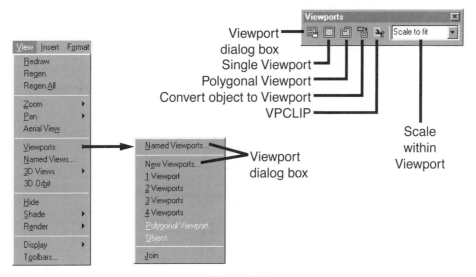

Figure 8.6

Though there is no limit to the number of floating viewports you can make, there is a limit to the number of active viewports you can have. (It is easy to confuse an active viewport with the current viewport. An active viewport is one that is visible, and displays a model space object. The current viewport is the one that contains the crosshair screen cursor.) If you create more than the maximum allowed number of active viewports, the extra viewports will not display anything. However, since the VPORTS and MVIEW commands allow you to switch active viewports by turning some on while turning others off, you can work with more than the number of floating viewports allowed. Moreover, the contents of all viewports, even inactive ones, will plot. The maximum allowed number of active viewports depends on your computer system.

THE VPORTS COMMAND

In paper space, AutoCAD displays the Viewports dialog box, shown in Figure 8.7. You can create up to four floating viewports at a time by selecting an option in the list box labeled Standard Viewports. The viewport arrangement for the option you select will be displayed in the Preview window.

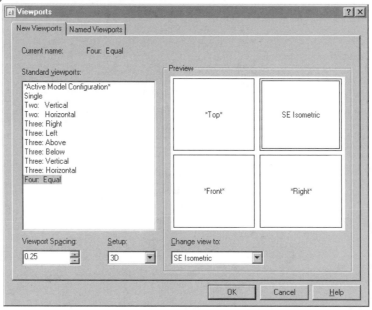

Figure 8.7

The drop-down list box labeled Setup has two options—2D and 3D. If you select 2D, the viewpoint within each newly created viewport will be the same as the viewpoint of the current model space tiled viewport. If you select 3D, AutoCAD will set an orthographic or isometric viewpoint in the new viewports. The default viewpoints are given in the Preview window, and you can change them by selecting another viewpoint from the drop-down list box labeled Change View To.

When the system variable Ucsortho is set to its default value of 1, the 3D option of

VPORTS will swivel the UCS in each viewport showing an orthographic view so that its XY plane directly faces the view's line of sight. If you do not want the UCS in each floating viewport to change, set Ucsortho to 0.

The Viewport Spacing edit box will create a gap between the viewports that is equal to the value in the edit box.

You can use the Named Viewports tab of the Viewports dialog box to create a set of floating viewports based on a named configuration of tiled viewports. You cannot, however, name and save a set of floating viewports.

When you click the OK button, AutoCAD will dismiss the dialog box and issue command line prompts for you to draw a rectangle that the new viewports are to fit within, or to specify that the viewports are to fit within the printable margins of the paper.

THE MVIEW COMMAND

The command line format of MVIEW (and VPORTS) is:

Command: MVIEW (or -VPORTS)

**Specify corner of viewport or
  [ON/OFF/Fit/Hideplot/Lock/Object/Polygonal/Restore/2/3/4] <Fit>:
  (Select an option, specify a point, or press ENTER.)**

SPECIFY CORNER POINT

Specifying a point location establishes one corner of a floating viewport, and AutoCAD will drag a rubberband rectangle to help you select its opposite corner.

ON

This option turns on an inactive floating viewport. The follow-up prompt is:

Select objects: (Select viewports using any object selection method.)

The inactive viewports you select will become active, displaying an image. If the number of floating viewports exceeds the maximum number of active viewports allowed, AutoCAD will automatically turn off other viewports as those that were selected are turned on.

OFF

This option turns off an active floating viewport. The follow-up prompt is:

Select objects: (Select viewports using any object selection method.)

The selected viewports will become inactive, and the images they display will disappear.

FIT

This option creates one floating viewport that fits the entire printable area of the current paper space layout. (refer back to Figure 8.7). It displays no additional prompts.

HIDEPLOT

Activates hidden line removal during plotting. The follow-up prompts are:

Hidden line removal for plotting [ON/OFF]: *(Enter ON or OFF.)*

Select objects: *(Select viewports using any object selection method.)*

- ON
- This option turns on hidden line removal during plotting. Hidden lines will not be shown.
- OFF
- This option turns off hidden line removal during plotting. All lines, hidden or visible, will be plotted.

LOCK

This option locks and unlocks the view, including the zoom level and view direction, within selected viewports. It issues the following command line prompts:

Viewport view locking [ON/OFF]: *(Enter ON or OFF.)*

Select objects: *(Select viewports using any object selection method.)*

The ZOOM, PAN, and DVIEW commands are disabled within locked floating viewports. The 3DORBIT command, however, is not.

OBJECT

You can transform a 2D or 3D polyline, ellipse, spline, region, or circle into a floating viewport with this option. 2D and 3D polylines must have been closed with the Close option. Splines, on the other hand, are acceptable even when they have been closed by picking an ending point that is on the starting point. Internal holes in a region create an opaque island in the viewport. The option issues the following prompt on the command line:

Select object to clip viewport: *(Select one object.)*

You can select just one object, and you must select it by picking a point on it. Examples of viewports based on objects are shown in Figure 8.8.

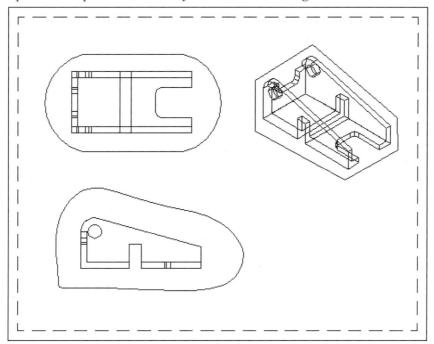

Figure 8.8

POLYGONAL

You can create a floating viewport that has a boundary composed of straight and arc-shaped 2D polyline like segments with this option. Command line prompts similar to those of the PLINE command will be issued for you to draw the boundary.

RESTORE

This option transforms a viewport configuration for tiled viewports into floating viewports. The number, arrangement, and views within the floating viewports will correspond to those of the tiled viewport configuration. The first follow-up prompt is:

Enter viewport configuration name or [?] <*Active>: (*Enter ?, a name, or press ENTER.*)

- *Active

- This, the default option, will use the tiled viewport configuration that is currently in effect in model space.

- ?

- Entering a question mark will bring up a list of all tiled viewport configurations that were saved through the VPORTS command.

- Enter Viewport Configuration Name
- When the name of a viewport configuration that was saved through the VPORTS command is entered, AutoCAD will translate those tiled viewports into floating viewports.

The second prompt is:

Specify first corner or [Fit] <Fit>: *(Enter F, specify a point, or press ENTER.)*

- First Corner
- The point specified will be one corner of a rectangle in which the viewports will fit. AutoCAD will drag a rubberband rectangle from this point to assist in locating the opposite corner. The length-to-height ratios of the floating viewports relative to the tiled viewports will be adjusted to fit within the rectangle.
- Fit
- The floating viewports will fit the printable area of the paper.

2

Makes two floating viewports placed either side by side, or one over the other (see Figure 8.9). The follow-up prompts are:

Enter viewport arrangement [Horizontal/Vertical] <Vertical>: *(Enter H or V, or press ENTER.)*

Specify first corner or [Fit] <Fit>: *(Enter F, specify a point, or press ENTER.)*

- First Corner
- The point specified will be one corner of a window in which the two viewports will fit. AutoCAD will drag a rubberband rectangle from this point to assist in locating the opposite corner.
- Fit
- The two floating viewports will fit the printable area of the layout's paper.

3

Makes three floating viewports in one of the arrangements shown in Figure 8.9. The follow-up prompts are:

Enter viewport arrangement

[Horizontal/Vertical/Above/Below/Left/Right] <Right>: (*Enter an option or press ENTER.*)

Specify first corner or [Fit] <Fit>: (*Enter F, specify a point, or press ENTER.*)

- First Corner
- The point specified will be one corner of a window in which the three viewports will fit. AutoCAD will drag a rubberband rectangle from this point to assist in locating the opposite corner.
- Fit
- The three floating viewports will fit the entire printable area of the layout's paper.

4

Divides a designated screen area horizontally and vertically into four equal-sized floating viewports (see Figure 8.9). The follow-up prompt is:

Specify first corner or [Fit] <Fit>: (*Enter F, specify a point, or press ENTER.*)

- First Corner
- The point specified will be one corner of a window in which the four viewports will fit. AutoCAD will drag a rubberband rectangle from this point to assist in locating the opposite corner.
- Fit
- The four floating viewports will fit the entire printable area of the layout's paper.

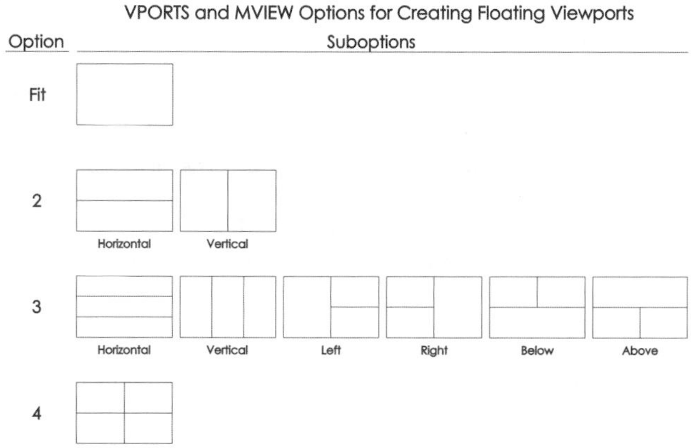

Figure 8.9

THE VPCLIP COMMAND

With this command, you can transform the boundary of an existing floating viewport into virtually any shape that can be drawn as a polyline, spline, circle, ellipse, or region. The format for VPCLIP is:

Command: VPCLIP

Select viewport to clip: (*Pick a point on the boundary of a viewport.*)

When you select a viewport that has not been modified by VPCLIP, the next prompt will be:

Select clipping object or [Polygonal]: (*Select a closed object, enter a P, or press ENTER.*)

POLYGONAL

Choose this option to draw a new viewport boundary composed of straight line and arc segments. Command line prompts similar to those of the PLINE command will be issued as you draw the new boundary.

CLIPPING OBJECT

Select an existing circle, ellipse, region, closed 2D or 3D polyline, or closed spline. Polylines must have been closed by their command's Close option.

When, in response to the Select Viewport to Clip prompt, you select a viewport that has been modified by VPCLIP, the next prompt will be:

Select clipping object or [Polygonal/Delete] <Polygonal>: (*Select a closed object, enter an option, or press ENTER.*)

DELETE

This option restores the original boundary of the floating viewport.

 Tips: Use a special layer for floating viewports. That layer can then be turned off to remove the border around the viewports. If you need to move, stretch, or perform any other operation on the viewport, however, the border is needed for selecting points.

The command line options of MVIEW and VPORTS that create multiple floating viewports and place the viewports touching one another. If you prefer a gap between them, you can move them after they are created.

After making one viewport, you can make copies of it for the other viewports you need. The copies will have the same properties as the original.

During plotting, hidden line removal will be done only in viewports that have hidden line removal turned on. The Hide Objects option of the PLOT command has no effect on objects in floating viewports.

Once you have established the scale and viewpoint you want within a floating viewport, you can ensure they are not inadvertently changed by locking the viewport.

If you forget which viewports have been selected for hidden line removal during plotting or have been locked to their current view and zoom level, the LIST command will tell you if a viewport will plot with hidden lines removed or if locking is on.

The options for creating nonrectangular viewport are most useful in 2D drawings of buildings and structures. On 3D models, they will occasionally be useful for making auxiliary views that are skewed to the X and Y axes.

Viewports that have been created by the Object or Polygonal options of VPORTS and MVIEW, or by VPCLIP can be rotated by the ROTATE command. The view within the viewport, however, will not rotate.

SWITCHING BETWEEN PAPER SPACE AND MODEL SPACE

You will often need to switch between paper space and model space as you prepare to make a drawing of your model. There are three ways to do this. One way is to type MS or PS in from the keyboard to enter floating model or paper space. A second way is to double click inside of any viewport to enter floating model space; double clicking outside of the viewport will switch back to paper space. A third way is to click the button labeled Model on the bottom edge of the AutoCAD graphics area to enter floating model space, and click one the Paper button to enter paper space.

Each time you reenter a space, it will be just as was the last time you worked in it, except for any model space changes that were made. When you work in model space, you can freely change the zoom level, viewpoint, and even the UCS without affecting any floating viewports you have set up. If you move or rotate the model, though, the views in floating viewports will be affected. Because you will generally not create any paper space layouts or viewports until you are finished constructing your 3D model, you will use the Tilemode method of switching between model and paper space only when you need to make significant modifications to your model.

The second way, which is the method you will use most often, is through a floating viewport. Once you enter a floating viewport, the paper space UCS icon will disappear, and the UCS icons in the floating viewports will reappear, as shown in Figure 8.10. Also, the screen crosshair cursor will be confined to one viewport. The border of the current viewport will be slightly wider than the borders of the other viewports. If several floating viewports are visible, you can change current viewports by moving the cursor to another viewport and clicking the pick button of your pointing device. You can also change viewports by simultaneously pressing the CTRL and R keys.

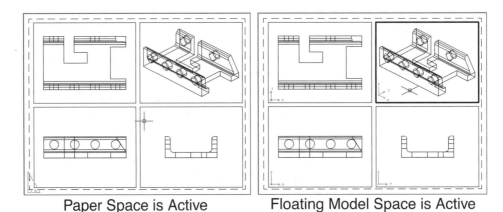

Paper Space is Active   Floating Model Space is Active

Figure 8.10

The MSPACE command is one of three ways to enter a floating viewport. This command has no options or prompts. It simply transfers control to the most recently used floating viewport. Conversely, you can invoke the PSPACE command to return to paper space.

In AutoCAD 2002, a second way to switch spaces is to double-click within the border of a floating viewport to enter model space and to double-click outside the border of a floating viewport to return to paper space.

The third way to switch spaces is with the rightmost button on the status bar located on the bottom edge of the AutoCAD window. When you are in paper space, the label on this button will be PAPER, and you can click it to enter the most recently used floating viewport. The label on the button will then change to MODEL, and you can click it to return to paper space.

 Tip: Model space zooms only affect the current viewport. Consequently, if the viewport is small, increasing the zoom level may not give you the magnification you desire. Therefore, you will often find it better to zoom in close within paper space before you switch to model space.

SETTING UP VIEWS OF THE 3D MODEL WITHIN FLOATING VIEWPORTS

After you have created the floating viewports for a multiview 2D drawing of your 3D model, you will need to:

1. Set the proper view direction within each viewport.

2. Set the proper scale for each viewport.

3. Align the views with each other.

Even though you can have AutoCAD perform the first two steps as viewports are created, we will explain in this section how you can manually perform them to ensure that you have alternate methods to set up viewports and to modify their setup. None of these steps require commands you do not already know. AutoCAD does, however, have a specialized command based on AutoLISP that provides some additional help in setting up floating viewports. We will describe that command after we go through these basic steps.

To set view directions, first use MSPACE or its equivalent to access model space and then set the viewpoint you want for each floating viewport. You can use VPOINT, DVIEW, 3DORBIT, or VIEW to set the viewpoint, whichever you prefer. Typically these will be orthographic views, looking straight at the model's top, front, or side, but you may want an isometric view also, because they are easy to set up when you have a 3D model. If the model has a slanted surface that calls for an auxiliary view, you may need to use the Twist option of DVIEW to rotate the view into the proper angle. Although you can rotate viewports (with the ROTATE command) made with the Object and Polygonal options of VPORTS and MVIEW, the images in them will not rotate.

After you have set the view directions, you will need to scale each viewport by using an XP scale factor with the ZOOM command options of Center, Left, or Scale. (Although the Left option is no longer shown in the command line prompt of ZOOM, AutoCAD still recognizes it.) If, for example, you wanted a view to be half-scale, you could use a zoom scale factor of 0.5XP; and if you wanted it to be twice true size, you could use a zoom scale factor of 2XP. If you forget what scale a viewport has, the LIST command will show you its scale relative to paper space.

The last step in setting up orthographic views is to align them so that points in the top view are exactly over corresponding points in the front view, points in the front view are horizontal with corresponding points in the side view, and so on. There are several ways to do this, but one good method involves moving the viewports. After selecting a viewport to move, use an object snap on a key point on the model for the base point, and use a snap to grid, a point filter (either .X or .Y), or typed-in coordinates for the destination point. Lining up auxiliary views, which are at odd angles, often requires ingenuity, and you may have to resort to temporary construction lines for aligning them.

Tip: It is considered good practice to organize all viewports created in paper space on a special layer named Viewports for example. Assign this layer the color gray for it to be easily distinguished from other colors. Finally assign the no plot state to this layer in the Layer Properties Manager dialog box. Now the layer will be visible on the display screen but will not plot out.

456

TRY IT! - SETTING UP PAPER SPACE VIEWPORTS

We will start a 2D orthographic drawing of the surface model shown in Figure 8.11. We started this model as a wireframe in Chapter 3, added most of its surfaces in Chapter 4, and finished surfacing it in Chapter 5. This model is for a mold of the enclosure of an electronic device, such as a cathode ray tube. Since it is for a mold, we need only one surface. Its front, which is pointed away in the figure, is open, and the back side has cutouts for switches and plugs. The overall size of the model is 7 by 7 by 6 inches.

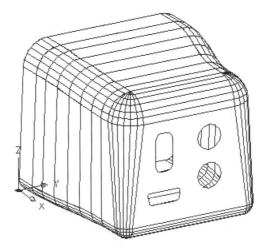

Figure 8.11

Take the following steps to set up the paper space drawing:

1. Open your computer file that contains this 3D model, or else open file 3d_ch8_01.dwg on the accompanying CD-ROM.

2. We intend to print the drawing on 22-by-17-inch paper, so you should make certain that you have a plotter configured that can handle that paper size. (If you don't, open the Plotting tab of the Options dialog box, select Add or Configure Plotter, and add a suitable plotter.) If you would rather use 11-by-8.5-inch paper, you will have to divide all of the distances and scales we will give for this drawing by 2.

3. Click the tab labeled Layout1 that is on the bottom side of the AutoCAD graphics window. If the Page Setup dialog box does not automatically open, enter PAGESETUP on the command line. Select the Plot Device tab, and select the plotter you intend to use for the drawing. Click the Layout settings tab of the dialog box, and set the paper size to 22 by 17 inches. If the plotter you selected cannot handle that size, select the size that most nearly matches it. Verify that the Plot Area is set to Layout, and that the Plot Scale is one-to-one. Then click the OK button.

4. The size of the paper in the layout will immediately adjust to the size you specified. The printable margins will be displayed as dashed lines, and the UCS origin will be on the lower left corner of the printable margins. If a floating viewport exists, erase it.

5. We will use three basic views for the drawing—a top view, a side view, and a back view. Don't let the view names we have used here confuse you. Although the front view is often located below the top view, what we refer to as the front of this particular model is of no interest to us. Therefore, we will position the side view of this model below the top view and position the back view to the right of the side view. Even though three full-size views will fit on the paper we intend to use, there would not be much room left for dimensions and notes. Therefore, we will make half-scale views and probably use an enlarged detail for the cutouts.

6. Create, if necessary, a layer named VPORT and make it the current layer. Use MVIEW or VPORTS to make four floating viewports. Do not use any options that set the viewpoint or scale as the viewports are created. Each viewport should be about 5 inches wide and 5 inches high. At this stage, neither the viewport sizes nor their locations need to be exact. The 3D model will appear in each viewport, with the same viewpoint it last had in model space.

7. Use the MSPACE command or its equivalent, to go into model space and set the view direction for each viewport. The VPOINT command is convenient here for setting the view directions shown in the Figure 8.12. The top view has the view direction coordinates of 0,0,1; the side view below it has the view direction coordinates of 0,-1,0; and the back view as well as the view for the cutout details has the coordinates of 1,0,0.

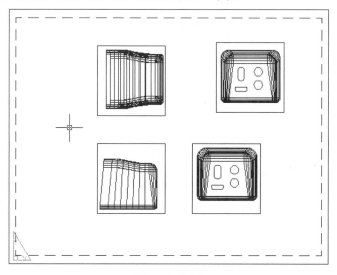

Figure 8.12

8. Next, set a scale in each viewport. Since we decided to use a half-size scale for the main views, go into those three viewports and ZOOM to a .5XP scale. Then scale the viewport for the detail 1XP. The model will be too large for the viewport, but that is not important.

9. Now that the view directions and the scales have been set up, we can align the three main views. Use PSPACE, or its equivalent, to go back into paper space. Then, with the Snap mode activated with a spacing of .25, move the viewports with the half-size views, one by one, until they are aligned as shown in Figure 8.13. This is easily done on this model by using object snaps on key points of the model to establish base points for the moves. The viewport borders do not line up, but that is of no consequence.

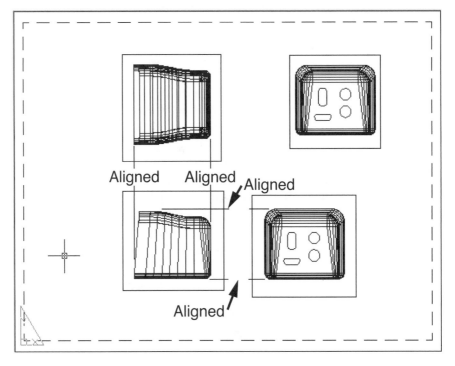

Figure 8.13

Our next exercise will be to add a border and a title block to the paper space drawing. Before we do that though, we will discuss an AutoCAD command that contains several different tools you can use in setting up paper space drawings, including one for inserting a predrawn border and title block. Therefore, you should save your drawing in its current state.

If you are not certain about the precision of your paper space drawing, you can compare your drawing to the one in file 3d_ch8_02.dwg.

THE MVSETUP COMMAND

MVSETUP is an AutoLISP program that serves as a front end for setting up a paper space drawing of a 3D model. It is automatically loaded and can be started from the command line. If Tilemode has not been set to 0, the program is able to do that and proceed with the paper space setup. MVSETUP can create floating viewports, set viewport scales, align views, and create or insert title blocks and borders. However, with one limited exception, the program is not able to set viewpoints within viewports.

In AutoCAD 2002 MVSETUP must be started by entering its name on the AutoCAD command line. The command line format for MVSETUP from model space is:

Command: MVSETUP

Enable paper space? [No/Yes]: <Y>: *(Enter a Y or press ENTER.)*

Enter an option [Align/Create/Scale viewports/Options/Title block/Undo]: *(Enter an option or press ENTER.)*

This menu reappears after an option has been selected and work on it is finished. Press ENTER to end the command.

ALIGN

This option is for aligning views between viewports. AutoCAD will ask for a base point in one viewport and a corresponding base point in a second viewport. It will then pan the image in the second viewport until the two points line up. Typically, you will use object snaps to find points on the model to serve as base points. This option can also rotate a view within a viewport. Its follow-up prompt is:

Enter an option [Angled/Horizontal/Vertical alignment/Rotate view/Undo]: *(Enter an option or press ENTER.)*

This prompt remains in place until you press ENTER to return the main menu.

- Angled pans the view in the second viewport in a specified direction and angle.

- Horizontal pans the second viewport in the Y direction until the base points line up.

- Vertical alignment pans the second viewport in the X direction until the base points line up.

- Rotate view rotates the view within a viewport about a specified base point through a specified angle.

- Undo reverses the most recent Align operation.

CREATE

This option is for creating floating viewports. The viewports will be in the current layer. The option can also delete objects. Its follow-up prompt is:

Enter option [Delete objects/Create viewports/Undo] <Create viewports>: *(Select an option or press ENTER.)*

This prompt remains until you press ENTER twice.

- Create viewports—the default option—uses a submenu that allows you to make one viewport, four viewports, or a set of viewports arranged in rows and columns. In each case AutoCAD will prompt for a boundary that the viewports will fit within. The option for creating four viewports, which is labeled Std. Engineering in the menu, will also set a specific viewpoint in each viewport. The upper left viewport will show the top view of the model, the lower left viewport will show its front, the lower right will show its right side, and the upper right viewport will show an isometric view of the model. The other options for creating viewports do not set viewpoints.

- Delete object is equivalent to the ERASE command.

- Undo reverses the last Create operation.

SCALE VIEWPORTS

The Scale viewports option will first ask you to select the viewports to be scaled and will then set their scale factors to be all the same, or allow you to set different scale factors for different viewports. In either case, AutoCAD will ask for two numbers in setting the scale factor:

Enter the number of paper space units <1.0>: *(Enter a value or press ENTER.)*

Enter the number of model space units <1.0>: *(Enter a value or press ENTER.)*

Pressing ENTER in response to both prompts will result in full-size (one-to-one) scale. Dividing the model space units into the paper space units sets the scale ratio. Therefore, if you wanted the viewport to show the model at one-fourth scale, you would enter 1 for paper space units and 4 for model space units.

OPTIONS

Sets preferences for the title block and for units. Its follow-up prompt is:

Enter an option [Layer/LImits/Units/Xref] <exit>: *(Select an option or*

press ENTER.)

This prompt repeats until ENTER is pressed.

- Layer permits you to specify a layer for the title block. If the layer does not exist, AutoCAD will create it, using the current color and linetype. This option has no effect on the layer for viewports. They are placed in the current layer.

- LImits will automatically set the drawing limits to correspond to the size of the drawing border. It has no practical effect if a border does not exist.

- Units specifies whether the paper space units are to be feet, inches, meters, or millimeters.

- Xref will insert the title block as an externally referenced object.

TITLE BLOCK

This option will insert a ready-made title block and border. It also can move the UCS origin and erase objects. Its follow-up prompt is:

Enter title block option [Delete objects/Origin/Undo/Insert] <Insert>: (*Select an option or press ENTER.*)

This prompt remains until you press ENTER twice.

- Insert will add a ready-made title block and border to the paper space drawing. Twelve standard drawing sheet sizes and styles, for both metric and inch dimensions, are offered, and you can add others to the list if you have them. When you select a title block, MVSETUP looks for a drawing file that matches it, and inserts it as a block if one is found. If a matching drawing is not found, MVSETUP uses data in a file named MVSETUP.DFS to draw the border and title block. For example: If you select the ANSI-D size title block, MVSETUP looks for a file named ANSI_D.dwg, and inserts it, if it is found. If a drawing named ANSI_D.DWG is not found, the program will use data in MVSETUP.DFS to draw the border and title block. This title block will consist of individual objects, rather than a block.

- Delete objects is equivalent to the ERASE command.

- Origin will move the paper space UCS to a new location.

- Undo reverses the previous Title block operation.

Tips: While the inability of MVSETUP to set viewpoints, except for the four viewport options of Create viewports, limits its usefulness, its functions for aligning viewports work very well. Sometimes, in viewports aligned with MVSETUP, the model will be panned partially out of the viewport, but it can be uncovered by stretching the viewport.

AutoCAD 2002 does not provide drawings for standard borders and title blocks. Consequently, the MVSETUP command will draw them from data in MVSETUP.DFS.

The data in MVSETUP.DFS uses the dimensions specified in the ANSI 14.1 drafting standard in drawing title blocks and borders. (ANSI is an acronym for American National Standards Institute.)

The MVSETUP options that set drawing limits and move the UCS are not relevant in the paper space layouts of AutoCAD 2002 and should not be used.

3D

E X E R C I S E S

TRY IT! - USING MVSETUP TO ADD A BORDER AND TITLE BLOCK

We will use MVSETUP to add a title block and border to our drawing of the display enclosure 3D model. They will be placed in a layer we will name BORDER. Open your drawing file of this model or file 3d_ch8_02.dwg. Then start the MVSETUP command.

Command: MVSETUP

Enter an option [Align/Create/Scale viewports/Options/Title block/Undo]: O

Enter an option [Layer/LImits/Units/Xref] <exit>: L

Enter layer name for title block or [. (*for current layer*)]: BORDER

Enter an option [Layer/LImits/Units/Xref] <exit>: (*ENTER*)

Enter an option [Align/Create/Scale viewports/Options/Title block/Undo]: T

Enter title block option [Delete objects/Origin/Undo/Insert] <Insert>: (*Press ENTER.*)

Available title blocks:

   0:   None

   1:   ISO A4 Size(*mm*)

   2:   ISO A3 Size(*mm*)

   3:   ISO A2 Size(*mm*)

   4:   ISO A1 Size(*mm*)

   5:   ISO A0 Size(*mm*)

6:     **ANSI-V Size(***in***)**

7:     **ANSI-A Size(***in***)**

8:     **ANSI-B Size(***in***)**

9:     **ANSI-C Size(***in***)**

10:   **ANSI-D Size(***in***)**

11:   **ANSI-E Size(***in***)**

12:   **Arch/Engineering (***24 x 36in***)**

13:   **Generic D size Sheet (***24 x 36in***)**

Enter number of title block to load or [Add/Delete/Redisplay]: 9

Create a drawing named ansi_c.dwg/ <Y>: N

Enter an option [Align/Create/Scale viewports/Options/Title block/Undo]: (*Press ENTER.***)**

As soon as the 9 key is pressed, signaling we want a C-sized (22-by-17 inch) border, AutoCAD inserts the drawing ansi_c.dwg as a block, or draws the border in place if it cannot find ansi_c.dwg. In AutoCAD 2002, ansi_c.dwg does not exist; therefore, the border and title block will be drawn and will consist of individual objects. If the border extends beyond the printable margins of the layout's paper, you can use the STRETCH command to make it fit within the margins.

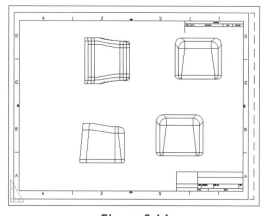

Figure 8.14

Our drawing with the title block in place is shown in Figure 8.14. In this figure, the layers that the surfaces are in have been turned off, along with the layer the viewports are in. It is beginning to look a little more like a standard, multiview engineering drawing.

The drawing at this stage is in file 3d_ch8_03.dwg.

Surface models often present a problem when making drawings because their mesh lines can be unsightly and confusing. We were able to turn off the layers with the surfaces on this model and use the underlying wireframe. How well this technique works depends on the geometry of the surface model; sometimes you simply have to accept the surface meshes. Solid models do not present these problems because you can turn off their mesh lines. Furthermore, AutoCAD has some special techniques and commands for making 2D drawings from 3D solids that we will discuss later in this chapter.

OBJECT VISIBILITY CONTROL

As you work with multiple floating viewports in paper space, it is easy to forget that each viewport is looking at the same 3D model. Any changes and additions you make to the model in one view are instantly shown in the other views (provided, of course, the area in which the changes occur are within a viewport's field of view). Most of the time this is desirable—in fact, it is one of the purposes of paper space—but, there are times when you do not want some model space objects to be seen in every floating viewport.

AutoCAD solves this problem through an object's layer. The VPLAYER command controls which layers are to be visible in each viewport. If there is some object in 3D space that you do not want shown in a particular floating viewport, you can place that object in a layer that will not be visible in that viewport. You will find this to be a useful tool for managing model space objects within paper space viewports and, as we will see later, it can be a necessary tool when adding dimensions.

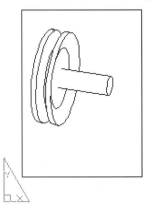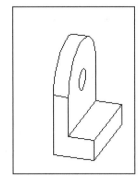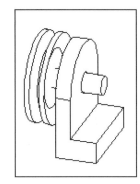

Figure 8.15

Figure 8.15 shows two different 3D solid objects within three floating viewports. The viewport on the right shows both objects, whereas in each of the other two viewports only a single 3D solid is shown. The other solid is in a layer frozen by VPLAYER in that viewport. Even though two objects are within each of the three viewports, each object is not visible in all of them.

THE VPLAYER COMMAND

VPLAYER (for Viewport Layer) controls layer visibility by freezing and thawing layers on a viewport-by-viewport basis. There is a subtle difference between layers frozen by the LAYER command and those frozen by the VPLAYER command, and if a layer has been frozen by the LAYER command, it must first be thawed by the LAYER command before VPLAYER can have any control over it. Also, because VPLAYER controls the relationship of layers to floating viewports—not to object creation—the current layer can be turned off, even in the current viewport.

In AutoCAD 2002, VPLAYER must be started by entering its name on the AutoCAD command line. As you would expect, VPLAYER can only be used in a paper space layout. However, you can be within a floating viewport invoke VPLAYER. Its command line format is:

Command: VPLAYER

Enter an option [?/Freeze/Thaw/Reset/Newfrz/Vpvisdflt]: (*Enter an option or press ENTER.***)**

This prompt reappears after each option is completed. Pressing ENTER ends the command.

?

The question mark option displays the names of layers frozen in specific viewports. A follow-up prompt asks you to select one viewport. AutoCAD will then show that viewport's ID number, along with a list of layers that are frozen in it. The list will not include layers that have been frozen globally through the LAYER command.

FREEZE

Freezes one or more layers in one or more viewports. Its follow-up prompts and options are:

Enter layer name(s) to freeze: (*Enter a name list.***)**

Enter an option [All/Select/Current] <Current>: (*Enter an option or press ENTER.***)**

You can freeze several layers at one time by entering their names separated by commas or by using wild card characters (? or *).

- Current
- This option freezes the selected layers in the current viewport. If VPLAYER was invoked from model space, the current viewport will be the floating viewport that currently contains the crosshair cursor. If VPLAYER was called from paper space, the current viewport is the main paper space viewport encompassing the entire layout.
- All
- The All option freezes the selected layers in all viewports, including the main paper space viewport.
- Select
- This option allows you to select specific floating viewports in which you want the selected layers frozen. If VPLAYER was called from model space, AutoCAD will temporarily switch to paper space to allow you to select the viewports.

THAW

Thaws one or more layers in one or more viewports. It will not thaw layers that were frozen through the LAYER command. Its follow-up prompts and options are:

Enter layer name(s) to thaw: *(Enter a name list.)*

Enter an option [All/Select/Current] <Current>: *(Enter an option or press ENTER.)*

You can thaw several layers at one time by entering their names separated by commas or by using wildcard characters (? or *).

- Current
- This option thaws the selected layers in the current viewport. If VPLAYER was invoked from model space, this will be the floating viewport that currently contains the crosshair cursor. If VPLAYER was called from paper space, the current viewport is the main paper space viewport encompassing the entire layout.
- All
- The All option thaws the selected layers in all viewports, including the main paper space viewport.
- Select
- This option allows you to select the floating viewports in which you want the

layers thawed. If VPLAYER was called from model space, AutoCAD will temporarily switch to paper space to allow you to select the viewports.

RESET

Resets one or more layers in one or more viewports to their default frozen/thawed condition established with the Vpvisdflt option. The follow-up prompts and options are:

Enter layer name(s) to reset: (*Enter a name list.*)

Enter an option [All/Select/Current] <Current>: (*Enter an option or press ENTER.*)

You can reset several layers at one time by entering their names separated by commas or by using wildcard characters (? or *).

- Current
- Resets the selected layers in the current viewport. If VPLAYER was invoked from model space, this will be the floating viewport that currently contains the crosshair cursor.
- All
- The All option resets the selected layers in all viewports, including the main paper space viewport.
- Select
- This option allows you to select the floating viewports in which you want the layers reset. If VPLAYER was called from model space, AutoCAD will temporarily switch to paper space to allow you to select the viewports.

NEWFRZ

This option creates new layers that will be automatically frozen in all viewports—both existing viewports (except for the main paper space viewport) and viewports that will be created later. The follow-up prompt is:

Enter name(s) of new layers frozen in all viewports: (*Enter a list of names.*)

You can enter several names, separated by commas, but not wildcard characters. This option creates new layers, but it does not allow you to specify their color or linetype.

VPVISDFLT

The name of this option stands for Viewport visibility default. It permits you to specify the layers that are to be frozen or thawed in subsequently created viewports. The follow-up prompts are:

Enter layer name(s) to change viewport visibility: (*Enter a list of names.*)

Enter a viewport visibility option [Frozen/Thawed] <Thawed>: (*Enter F or T, or press ENTER.*)

You can name several layers at one time by entering their names separated by commas or by using wildcard characters (? or *).

- Thawed

- The selected layers will be in a thawed condition in subsequent viewports. This option can override the effect of the Newfrz option.

- Frozen

- The selected layers will be frozen in subsequent viewports.

FREEZING LAYERS IN VIEWPORTS THROUGH THE LAYER PROPERTIES MANAGER DIALOG BOX

The AutoCAD 2002 Layer Properties Manager dialog box of the LAYER command can be used to set some of the VPLAYER options. First, you must make a viewport active while in Paper Space. Then, activate the Layer Properties Manager dialog box (you may have to stretch the dialog box slightly to the right to expose the viewport freeze tools.) Once in the dialog box, the column headed Active VP Freeze and the checkbox labeled Freeze in Active Viewport is equivalent to VPLAYER's Freeze option. The column headed New VP Freeze and the checkbox labeled Freeze in New Viewports is equivalent to VPLAYER's Newfrz option.

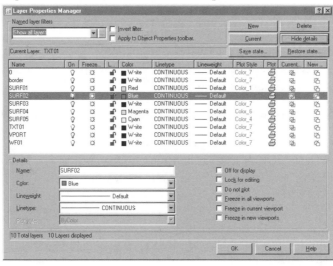

Figure 8.16

Much of the time, though, you will prefer to use the command line options of VPLAYER, not only because there are more options but also because it allows you to control the visibility of several different layers in several different viewports at one time.

ANNOTATING AND DIMENSIONING 3D MODELS

Title blocks, borders, notes, and other annotations belong in paper space. The only exceptions would be for notes and symbols pertaining to a particular feature, or points, on the model—such as a surface roughness symbol—that you may prefer to keep closely tied to the model.

You add these items just as you have always done in 2D drafting with AutoCAD, except you do not have to compensate for drawing scale. The drawing will be plotted in one-to-one scale, and your paper space text and symbols will also be one-to-one. The scale of the model, as we have seen, is handled by viewport zoom levels.

A potentially undesirable side effect from viewports having different scales is that linetypes having broken segments, such as center lines and dashed lines, will have shorter line segments in some viewports than in others, depending on each viewport's zoom scale factor. To prevent this from happening, the system variable Psltscale (Paper Space Linetype Scale) can be set to 1 to force all linetypes to be scaled to each viewport's zoom scale factor. Then, all center lines and all dashed lines will have the same appearance in all floating viewports, even if the scale of the model varies between viewports.

In Figure 8.17, the center line in the detail view, which has a 2XP zoom scale, would appear to have line segments twice as long as it does in the other viewports if Psltscale were set to 0. When it is set to 1, the center line has the same look in all viewports.

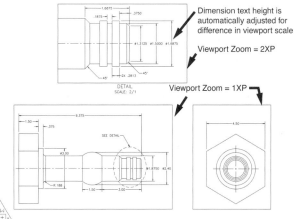

Figure 8.17

Dimensions can be added using either of two fundamentally different approaches. One is to add the dimensions in paper space, while the other is to add dimensions in model space. Each approach has advantages and disadvantages when compared, and both have some implementation problems.

The preferred approach, though, is to add dimensions in model space through floating viewports. The dimensions are closer to the model and are more likely to be changed when changes are made to the model than dimensions in paper space. In fact, associative dimensions have no meaning when they are in paper space while the object of the dimensions is in model space. Also, since AutoCAD does not fully recognize model space objects from paper space, you must establish dimension points by object snaps—especially endpoint snaps. In addition, model space arcs and circles cannot be dimensioned directly from paper space.

On the other hand, AutoCAD 2002 allows for dimensions to be easily added in paper space. You generally have all of the views in front of you; which view should receive a dimension, as well as where the dimension should be located, is often more apparent than when you are in model space. Furthermore, you are not concerned with the visibility of dimensions, so you can use a simpler layer setup.

We will explore both the model space and the paper space approaches in adding dimensions to the drawing of the mold for the electronic case we've been working on. If you would like to add dimensions this model yourself, you can use the version in file 3d_ch8_04.dwg, so that your layer names and visibility will be consistent with our descriptions. Although we will not describe the dimensioning process step by step, you will still be able to follow the process.

We have globally frozen the layers of the surfaces so that their mesh lines will not show, and we have used VPLAYER to selectively freeze wireframe objects so that they show only in the appropriate view. After adding some notes and completing the title block the drawing looks like the one in Figure 8–18, and it is ready for dimensions.

DIMENSIONING IN MODEL SPACE

The first step in adding dimensions in model space is to set the Dimscale dimension variable to 0. You can do this directly from the command line, but you will probably use the dialog box of the DDIM command. Select the Fit tab, and select the checkbox labeled Scale Dimensions to Layout (Paperspace).

When Dimscale is set to 0, AutoCAD will automatically adjust the size of dimension text, arrow heads, dimension line offsets, and other dimension features according to each floating viewport's zoom scale factor. AutoCAD makes these adjustments by dividing overall paper space scale by the viewport's paper space zoom scale factor. Because overall paper space scale is virtually always 1, the size adjustment scale factor is, in effect, the reciprocal of the viewport's zoom scale relative to paper space.

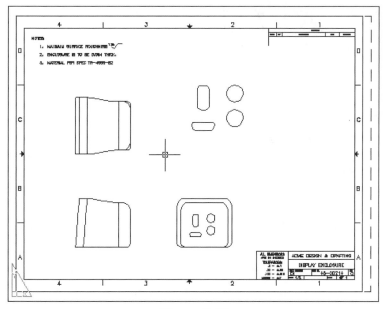

Figure 8.18

For example: The image in the two lower viewports of the paper space drawing in Figure 8.17 has been zoomed to 1XP (full scale relative to paper space), and the image in the upper right viewport has been zoomed to 2XP (magnified by 2 relative to paper space) to show some details on the model. Since Dimscale was set to 0, dimension features, such as text and arrow heads, are the same size in all viewports. Even though the dimension features (not their values) in the detail view are actually half their image size, no calculations or adjustments were needed from the user. AutoCAD did them all.

Dimensioning is a 2D operation, with dimension features—text, extension lines, dimension lines, and so forth—always being placed on the XY plane by AutoCAD. Horizontal dimensions are parallel with the X axis, and vertical dimensions are always parallel to the Y axis. Moreover, the object being dimensioned must be parallel to the current XY plane. If it not, AutoCAD will project the length of the dimension onto the XY plane, resulting in a dimensioned length that is too short.

Consequently, you must properly orient the UCS before adding any dimensions. One way to do this is to use MSPACE or its equivalent to activate model space, go to the viewport you intend to add the dimensions to, and use the View option of the UCS command.

One more consideration when adding dimensions in model space is that every dimension is likely to show up in every viewport. It may be backward, or it may look like a line, but it will be there (unless it is beyond the viewport's border). Therefore, you must have a separate layer for dimensions in each viewport, and then use VPLAYER to freeze all but the applicable layer in each viewport. As the drawing of the mold we've been working on has four viewports, we'll use four different layers for its dimensions—DIM_TOP for the top view, DIM_SIDE for the side view, DIM_BACK for the back view, and DIM_DETAIL for the detail view. In the side view, for example, only the DIM_SIDE layer for dimensions will be thawed, with the other three views handled similarly.

Figure 8.19 shows the side view of our mold model in model space from slightly off-center (the view direction is -95° from the X axis, and 5° from the XY plane), and with just the side view dimensions. Notice that all of the dimensions, even the radii of the arcs, have been projected onto the UCS XY plane. AutoCAD makes these projections regardless of where or how the dimension points were selected. Notice also that the spline curves have not been dimensioned. Radius and diameter dimensions have no meaning on 3D curves such as these.

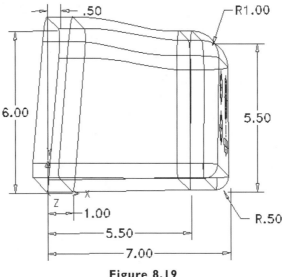

Figure 8.19

Usually, simply using the View option of the UCS command will orient the UCS properly for dimensioning within a viewport, but sometimes you will need to move the UCS to the front part of the view. Otherwise, dimensions in back of the model may be all, or partially, hidden by the model itself. Even though this isn't a problem with this model because we're dimensioning the model without its surfaces, we moved the UCS to the back of the model (which is in the front of the viewport) before adding the back view and detail view dimensions. Figure 8.20 shows just the back view dimensions in model space. The view direction is -5° from the X axis and 5° from the XY plane.

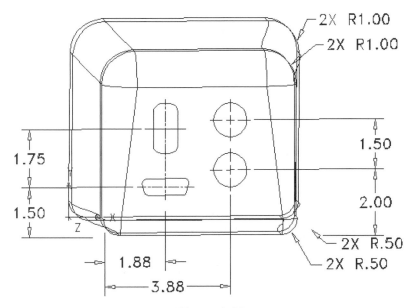

Figure 8.20

Most of the same techniques used in dimensioning 2D objects can be used to add dimensions to 3D models, especially if the model is made of wireframe objects like this one. Often dimensions will extend beyond the viewport border, but you can stretch the viewport to uncover them. The completely dimensioned drawing is shown in Figure 8.21 with the viewport layer turned off. Notice that the dimension features, such as arrow heads and text, are the same size in the detail view as in the other views, even though its scale is twice as large.

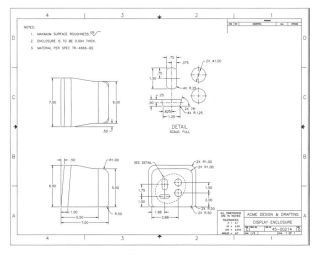

Figure 8.21

The completely dimensioned drawing is in file 3d_ch8_05.dwg.

DIMENSIONING IN PAPER SPACE

A major problem with dimensioning in paper space, aside from the separation of dimensions from the objects they represent, is that AutoCAD does not recognize model space objects from paper space. AutoCAD can find object snap points—including centers, midpoints, and quadrants—but not lines or circles. As a result, you must set extension line origins by pointing to endpoints, not just by picking an entity. Diameters, radii, and angles can be especially bothersome to dimension because AutoCAD cannot recognize circles, arcs, or lines from paper space.

The dimension variable Dimscale, which we set to 0 when dimensioning in model space, can be set to either 0 or 1 when dimensions are added in paper space. When it is set to 0, AutoCAD automatically uses a scale factor of 1 when dimensions are added within paper space. This is to be expected because paper space is intended to have a one-to-one relationship between drawing size and plot size, so there is no need to adjust the size of arrowheads, text, extension line offset, or other dimensioning features having a size.

When the paper space drawing has floating viewports with images that are not scaled one-to-one relative to paper space, as are the top, side, and back views of our drawing, compensation must be made for the differences in scale. This is done through the Dimlfac (Dimension Length Factor) dimension variable. AutoCAD multiplies the measured dimension distances by the value of Dimlfac to come up with a dimension length. In our drawing, dimensioned distances in the top, side, and back views, which are half-scale, will be correct when Dimlfac is set to 2. Since the detail view is full scale, no compensation is needed.

AutoCAD can help you set Dimlfac, although it is not as automatic as with the Dimscale dimension variable. When you set Dimlfac from the dimension mode, AutoCAD will give you the option of entering a new value by typing in a number or selecting a viewport. If you use the option to select a viewport, AutoCAD will use that viewport's zoom scale factor relative to paper space to set Dimlfac. You do not get this option when Dimlfac is accessed directly, or as a system variable. This method works even in AutoCAD 2002, which does not need but still accepts the dimension mode.

If you used this method to set Dimlfac for the either the top, side, or back views of our drawing, the command line sequence of prompts and input would be:

Command: DIM1 (*or DIM.*)

Dim: DIMLFAC

Enter new value dimension variable, or Viewport <1.0000>: V

Select viewport to set scale: (*Select the viewport you wish to dimension.*)

DIMLFAC set to -2.0000

As a result, all dimension lengths in that viewport will be multiplied by 2. The minus sign in Dimlfac is a paper space signal to AutoCAD.

Once Dimlfac has been set, you can begin adding dimensions as you choose. It is easy to skip from one view to another—adding, moving, and changing dimensions at will—as long as the viewports are the same zoom level. While associative dimensions are possible, there is little reason to use them because the dimensions and the model are in different spaces.

The drawing of the mold we've been working on would look virtually the same when dimensioned from paper space as when dimensioned in model space.

PLOTTING PAPER SPACE DRAWINGS

Plotting is straightforward because the floating viewports have already taken care of scales. All plots of the complete paper space drawing are at one-to-one scale, with the plot area based on the layout. Also, since hidden lines are also taken care of by viewports, you do not need to hide lines during plotting. In fact, that plotting option has no effect on paper space drawings.

You can turn off, or freeze, the layer in which the viewports reside to eliminate the border around the viewport. Just their border will disappear, while the viewport contents remain. You must do this through the LAYER command, not through VPLAYER.

Orthographic views of 3D models will almost always have lines on top of other lines, which will cause pen plotters to redraw a line several times, often with undesirable results. To avoid this, you can set AutoCAD's plotter pen optimization to eliminate plotting of overlapping vectors. To do this, select the Plot Device tab of the Page Setup dialog box and then select Properties to open the Plotter Configuration Editor dialog box (see Figure 8.22). Select Pen Configuration from the outline list of plotter properties, and set the Pen Optimization Level to Adds Elimination of Overlapping Diagonal Vectors.

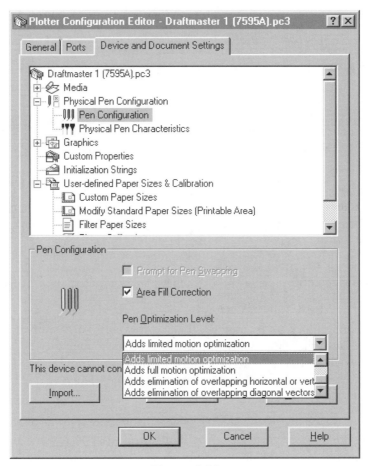

Figure 8.22

Plotting overlapping vectors is not a problem with laser and inkjet printers. Also, the plot optimization levels that eliminate overlapping vectors are not efficient for plotting other types of drawings.

SOLID MODELS IN PAPER SPACE

AutoCAD has three specialized commands for working with 3D solids in paper space—SOLPROF, SOLVIEW, and SOLDRAW. These commands create an intermediary object from the 3D solid for viewing and dimensioning rather than using the model itself, as is done with 3D surface and wireframe models. These intermediary objects are made of wireframe entities—such as lines, arcs, circles, and ellipses—that are easier to dimension than 3D solids, and they allow hidden edges to be displayed as dashed lines. Generally, they are projected on a plane that is perpendicular to the line of sight, although SOLPROF can make a 3D wireframe.

These three commands are in an external program named SOLIDS.ARX, which is loaded automatically the first time any of the three commands are invoked. The menus and toolbar buttons for initiating these commands are shown in Figure 8.23.

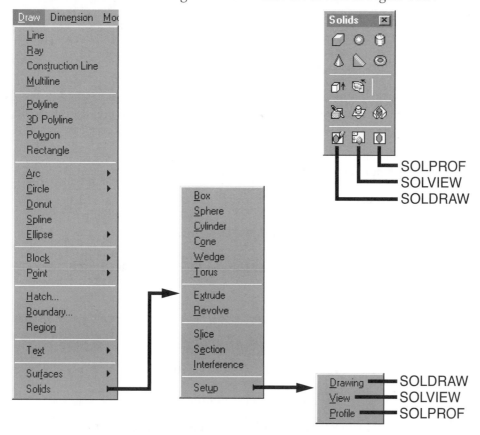

Figure 8.23

THE SOLPROF COMMAND

The SOLPROF command creates blocks made of wireframe objects based on the profile and edges of one or more 3D solids. Although the blocks are located in model space, the command must be invoked within a paper space floating viewport. All edges, whether they are visible or hidden (except those seen head on), are included in the blocks. However, the user does have the option of placing the visible and hidden edges in separate blocks.

AutoCAD uses a layer named PV-handle for the blocks showing the solid's visible edges, and a layer named PH-handle for hidden edges (provided the user specifies hidden edges to be separated from visible edges). Handle in these layer names is the handle of the floating viewport. For example: If the floating viewport's handle is 7A, the names of the layers will be PV-7A and PH-7A. (A floating viewport's handle is a hexa-decimal number assigned to it by AutoCAD. You can see a viewport's handle through the LIST command.)

If these layers do not exist, AutoCAD will automatically create them, assigning the HIDDEN linetype to the PH-handle layer, if that linetype is loaded. Otherwise, its linetype will be CONTINUOUS.

If the PH-handle layer is not used, the blocks for hidden edges are placed in the PV-handle layer—the same layer as the visible edges. Also, you can choose whether the blocks will be planar or 3D, and you can choose whether or not to display tangent lines that are at the edges of curved surfaces.

The command line format for SOLPROF is:

Command: SOLPROF

Select objects: (*Select one or more 3D solids.*)

Display hidden profile lines on separate layer? [Yes/No] <Y>: (*Enter Y or N, or press ENTER.*)

Project profile lines onto a plane? [Yes/No] <Y>: (*Enter Y or N, or press ENTER.*)

Delete tangential edges? [Yes/No] <Y>: (*Enter Y or n, or press ENTER.*)

Following is a description of each of these options:

DISPLAY HIDDEN PROFILE LINES ON SEPARATE LAYER?

If you respond by entering Y or pressing ENTER, AutoCAD will place visible edges in one block in the PV-handle layer, and hidden edges in a second block in the PH-handle layer. If more than one solid was selected, edges on solids hidden by other solids will be placed in the block located in the PH-handle layer. Only two blocks will be created, regardless of how many solids were selected.

If you respond by pressing N, AutoCAD will place both the visible and hidden edges of each solid in a single block in the PV-handle layer. One block will be created for each solid, with all edges shown on all solids, including those that are behind other solids as well as edges hidden within a single solid.

PROJECT PROFILE LINES ONTO A PLANE?

A positive response will result in 2D blocks placed on a plane that passes through the origin of the UCS and is perpendicular to the viewing direction.

A negative response will make 3D blocks that are located on the selected solids. However, these blocks are not necessarily complete copies of the edges of the solids. For instance, if the view of a solid box is directly toward one side of the box, so that the box looks like a rectangle, the resulting block will not include the four edges seen head on.

DELETE TANGENTIAL EDGES?

Tangential edges occur where curved surfaces on the solid are tangent with an adjacent surface, such as where a fillet blends with adjacent faces on a solid. If you answer Y or press ENTER, to this prompt, AutoCAD will not show tangential edges. If you answer N, they will be shown.

Tips: Results from the SOLPROF command are not immediately apparent because the blocks containing the edges are exactly over corresponding edges of the solid. To see the blocks made by SOLPROF, you will have to move or erase the solid or (a better method) turn off or freeze the layer the solid is in.

The blocks created by SOLPROF are anonymous blocks with no name. If you want to export one of them to another drawing, you can make a named block from it with either the BLOCK or the WBLOCK command. If the SOLPROF blocks are 2D, be certain to first set the UCS XY plane on the plane of the blocks.

Although you can make orthographic views of 3D solids using SOLPROF, the SOLVIEW and SOLDRAW commands are more convenient. SOLPROF is best used for making unique views of 3D solids, perhaps for use in another drawing or even another program.

Often, you will not want tangential edges to be shown. However, on certain geometric shapes they are needed, as shown in Figure 8.24. Suppose, for example, you have a flat plate with a recessed area on one of its surfaces. If all of the edges of that recessed area have been filleted, it will not even show up when tangential edges are eliminated.

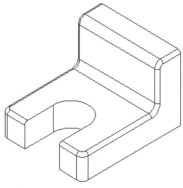
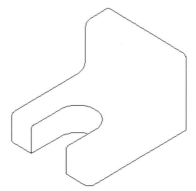

Tangential Edges Shown   Tangential Edges Not Shown

Figure 8.24

TRY IT! - USING THE SOLPROF COMMAND

Figure 8.25 shows a 3D solid in its wireframe mode, and Figure 8.26 shows the same solid with HIDE activated. The Dispsilh system variable has been set to 1 to eliminate polygon meshes from being displayed when HIDE is used.

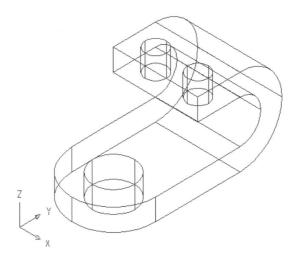

Figure 8.25

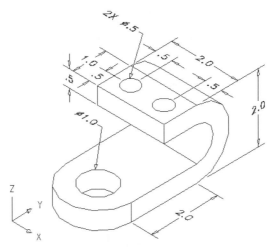

Figure 8.26

Construct this 3D solid using the dimensions shown in Figure 8.26, and set the view direction to 1,-1,1. (An SE isometric view.) Make certain that AutoCAD's HIDDEN linetype is loaded. Then, open an AutoCAD Layout tab, and if one does not exist, make one floating viewport that is about 4 units square. Use MSPACE or any of its equivalents to enter model space, and start SOLPROF.

Command: SOLPROF

Select objects: (*Select the 3D solid.*)

Display hidden profile lines on separate layer? [Yes/No] <Y>: (*Enter Y or press ENTER.*)

Project profile lines onto a plane? [Yes/No] <Y>: (*Enter Y or press ENTER.*)

Delete tangential edges? [Yes/No] <Y>: (*Enter Y or press ENTER.*)

After you turn off the layer the solid is in, the view should look like the one in Figure 8.27. This 3D model and the profile blocks are in file 3d_ch8_06.dwg. Although it looks very similar to the 3D solid, it is actually two 2D blocks made up of lines and ellipses. These profile blocks would not be suitable for dimensioning because their lines are 81.65 percent of their true length. It could, however, be used in an assembly drawing; if it were scaled 1.2247 times, it would be equivalent to an AutoCAD 2D drawing made in the isometric snap mode. Notice that the tangential edges, which AutoCAD always shows on solids, are missing in the profile blocks.

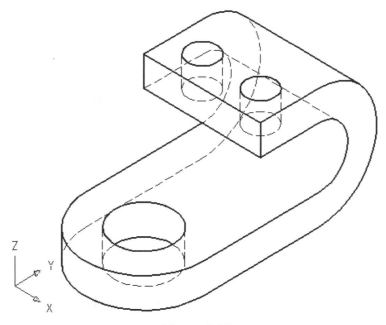

Figure 8.27

THE SOLVIEW COMMAND

SOLVIEW sets up floating viewports for making multiview production drawings from 3D solid models. It is intended to be used with the SOLDRAW command. SOLVIEW sets up the viewports, and SOLDRAW completes them. Specifically, SOLVIEW:

- Creates and aligns floating viewports for the principal orthographic views, auxiliary views, and section views.

- Sets the proper view direction and scale of the model within these viewports.

- Creates layers in each viewport for visible lines, hidden lines, dimensions, and hatches.

- Places the viewports in a layer named VPORTS. If this layer does not exist, AutoCAD will automatically create it.

The layers AutoCAD creates for visible and hidden lines, and for hatches are used by the SOLDRAW command. The layers for dimensions, which are included as a convenience for the user, are thawed only in the applicable viewport, and frozen in all of the others. These layers are:

Layer Name	Items to be Placed in the Layer
View_name-VIS	Visible object lines and edges
View_name-HID	Hidden object lines and edges
View_name-DIM	Dimension objects
View_name-HAT	Section hatches

View_name in these layer names is the name you assign to the view in each viewport as you are using SOLVIEW. For example, if you named the viewport's view SEC, AutoCAD would name the layers SEC-VIS, SEC-HID, SEC-DIM, and SEC-HAT. The View_name-HAT layer is created only if the Section option of SOLVIEW is used.

SOLVIEW is an interactive command—it prompts you for locations, sizes, scale, and view names. If the Tilemode system variable has not been set to 0, AutoCAD will set it for you and proceed with the command. The command line format of SOLVIEW is:

Command: SOLVIEW

Enter an option [Ucs/Ortho/Auxiliary/Section]: (*Enter an option or press ENTER.*)

This prompt repeats after an option is complete. Press ENTER to end the command. All of the options, except Ucs, require an existing floating viewport. Therefore, the first option you select must be Ucs. Examples of these four view types are shown in

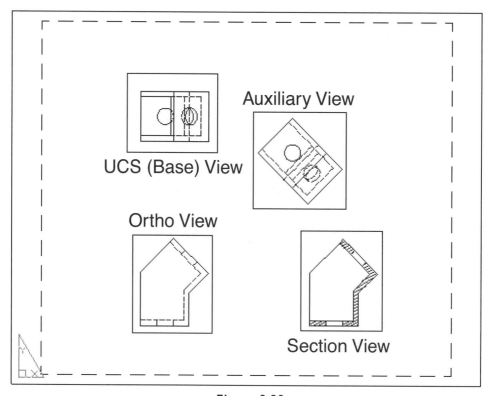

Figure 8.28

UCS

This option creates a floating viewport with a view direction based on the plan view of a UCS. Its follow-up prompt is:

Enter an option [Named/World/?/Current] <Current>: (*Enter an option or press ENTER.*)

- Named

- This option sets the viewpoint to the plan view of a named user coordinate system. AutoCAD will prompt for the name of a UCS.

Enter name of UCS to restore: (*Enter the name of a UCS.*)

- The UCS name you specify must be one that has been saved in the UCS command.

- World

- Sets the viewpoint of the new viewport to the plan view of the WCS.

- ?
- Displays the names of existing named user coordinate systems. AutoCAD displays a prompt that enables you to filter the list of names.

Enter UCS names to list <*>: *(Enter a name list or press ENTER.)*

- Press ENTER to see a list of all UCS names, or use wildcard characters (* or ?) to see a filtered list of UCS names. After listing the names AutoCAD will redisplay the "Named/World/?/Current" prompt.
- Current
- This option sets the viewpoint of the new viewport to the plan view of the current UCS.

Once a UCS has been specified for setting the view direction of the new floating viewport, AutoCAD prompts for the scale of the viewport.

Enter view scale <1.0000>: *(Enter a value or press ENTER.)*

The value you enter will set the viewport's scale relative to paper space. Then AutoCAD will prompt for the location of the viewport.

Specify view center: *(Specify a point.)*

As soon as you specify a point for the viewport's center, AutoCAD will display the model centered at that point, in the specified scale, and will display the prompt:

Specify view center <specify viewport>: *(Specify a point or press ENTER.)*

If you are satisfied with the location, press ENTER. Otherwise, continue to specify other points until you find the location you want, and press ENTER to select it. AutoCAD will then prompt for the size of the viewport.

Specify first corner of viewport: *(Specify a point.)*

Specify opposite corner of viewport: *(Specify a point.)*

These two points establish the opposite corners of the floating viewport's boundary. AutoCAD will continue to display the model and drag a rubberband rectangle from the first corner to help you size the viewport.

Finally, AutoCAD will prompt for the name of the view.

Enter view name: *(Enter a view name.)*

This view name is the same as those assigned and used by the VIEW command. It cannot be the name of an existing view. AutoCAD uses this view name in the names of the layers created for each viewport.

ORTHO

This option creates a floating viewport containing a view that is folded 90° from the view within an existing viewport. AutoCAD will first prompt you to pick one side of the viewport you want the new viewport to be based on.

Specify side of viewport to project: (*Select the edge of a viewport.***)**

AutoCAD will turn ORTHO on, anchor a rubberband line at the middle of the viewport edge, and prompt for the center of the new viewport.

Specify view center: (*Specify a point.***)**

When you select a point AutoCAD will display the model centered at that point in the same scale as the base viewport, and with a view direction rotated 90° from the view direction of the base viewport. Then you will be prompted to move or accept the viewport location.

Specify view center <specify viewport>: (*Specify a point or press ENTER.***)**

You can try any number of center locations until you find the one you want. Signal acceptance by pressing ENTER. Then AutoCAD will prompt for opposite corners of the viewport and for a view name.

Specify first corner of viewport: (*Specify a point.***)**

Specify opposite corner of viewport: (*Specify a point.***)**

Enter view name: (*Enter a view name.***)**

As soon as the viewport's first corner is selected, AutoCAD will anchor a rubberband rectangle to it to help you locate the opposite corner. The view name will be used in the layer names for the viewport.

AUXILIARY

Views that are not parallel to any of the principal orthographic projection planes are made with this option. It creates a viewport containing a view that is perpendicular to a plane on the model and shows the true size of that plane. The option starts by asking you to establish the angle of projection by selecting two points on the edge of the plane.

Specify first point of Inclined planes: (*Specify a point.*)

Specify second point of Inclined plane: (*Specify a point.*)

Specify side to view from: (*Specify a point.*)

The two points for identifying the plane must be in the same viewport, and the plane must be seen edge on—that is, the plane must be parallel to the viewport's view direction. Otherwise, the plane will not be shown true-size in the auxiliary viewport. The point specifying the side to view the plane from must also be in the same viewport, and it must be off the line between the two points for the inclined plane. AutoCAD will then rotate the paper space UCS so that the X axis is perpendicular to the plane, extend a rubberband line from it, turn on the Ortho mode, and prompt for a viewport center.

Specify view center: (*Specify a point or press ENTER.*)

When you select a point, AutoCAD will display the model centered at that point, in the same scale as the base viewport, and with a view direction looking directly toward the plane selected in the base viewport. It will then prompt:

Specify view center <specify viewport>: (*Specify a point or press ENTER.*)

When you find a viewport center location that suits you, press ENTER to signal acceptance. AutoCAD will then prompt for opposite corners of the viewport and for a view name.

Specify first corner of viewport: (*Specify a point.*)

Specify opposite corner of viewport: (*Specify a point.*)

Enter view name: (*Enter a view name.*)

AutoCAD will anchor a rubberband rectangle to the first viewport corner selected to help you locate the opposite corner. Even though the view within the viewport will be tilted, the borders of the viewport will be parallel to the paper space X and Y axes. The view name you specify will be used in the layer names for the viewport.

SECTION

This option sets up a section view of the model. The final view will consist of a cross-hatched section plus edges on the far side of the section that would be visible from the section plane. The objects for the section view, however, are not made until the

SOLDRAW command is used. This option first prompts for two points to define the cutting plane, and then for a direction from which to view the section, plus a scale for the new viewport.

Specify first point of cutting plane: (*Specify a point.*)

Specify second point of cutting plane: (*Specify a point.*)

Specify side to view from: (*Specify a point.*)

Enter view scale <current>: (*Enter a value.*)

The section will be perpendicular to the two cutting plane points, and the "side to view from" point must be off the cutting plane. Also, these three points must be in the same viewport. The view scale will set the viewport's zoom scale relative to paper space. As soon as you enter a view scale AutoCAD will anchor a rubberband line on the cutting plane, turn on the Ortho mode, and prompt for a viewport center.

Specify view center: (*Specify a point or press ENTER.*)

When you select a point, AutoCAD will display the model, centered at that point, in the specified scale, and with a view direction looking directly toward the cutting plane selected in the base viewport. AutoCAD will issue a prompt for you to select another viewport location or accept the current one.

Specify view center <specify viewport>: (*Specify a point or press ENTER.*)

You can try number of center locations until you press ENTER. The view center must be perpendicular to the cutting plane specified in the base viewport for the option to work correctly. After you specify the viewport location, AutoCAD will prompt for opposite corners of the viewport, and for a view name.

Specify first corner of viewport: (*Specify a point.*)

Specify opposite corner of viewport: (*Specify a point.*)

Enter view name: (*Enter a view name.*)

AutoCAD will drag a rubberband rectangle from the first viewport corner to help you locate the opposite corner. The view name will be used in the layer names for the viewport.

Tips: Before you invoke SOLVIEW, you should make certain that the layout you are going to use is set to the paper size you intend to use for the drawing and that no floating viewports exist in the layout.

By default, AutoCAD will place the paper space UCS origin on the lower left corner of the printable paper margin. You should not move the UCS origin from that location.

The Ucs option of SOLVIEW must be the first option you select. This creates a base viewport that enables you to use the Ortho, Auxiliary, and Section options to create the other viewports you need.

Do not place any objects in the layers SOLVIEW makes for each viewport. Objects in some of these layers are deleted whenever the SOLDRAW command is used.

After using SOLVIEW you should examine the layers that were created and set their colors and linetypes to your preferences.

SOLVIEW performs many separate operations, which makes it difficult to undo. Therefore, you should save the drawing and set an UNDO mark just prior to using SOLVIEW.

Although viewports made with the Section option must initially be perpendicular to their cutting plane, they can be moved later.

TRY IT! - MAKING A SECTION VIEW WITH SOLVIEW

In this exercise you will create a 3D solid and then use SOLVIEW to create a two-view drawing of it. The initial setup of the drawing is not important. Draw the center line and the profile shown in Figure 8.29 on the XZ plane of the WCS. The centerline should be on the WCS Z axis so that the coordinates we will use during SOLVIEW will be correct.

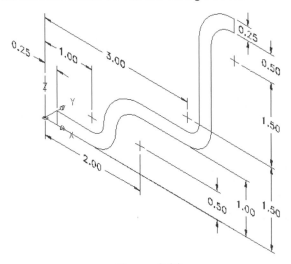

Figure 8.29

Turn the lines and arcs of the profile into a region or 2D polyline and use REVOLVE with the centerline as an axis to make a 3D solid. Your revolved solid, with HIDE activated and Dispsilh set to 1, should look similar to the one shown in Figure 8.30.

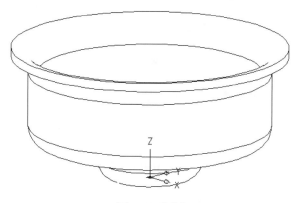

Figure 8.30

Restore the WCS. Then open a layout tab, set the paper size to 11 by 8.5 inches, and erase any floating that may be on the layout page. Now, we will use SOLVIEW to set up a top view and a section view through the center of the solid. We will use a scale of one-half for each view. The command line sequence of prompts and input to create these views is:

Command: SOLVIEW

Enter an option [Ucs/Ortho/Auxiliary/Section]: Ucs

Enter an option [Named/World/?/Current] <Current>: World

Enter view scale <1.0000>: .5

Specify view center: 5.0,5.5

Specify view center <Specify viewport>: (*Press ENTER.*)

Specify first corner of viewport: 2.5,3.0

Specify opposite corner of viewport: 7.5,8.0

View name: top

Enter an option [Ucs/Ortho/Auxiliary/Section]: Section

(AutoCAD will enter model space through the newly created viewport.)

Specify first point on cutting plane: -4,0

Specify second point on cutting plane: 4,0

Specify side to view from: 0,-1

Enter view scale <.5000>: (*Press ENTER.*)

(*AutoCAD will return to Paper space.*)

Specify view center: 4.0<270

Specify view center <Specify viewport>: (*Press ENTER.*)

Specify first corner of viewport: 2.5,0.25

Specify opposite corner of viewport: 7.5,2.5

View name: sec

Enter an option [Ucs/Ortho/Auxiliary/Section]: (*Press ENTER.*)

Your paper space drawing should now look similar to the one shown in Figure 8.31. Also, your drawing file will contain the following new layers: SEC-DIM, SEC-HAT, SEC-HID, SEC-VIS, TOP-DIM, TOP-HID, TOP-VIS, and VPORTS. The layers for the section view are frozen by VPLAYER in the viewport for the top view, and the layers for the top view are frozen in the viewport for the section view. The SEC-HID layer is frozen in the section viewport as well, since hidden lines are not normally shown in section views.

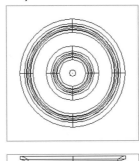

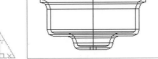

Figure 8.31

Although SOLVIEW has created new viewports and new layers, no new objects have been drawn. The only object in the viewports is your original 3D solid. SOLDRAW will be used in the next exercise to complete the drawing.

THE SOLDRAW COMMAND

SOLDRAW is the companion command of SOLVIEW. SOLVIEW sets up the viewports, while SOLDRAW draws the outlines and edges (both visible and hidden), plus sections and hatches in those viewports. It does a lot of work, but all you have to do is tell it which viewports to work in. The command line format for SOLDRAW is:

Command: SOLDRAW

Select viewports to draw:..

Select objects: (*Select one or more viewports set up by SOLVIEW.*)

The first response by the command is not a prompt—it is a message reminding you what type of objects to select. SOLDRAW makes the edges from individual lines, circles, and arcs, rather than blocks, as SOLPROF does. The entities for the orthographic views are projected onto planes parallel to the XY, XZ, and YZ planes, while entities for auxiliary views and sections are projected onto the planes selected during SOLVIEW. Hatch patterns used for sections are based on the current settings of the Hpname, Hpscale, and Hpang system variables.

In each viewport, layers not needed to display edges and sections, including the original 3D solid's layer, are frozen via the VPLAYER command. SOLDRAW will not work if the viewports were not prepared by SOLVIEW, or if the 3D solid is a block.

 Tip: SOLDRAW does a surprising amount of work, and goes through numerous commands. Consequently, you should save your drawing file and set a mark with the UNDO command before using SOLDRAW.

TRY IT! - MAKING STANDARD PRODUCTION DRAWINGS

First, we will use SOLDRAW to finish drawing our revolved solid. Before you begin, be sure to load the HIDDEN linetype and assign it to the SEC-HID and TOP-HID layers if this linetype wasn't loaded before you activated the SOLVIEW command.

Command: SOLDRAW

Select viewports to draw:..

Select objects: *(Select the two viewports set up with SOLVIEW.)*

You will see much screen flickering and other evidence of activity as SOLDRAW does its work. The results from your drawing, after turning off the VPORTS layer, should be similar to those in Figure 8.32.

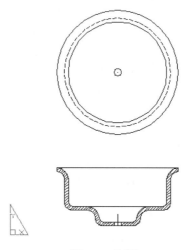

Figure 8.32

 File 3d_ch8_07.dwg contains the 3D solid model (its layer is frozen) along with the SOLDRAW objects.

We will now use SOLVIEAW and SOLDRAW to make a standard production drawing of the bracket we built as a 3D solid in Chapter 5. This model is shown in Figure 8.33, as it appears when HIDE has been used, and with Dispsilh set to 1.

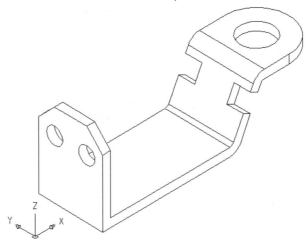

Figure 8.33

Although you could use the model directly in making the drawing, you could not show any hidden edges, and tangential edges would always show. Furthermore, adding dimensions directly to solids is not straightforward. AutoCAD dimensions are intended for wireframe objects—lines, circles, arcs, and so on. The edges on a solid may look like lines, circles, and arcs, but they are not. Consequently, AutoCAD does not always dimension them correctly—especially circular and arc-shaped edges of solids. Therefore, we will use the 2D objects made by SOLVIEW and SOLDRAW for dimensions.

We will use four views in the drawing—a top view, a front view, a left-side view, and an auxiliary view of the slanted region. Also, we will show the model at full scale and use 22-by-17-inch paper. You should also make sure that the HIDDEN linetype is loaded.

Open a layout tab, and select a paper size of 22 by 17 inches, or as close to that size as your plotting device can handle. If any floating viewports exist, erase them. Then use SOLVIEW to make the three orthographic views.

Command: SOLVIEW

Enter an option [Ucs/Ortho/Auxiliary/Section]: Ucs

Enter an option [Named/World/?/Current] <Current>: World

Enter view scale <1.0000>: (*Press ENTER.*)

Specify view center: 11,13

Specify view center <specify viewport>: (*Press ENTER.*)

Specify first corner of viewport: 7,11

Specify opposite corner of viewport: 15,15

Enter view name: top

Enter an option [Ucs/Ortho/Auxiliary/Section]: Ortho

Select side of viewport to project: (*Select the bottom edge of the existing viewport.*)

Specify view center: @3<270

Specify view center <specify viewport>: (*Press ENTER.*)

Specify first corner of viewport: 7,6

Specify opposite corner of viewport: 15,10

Enter view name: front

Enter an option [Ucs/Ortho/Auxiliary/Section]: Ortho

Select side of viewport to project: (*Select the left edge of the previous viewport.*)

Specify view center: @3<180

Specify view center <specify viewport>: (*Press ENTER.*)

Specify first corner of viewport: 2,6

Specify opposite corner of viewport: 6,10

Enter view name: side

Enter an option [Ucs/Ortho/Auxiliary/Section]: (*Press ENTER.*)

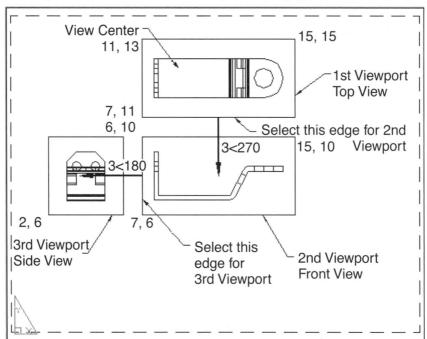

Figure 8.34

These locations are shown in Figure 8.34. Actually, the coordinates given are not critical. What is important is the relative location of the viewports and that they are aligned. To help you align them, AutoCAD turns on the Ortho mode when you are selecting view centers. Now, we will add the viewport for the auxiliary view.

Command: SOLVIEW

Enter an option [Ucs/Ortho/Auxiliary/Section]: Auxiliary

Specify first point of inclined Plane: (*Use an object endpoint snap.*)

Specify second point of inclined Plane: (*Use an object endpoint snap.*)

Specify side to view from: 11,8

(*AutoCAD switches to paper space, rotates the UCS so that its X axis is parallel to the two points, and turns on the Ortho mode.*)

Specify view center: @5<270

Specify view center <specify viewport>: (*Press ENTER.*)

Specify first corner of viewport: 15,4

Specify opposite corner of viewport: 19,8

View name: aux

Enter an option [Ucs/Ortho/Auxiliary/Section]: (*Press ENTER.*)

3D

EXERCISES

The locations of the two points for the inclined plane and the resulting auxiliary view are shown in Figure 8.35. Notice that the viewpoint for the auxiliary view is determined by the direction of the viewport center relative to the inclined plane, rather the side to view from point. Aligning this auxiliary view would have been difficult without SOLVIEW, probably requiring some temporary construction lines.

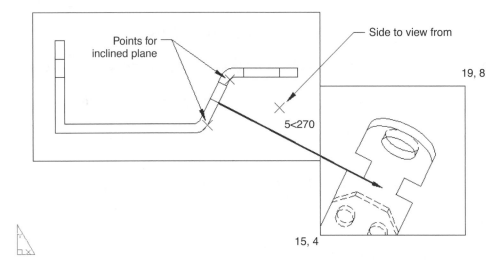

Figure 8.35

If you look at a list of the layers in your drawing, you will see there are 13 new ones. The names of three of these new layers begin with TOP, three begin with FRONT, three with SIDE, and three with AUX. The SOLDRAW command will use the layers having names that end in VIS and HID to draw the intermediary 2D objects, whereas those ending in DIM are for you to use when you add dimensions. The thirteenth layer is VPORTS, which is the layer the floating viewports are in. Now is a good time to set these layers' colors to your preferences and make sure that the layers for hidden lines use a HIDDEN linetype.

Next, we will use SOLDRAW to draw 2D intermediary objects representing the edges of the model.

Command: SOLDRAW

Select viewports to draw:..

Select objects: (*Select the four floating viewports.*)

Then relax while AutoCAD draws the arcs, circles, ellipses, and lines representing the edges of the bracket. When SOLDRAW is finished, your drawing should look like the one in Figure 8.36.

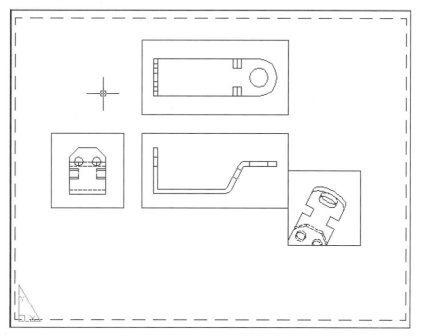

Figure 8.36

Now the drawing is ready to be dimensioned and annotated. As all of the objects are 2D (SOLDRAW froze the original 3D solid's layer), dimensioning can proceed almost as in any 2D drawing. You will want, however, to locate the dimensions in model space, and you will need to use the appropriate layer for dimensions in each viewport. Dimensions in the top view, for instance, will be in the DIM-TOP layer.

Generally, you will add dimensions to one viewport at time using these steps:

1. In paper space, zoom to a comfortable distance to the viewport you intend to dimension.

2. Set the current layer to match the appropriate dimension layer for the viewport.

3. Use MSPACE or its equivalent to switch to model space.

4. For orthographic views, set the UCS to World, and then to View. This ensures that all of the dimensions will be placed on the three principal planes. For auxiliary views, set the origin of the UCS on the object, and then set the UCS to View. This will cause those dimensions to be placed on the auxiliary view's plane.

5. Dimension the objects in the viewport as you would any 2D object.

Two easy-to-make mistakes as you add dimensions are (1) forgetting to switch to model space, and (2) forgetting to set the appropriate layer. You will probably need to stretch some of the viewports to fit the dimensions in and you may need to move the viewports a little. Make sure viewports remain aligned if you do move them. You should also use MVIEW or VPORTS to lock the scale of the viewports to ensure that you do not inadvertently zoom within a floating viewport.

Our finished drawing is shown in Figure 8.37. Your drawing should be similar. We used a title block and border from MVSETUP and added some general notes in paper space. In model space we added two phantom lines in the top view to indicate the bend lines; in all views we erased hidden lines that were not appropriate to standard drafting practices.

This completed drawing is in file 3d_ch8_08.dwg.

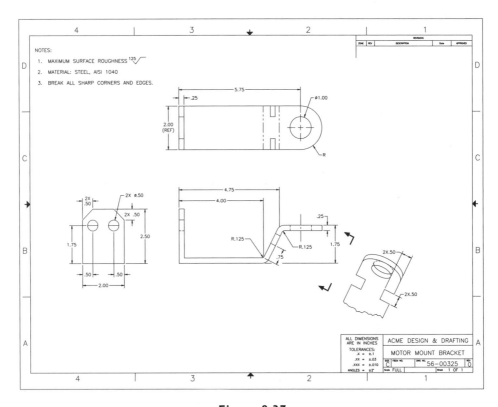

Figure 8.37

TRY IT! - SUPPLEMENTAL SOLVIEW AND SOLDRAW

To gain more experience in using SOLVIEW and SOLDRAW, make a 2D multiview drawing from the 3D solid model shown in Figure 8.38. The drawing for this model will have a top view, a front view, an auxiliary view, and a section view, but it will have relatively few dimensions. You should have no trouble making the drawing.

 The model for this exercise is found in file 3d_ch8_09.dwg.

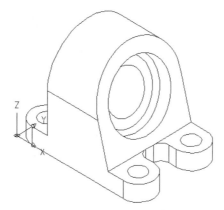

Figure 8.38

Although we will not list the specific steps to create the drawing, the following suggestions may be helpful:

- If you use a layout paper size that is about 21 inches long by 16 inches high (to fit on standard C-size paper), the four views will fit nicely when a 2:1 drawing scale is used.

- Before you start SOLVIEW, make certain that the HIDDEN linetype is loaded.

- The floating viewports can be resized and moved after you are finished with SOLVIEW, but you must be certain that you maintain their alignment, viewpoint, and scale.

- After you have used SOLVIEW to the set the viewports up, use SOLDRAW to create 2D objects in the viewports.

- Dimension the 2D objects on a viewport-by-viewport basis. Go into a viewport (with MSPACE or its equivalent), switch to the layer SOLVIEWcreated for dimensioning in that viewport, and add the dimensions as you would in a 2D drawing.

Figure 8.39 shows the SOLVIEW views we used, with some notes on how they were set up and the names we used for the views.

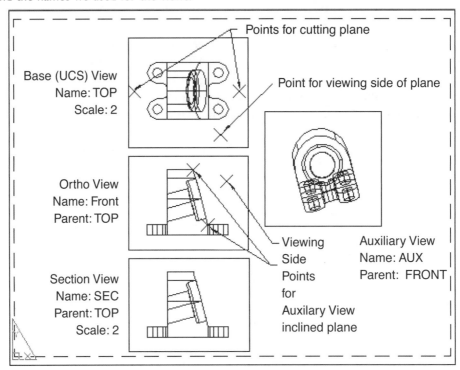

Figure 8.39

Our version of the completed 2D multiview drawing is shown in Figure 8.40. Of course, your arrangement of the dimensions and even your choice of views may be different. The drawing border and title blocks are from MVSETUP, and the center lines in the front view and in the section view were added manually.

This completed version of the drawing is in file 3d_ch8_10.dwg.

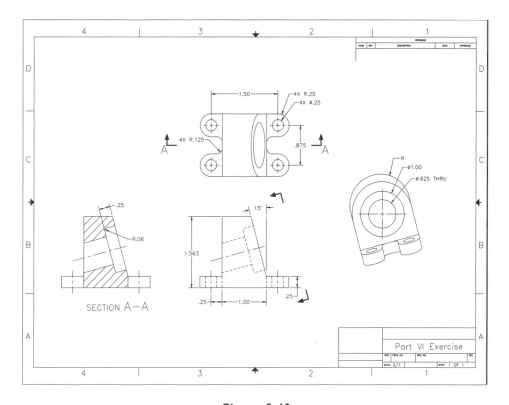

Figure 8.40

COMMAND REVIEW

SOLVIEW

This command sets up floating viewports for **SOLDRAW.**

SOLDRAW

This command draws objects in floating viewports set up by **SOLVIEW.**

SYSTEM VARIABLE REVIEW

CVPORT

This variable contains the viewport number of the current viewport (the viewport which currently has the screen crosshair cursor). The main paper space viewport is always viewport number 1.

DISPSILH

This variable controls the display of 3D solid edges. When this variable is set to 1, AutoCAD will display only the edges of solids during the **HIDE** command. Unlike **SOLPROF,** no new object is created, hidden edges are not shown, and tangential edges are shown.

MAXACTVP

This variable determines the maximum number of active viewports allowed. The default value in Maxactvp is 64, but it can be set to any integer value between 2 and 64. Because the main paper space screen counts as a viewport, the total number of active floating viewports is one less than the value of Maxactvp. You are unlikely to ever need to exceed the number of active viewports allowed by this system variable.

TILEMODE

This variable controls whether floating or tiled viewports are used. When it is set to 1, tiled viewports are used; when it is set to 0, floating viewports are used.

UCSORTHO

When Ucsortho is set to 1, the **UCS** of the current viewport will swivel so that its **XY** plane is perpendicular to the view direction when an orthographic view is set by the **VIEW** command. When Ucsortho is set to 0, the viewport's **UCS** will remain unchanged when an orthographic view is set by **VIEW.** This system variable also affects the **UCS** in viewports created by the 3D Setup option of the **VPORTS** command.

CHAPTER REVIEW

Directions: Answer the following questions.

1. List the problems in making 2D multiview paper drawings of 3D models from model space (that is, without using paper space at all).

2. Name the AutoCAD system variable that controls whether model space or paper space is active.

3. How can you control whether hidden lines will or will not be plotted within a floating viewport?

4. How do you set a view direction within each floating viewport? How do you set the scale within each floating viewport?

5. Suppose a model is only partly visible within a floating viewport. How can you make it entirely visible?

6. Once you are in model space through a paper space viewport, how do you know which viewpoint is the current one? How do you move from one floating viewport to another?

7. What is the object property that controls whether or not an object will be visible within a floating viewport?

8. What happens when the system variable Dimscale is set to 0, or when the Scale Dimensions to Layout button in the Fit tab of the Dimension Style Manager dialog box is checked?

9. What is the relationship of the XY plane with dimensions?

10. List some disadvantages in dimensioning a model from paper space.

11. What is the relationship between the SOLVIEW and SOLDRAW commands?

12. What does the Lock option of MVIEW and VPORTS do.?

13. List the two options of MVIEW and VPORTS that can create nonrectangular floating viewports, and explain their differences. List the object types that can be used to define the boundary of a floating viewport.

14. Describe the function of the VPCLIP command.

15. Name the system variable that sets the scale of the model within new view ports.

16. By default, AutoCAD creates a floating viewport in every new layout. How can you turn that default off?

17. What is the function of the Psltscale system variable?

Directions: Circle the correct response(s) in each of the following.

18. Which of the following statements about paper space and about floating view ports are true?

   a. Floating viewports can be copied.

   b. Floating viewports can overlap and can even hide one another.

   c. Once paper space floating viewports have been established, it is not possible to change the 3D model.

   d. The border of floating viewports is always visible.

   e. You can draw 3D objects within paper space.

   f. You can set viewpoints from any point in space within paper space.

19. Which of the following statements about SOLPROF are false?

   a. SOLPROF does not require the use of paper space.

   b. SOLPROF is useful for creating isometric views of 3D solids.

   c. SOLPROF works on both surface and solid models.

   d. Tangential edges are always displayed.

20. Which of the following statements about SOLVIEW and SOLDRAW are true?

   a. Hidden lines are drawn in the HIDDEN linetype, provided that linetype is loaded.

   b. Section views may have a scale different than their parent views.

   c. SOLDRAW automatically adds dimensions to the views.

   d. SOLDRAW creates 2D objects that represent the 3D solid and freezes the layer of the 3D solid.

   e. SOLVIEW is able to set up auxiliary views.

   f. The first viewport created must be for a plan view.

21. Match a command in the list on the left with a function or result from the list on the right.

   _____ a. MSPACE   1.   Controls the visibility of layers on a view port-by-viewport basis.

   _____ b. MVIEW   2.   Creates floating viewports.

   _____ c. MVSETUP   3.   Helps in creating and aligning floating view ports.

   _____ d. PSPACE   4.   Switches from model space to paper space.

   _____ e. VPLAYER   5.   Switches from paper space to model space.

CHAPTER 9

Rendering

LEARNING OBJECTIVES

You will learn how to use the features of AutoCAD's rendering module in this chapter. When you have finished, you will:

- Know how to set up various parameters to efficiently make renderings, know how to control the output of renderings, and be able create special backgrounds for renderings.

- Be familiar with the properties of the different types of lights for renderings, and be able to install them.

- Be able to control shadows in renderings.

- Know how to create materials that have color, transparency, and reflectivity.

- Be able to attach bitmap images to objects in creating realistic renderings.

- Know how to enhance realism by installing images of objects, particularly landscape objects, in rendering.

- Know how to emphasize distance by fading or shading objects.

WHAT ARE RENDERINGS

Engineering and architectural drawings are able to pack a vast amount of information into a 2D outline drawing supplemented with dimensions, some symbols, and a few terse notes. However, training, experience, and sometimes imagination are required to interpret them, and many people would rather see a realistic picture of the object. Actually, realistic pictures of a 3D model are more than just a visual aid for the untrained. They can help everyone visualize and appreciate a design and can sometimes reveal design flaws and errors.

Shaded, realistic pictures of 3D models are called renderings. Until recently, they were made with colored pencils and pens or with paint brushes and airbrushes. Now they are often made with computers, and AutoCAD comes with a rendering program that is automatically installed by the typical AutoCAD installation and is ready for use. Figure 9.1 shows, for comparison, the surface model of a cone in its wireframe form, as it looks when the HIDE command have been invoked, and when it is rendered.

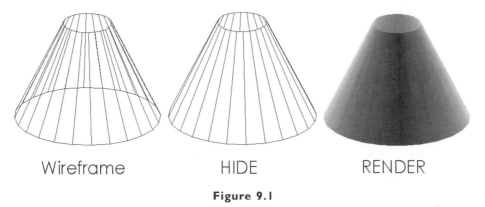

Wireframe   HIDE   RENDER

Figure 9.1

For a quick demonstration of a rendering, make a sphere from the AutoCAD command line by typing in SPHERE for a solid or AI_SPHERE for a surface model. The diameter of the sphere is not critical, but it should be large enough to be easily visible. The viewpoint is not important. Give the sphere a color that gives good contrast with your graphics screen background. Then, on the command line, type in RENDER, and press ENTER to render the current viewport with the default render settings. After performing some calculations, AutoCAD will replace the wireframe sphere with a shaded, colored, round ball, similar to the one shown on the right in Figure 9.2. That is a rendering. REDRAW will restore the wireframe image.

Figure 9.2

AutoCAD's rendering module is based on a file named acrender.arx, which is automatically loaded whenever a rendering command is invoked. This module has capabilities that equal or exceed those of add-on renderers that cost hundreds of dollars a few years ago. AutoCAD's renderer:

- Supports three types of lights—spot lights, point lights, and distant lights—that can have colors and can cast shadows.

- Supports transparent and reflective materials.

- Can add bitmap images to surfaces to help create realistic renderings.

- Can add bitmap images of people, trees, and other items to renderings.

- Gives you full control over the background of the rendering.

- Can fade distant objects to emphasize distance.

BITMAP FILES

When you work with renderings you work with bitmap files because AutoCAD uses them for enhancing renderings, as well as for saving rendered images. Bitmap files are based on pixels, the rows of tiny colored dots on your computer screen, rather than vector graphics. Vector graphics, which AutoCAD normally uses, depend on coordinate systems and equations to draw objects on a computer screen. Bitmapped graphics, on the other hand, use lists that tell the computer how each pixel on the screen is to be colored. They are analogous to those masses of people in football stadiums holding up colored cards to make giant pictures. Sometimes bitmap graphics are referred to as raster graphics.

As you would surmise, bitmap files use binary numbers, called bits, to specify colors. The maximum number of colors available depends on the number of bits available, which in turn depends on the bitmap file format and on the computer's video system. Virtually all computer video systems capable of running AutoCAD will support at least 8 bits for each pixel, and some will support 24 bits. Occasionally even 32 bits are available, but from a practical standpoint, 8-bit bitmap files are generally all you need when you are working with AutoCAD renderings.

Often the dialog boxes used by AutoCAD for bitmap files will use the term color depth when referring to colors and will offer options for 8-bit, 16-bit, and 24-bit output. The maximum number of colors that an 8-bit system can handle is 256 (2 to the 8th power). 16-bit systems are capable of 65,536 colors, and 24-bit systems can have over 16 million colors.

When many pixels in a row are to have the same color, most bitmap file formats use an internal code to signal that X number of pixels are to have color number Y, instead of designating a color number for each pixel. This feature, which is called compression, is available for some bitmap formats when saving AutoCAD images.

Paint programs were the type of computer program that first began using bitmap files.

Each program developed its own file format, and because no program became dominate, no single file format became the standard. Consequently, there are currently over two dozen different bitmap file formats, identified by their three-character file name extension. The file types you will most often encounter as you work with AutoCAD renderings are listed in Table 9.1.

Table 9.1 Bitmap File Extensions

Filename

Extension	Remarks
BMP	A format created by Microsoft for the Windows operating system. The files tend to be large because they are uncompressed.
GIF	This format, the Graphics Interchange Format, was developed by the CompuServe Information Service to support bitmap files on different computer platforms. Generally, it supports a maximum of 256 colors.
JPG	This is a graphics compression format developed by an international commit tee—The Joint Photographic Experts Group (JPEG)—for digitizing phot ographs. It is widely used for transferring files on the Internet because it is able to squeeze many colors and high-resolution images into relatively small files.
PCX	This format originated from an early paint program, PC Paintbrush by Z-So ft. Although it has undergone many revisions—and not all revisions are compat ible with others—it is widely used by programs running on a variety of operat ing systems.
TGA	A format developed by Truevision, a manufacturer of high-end graphics hard ware (Targa graphic adapters) and software. AutoCAD uses this format for most of its bitmap rendering files.
TIF	The Tagged Image File Format (TIFF), designed for high-end graphics on a variety of computer platforms. TIFF has several variants, which sometimes cause compatibility problems.

 Tip: If you selected Typical during the installation of AutoCAD, you will probably have only three bitmap files available for renderings in your computer. They will be in the Acad2002\Textures folder. To install all of the bitmap files for renderings that come with AutoCAD, open the Acad\Textures folder of your AutoCAD 2002 installation CD-ROM, and copy all of its files to the Acad2002\Textures folder of your computer. There will be about 140 TGA files to copy.

MAKING RENDERINGS

Renderings are commonly done in a single viewport on the AutoCAD graphics screen. You can, however, direct the rendering output to a file or to a hard-copy printing device. Although rendering is not allowed in paper space, you can render a paper space floating viewport (after invoking MSPACE or its equivalent). Then you can return to paper space (via PSPACE or its equivalent), with the rendered image remaining in the viewport. If no lights have been installed, AutoCAD uses a single distant light, pointed in the viewing direction, for the rendering.

Although the bulk of your work and time will be spent setting up lights and materials for renderings, we will first discuss the command that performs renderings. Then, we will discuss some commands that set parameters and direct the output for AutoCAD's renderer and the command that reports statistics about renderings. The toolbar buttons and the screen pull-down menus for initiating these commands are shown in Figure 9.3.

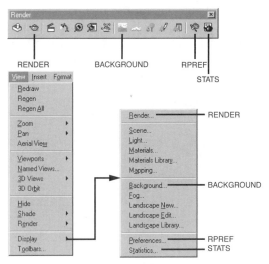

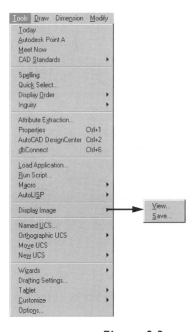

Figure 9.3

THE RENDER COMMAND

The RENDER command gives you control over what will be rendered, how it will be rendered, where the rendering output will be directed, and how the rendering background will appear; plus it allows you to fade selected areas within renderings. Normally, the command uses the Render dialog box, which is shown in Figure 9.4. This dialog box is virtually identical with the Rendering Preferences dialog box that the RPREF command uses. If the Skip Render Dialog checkbox in either of these dialog boxes has been turned off (deselected), the Render dialog box will not be displayed. AutoCAD will then either render the viewport or prompt for objects to render, depending on settings in the Rendering Preferences dialog box.

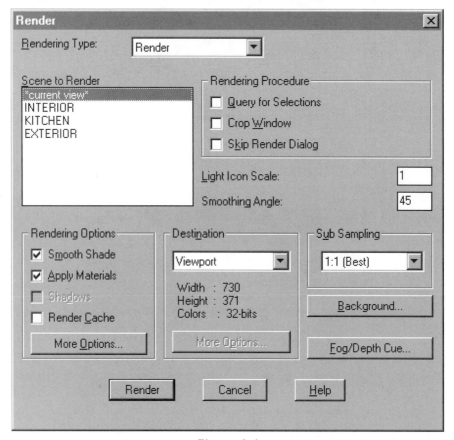

Figure 9.4

RENDERING TYPE

This drop-down list box, located near the top of the Render dialog box, gives you a choice of three rendering types.

- Render

- This rendering type does not support shadows, bitmaps, transparent materials, or other advanced rendering features. You will use it only for quick, preliminary renderings. An example is shown in Figure 9.5.

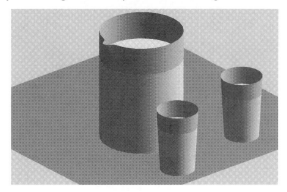

Figure 9.5

- Photo Real

- Most of the advanced rendering features, including shadows, bitmaps, transparent materials, and landscape objects, are supported by this rendering type. See Figure 9.6 for an example.

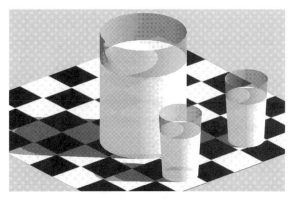

Figure 9.6

- Photo Raytrace

- Raytrace renderings are based on the paths that individual light beams traverse from the light source to the viewer. This type of rendering can do everything that the Photo Real type can, plus it can generate reflected images and handle light beam refraction. The price of these extra capabilities is in rendering time—Photo Raytrace renderings are generally much slower than

Photo Real renderings. Figure 9.7 for an example.

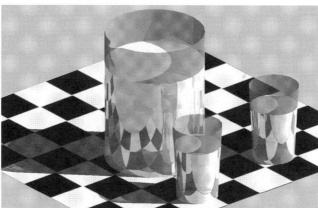

Figure 9.7

SCENE TO RENDER

A scene is a specific view combined with an assortment of lights that has been assigned a name. The command SCENE, which we will describe when we discuss lights, creates and manages scenes. There will always be one scene labeled *current view* in this list box. When several scenes are listed, one scene is to be selected for rendering.

RENDERING PROCEDURE

This cluster of three checkboxes allow you to control which objects will be rendered and whether or not the Render dialog box will be displayed when the RENDER command is initiated.

- Query for Selections

- When this checkbox is selected, AutoCAD will prompt for the objects that are to be rendered. If the Render dialog box is displayed, this prompt will occur after the Render button has been clicked. If the Render dialog box is not displayed, the prompt will occur when the RENDER command is invoked. Objects that are not included in the selection set will completely disappear during the rendering (unless, as we will describe shortly, the Merge Background mode is active).

- Crop Window

- Use this option when you want to render just a certain area of the screen. When this checkbox is selected, AutoCAD will ask you to designate a window. Objects within the crop window will be rendered and those outside the window will disappear. The prompt for a crop window will be issued after the Render button in the Render dialog box has been clicked, or when the RENDER command is initiated if the Render dialog box is not displayed.

- Skip Render Dialog
- The Render dialog box will not be displayed when the RENDER command is initiated if this checkbox is selected. You must use the RPREF command to turn this button off and enable the Render dialog box.

LIGHT ICON SCALE

This edit box controls the relative size of the icon blocks that AutoCAD inserts at light locations. (These blocks will be described when we discuss the LIGHT command.) The size of all currently inserted light icon blocks is affected, as well as the size of blocks to be inserted.

SMOOTHING ANGLE

This option sets the blending angle for polygon mesh surfaces. It has no effect on other types of objects. When smooth shading is turned on, AutoCAD uses angles to decide whether an individual face is part of a rounded surface, or if it is an individual flat surface that should not be blended with the adjacent face. The default angle is 45°. Faces at angles less than 45° are blended, and faces at angles greater than 45° with one another are not blended. See Figure 9.8 for examples.

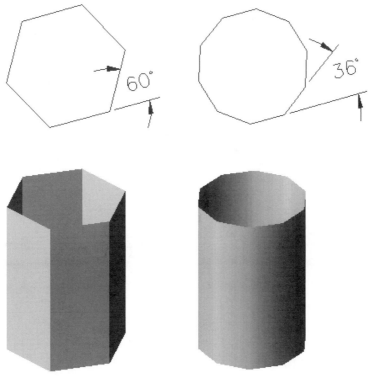

Figure 9.8

RENDERING OPTIONS

- Smooth Shade

- When Smooth Shade is deselected, shading on each individual face is uniform, which gives a faceted appearance to rounded surfaces. When this option is selected (the default option), shading is blended across faces, which makes rounded surfaces more realistic, as shown in Figure 9–9.

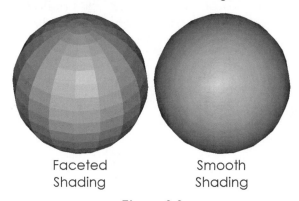

Faceted
Shading

Smooth
Shading

Figure 9.9

- Apply Materials

- This option applies materials that were created and attached to objects through the RMAT command. If this checkbox is not selected, AutoCAD ignores attached materials and uses object colors for renderings.

- Shadows

- Shadows are enabled when this checkbox is selected. See Figure 9.10 for examples. The checkbox will be grayed out and unavailable when Rendering Type is set to Render.

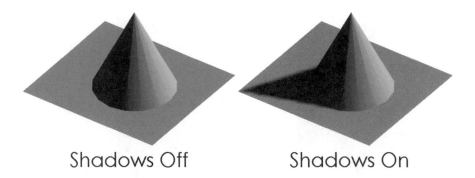

Shadows Off   Shadows On

Figure 9.10

- Render Cache

- When this checkbox is selected, AutoCAD stores some of the data it needs for making renderings of the specified view. This can speed up subsequent renderings, especially when 3D solid models are being rendered.

- More Options

- This button leads to one of three dialog boxes, each appropriate for the selected rendering type. Each of these dialog boxes has two options related to back faces. Faces on surface meshes have front and back sides according to how the surface was drawn. (AutoCAD often uses the word normal when referring to surface faces. A normal is an imaginary line, positioned at the geometric center of a surface, that points in the direction the surface is facing.) Normally, AutoCAD renders both sides of a surface, but you can activate an option in these dialog boxes to have AutoCAD render only the front side of surfaces. Another option related to back faces in these dialog boxes enables you switch the front and back sides of surface meshes. Figure 9.11 shown examples of discarded back faces and reversed normals. You are not likely to ever need to discard back faces in a rendering or reverse face directions.

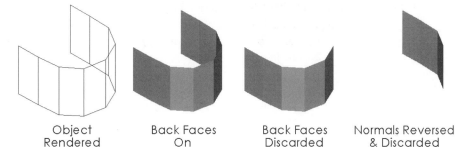

Object
Rendered

Back Faces
On

Back Faces
Discarded

Normals Reversed
& Discarded

Figure 9.11

- The dialog boxes for More Options for Photo Real and Photo Raytrace renderings contain options for shadow map controls that we will describe later in this chapter when we discuss the LIGHT command. The dialog boxes for these rendering types also contain options for setting anti-aliasing to one of four levels. This controls the representation of slanted lines during renderings, with a setting of Minimal resulting in relatively jagged slanted lines. For the smoothest lines, select High. You are unlikely to ever change the setting for texture map sampling that is in this dialog box. Also, the dialog box of additional rendering options for Photo Raytrace rendering contains an option for tracing light rays that you are not likely ever to change.

DESTINATION

- This drop-down list offers several options for directing render output.

- Viewport

- This, the default option, renders to the current viewport. Data about the current viewport's size (in pixels) and the number of available rendering colors is shown below the list box.

- Render Window

- This option sends the rendering to a Microsoft-style window. You can open this window by clicking the Render button on the Microsoft toolbar. From there, you can save the rendering to a BMP file, send it to the Windows clipboard, or print it using the current Windows printer. Menus and toolbar buttons in the render window give you control over the rendering's resolution (in pixels) and its number of colors.

- File

- When this option is selected, the button labeled More Options located below the drop-down list will be activated. Clicking that button will allow you to save the file in either the BMP, PCX, PostScript, TGA, or TIF format, as well as specify the rendering's resolution (in pixels) and its number of colors. You can even save the file using a resolution and color depth beyond the capabilities of your system.

SUB SAMPLING

This option allows you to reduce rendering time—and still retain shadows, transparency, and other special effects—by rendering only a fraction of the screen's pixels. The option's drop-down list contains ratios ranging from 1:1 to 8:1. The default setting of 1:1 gives the highest-quality renderings; as the ratio increases, the quality of the renderings decreases and rendering speed increases. Figure 9.12 shows the same scene shown in Figure 9–6 as an example of Photo Real rendering. The rendering for Figure 9.6 used a 1:1 subsampling ratio, and the rendering for Figure 9.12 used an 8:1 ratio.

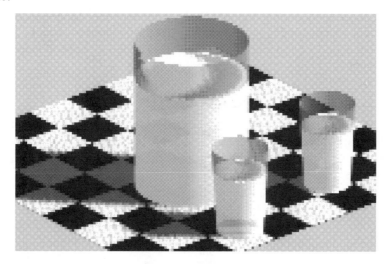

Figure 9.12

BACKGROUND

This button brings up the same dialog used by the BACKGROUND command. We will describe its contents when we discuss the BACKGROUND command.

FOG/DEPTH CUE

This button brings up a dialog box that allows you to give an illusion of distance to renderings by fading selected images and objects. This dialog box is also used by the FOG command, and we will describe its contents when we discuss the FOG command later in this chapter.

RENDER

Clicking this button, which is the default response for the Render dialog box, renders the selected scene using the specified parameters. The length of time required for the rendering depends on the complexity of the objects being rendered, the destination of the rendering, and the type of rendering. AutoCAD will display the progress of its rendering calculations on the command line.

Tips: The Query for Selections option of the RENDER command is useful when you are installing lights or creating materials and want to test your setup without rendering the entire scene.

Screen size and resolution is a major factor in rendering speed when rendering to a viewport. Therefore, when you are setting up a rendering you will find it helpful to divide the AutoCAD graphics area into multiple viewports. This will not only speed up renderings but also enable you to compare the effects of different settings and parameters.

AutoCAD continues to display the UCS icon in renderings to the viewport. Therefore, you may want to turn off the icon. This can be done, you will recall, with the UCSICON command. The UCS icon is not displayed, however, in renderings to a file or to the render window.

THE BACKGROUND COMMAND

This command gives you control over the appearance of rendering backgrounds—those areas within the rendered viewport that not covered by objects or images. Backgrounds can be colored in with the same color used for the AutoCAD graphics screen, with a specified color that is different than that of the graphics screen, or with a gradient of colors blending across the background. You can also use a bitmap image for the background. The dialog box used by the BACKGROUND command can also be accessed by the Background button in the dialog boxes used by the RENDER and RPREF commands. This dialog box is shown in Figure 9.13.

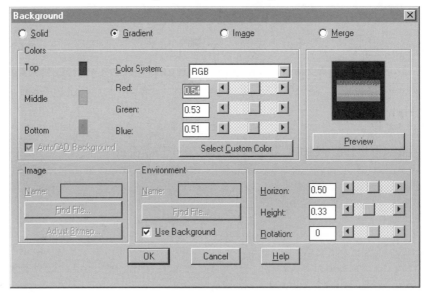

Figure 9.13

SOLID/GRADIENT/IMAGE/MERGE

These four radio buttons are for selecting a rendering background type. When you select one of these buttons, other buttons and edit boxes appropriate for the background type will become available for setting additional parameters.

- Solid

- This option allows you to specify one color for the background. The Colors section of the dialog box will be activated so that you can specify the color. This radio button also activates the AutoCAD Background checkbox in the Colors section of the dialog box.

- Gradient

- When this option is selected, you can have either two or three colors blending across the background, as shown in Figure 9.14. The Colors section of the dialog box will be activated so that you can select the colors, and the Horizon, Height, and Rotation edit boxes will be activated for your use in placing the colors.

Figure 9.14

- Image

- With this option, you can fill the background with a bitmap image. When it is used, buttons and list boxes in the Image and in the Environment sections of the Background dialog box will be available for you to specify a bitmap file and set its parameters. An example of a rendering with a background image is shown in Figure 9.15.

Figure 9.15

- Merge
- Normally AutoCAD clears the current viewport when it makes a rendering. However, when this radio button is selected, AutoCAD will leave the screen intact and will change only what needs to be changed to reflect changes within the rendering.

COLORS

This area of the dialog box is used for setting the colors for solid and gradient backgrounds.

- AutoCAD Background
- This checkbox is available when the Solid background type has been selected. No other options in the Background dialog box will be available, and the current color of the AutoCAD graphics area will be used for the rendering background.
- Top, Middle, Bottom
- These three buttons are for specifying the background area in which a color will be applied. Only the Top color button will be available for a solid back-

ground. The other two buttons are for setting gradient colors.

- Color System

- These edit boxes and sliders are for setting the color of the specified background area. You can use either the Red-Green-Blue (RGB) or the Hue-Lightness-Saturation (HLS) color systems in choosing a color. Clicking the Select Custom Color button will bring up a supplementary dialog box that displays the available colors and allows you select a standard color or a custom color.

PREVIEW
Clicking this button enables you to see the background in effect, including gradient colors and bitmap images.

IMAGE
This area of the Background dialog box is available when the Image radio button is selected. Enter the name of the image file in the edit box labeled Name, or click the Find File button to browse for it. Seven different bitmap file formats are supported: BMP, GIF, JPG, PNG, PCX, TGA, and TIF. The Adjust Bitmap button brings up a dialog box for adjusting the bitmap image's scale and location. This dialog box is essentially the same as the Adjust Object Bitmap Placement dialog box used by the SETUV command, which we will discuss later.

ENVIRONMENT
You can control the interaction of background images and objects through this section of the Background dialog box. However, the objects must have a mirror-like material attached. (We will discuss mirror-like materials later.) If objects do not have a mirror-like material attached, the background will have no effect on them, as shown in Figure 9.16.

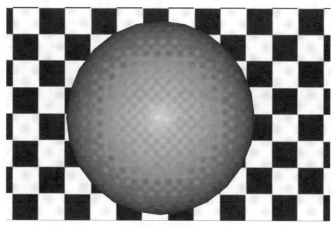

Figure 9.16

If, on the other hand, objects do have a mirror-like material attached, they will reflect the background, as shown in Figure 9.17. This occurs with both Photo Real and Photo Raytrace rendering types.

Figure 9.17

You can change this effect by turning off the Use Background checkbox. This will activate the Name edit box for you to specify a bitmap file to be used in reflections instead of the background image. Figure 9.18 shows a rendering of the same sphere used in Figure 9.17 with the same background image, but a separate bitmap file has been specified for the environment image.

Figure 9.18

HORIZON, HEIGHT, ROTATION

These three edit boxes and their corresponding slider bars control background color gradients. Horizon, which accepts a value from 0.0 to 1.00, establishes the center of the gradient. Height, which also accepts a value from 0.00 to 1.00, sets the start of the second color in a three-color gradient. If you want to use only two colors, set Height to 0. Then AutoCAD will ignore the middle color, and use only the top and bottom colors in the gradient.

Colors are blended vertically in the default option, but you can specify a different direction through the Rotation edit box and slider. You may specify any angle between 90° and -90°. A 90° rotation angle will blend the colors from left to right.

 Tip: When you are establishing the parameters of a complicated rendering, the Merge background option can be helpful when used in conjunction with the Query for Selections option in the Render dialog box. You can select only the objects you have changed for a revised rendering; the unaffected sections of the rendering will remain as they were. If Merge is not in effect, the viewport will be cleared and only the selected objects will be shown in the new rendering.

THE SAVEIMG COMMAND

The SAVEIMG command gives you an additional tool for saving a rendering to a bitmap file. With this command, you can save the image in the current viewport to a file using a BMP, TGA, or TIF bitmap format. The command will save everything shown in the current viewport, including rendered objects, wireframe objects, and shaded objects (those shaded with the SHADEMODE command). It uses a dialog box for you to specify the file's format and name. Because this dialog box is uncomplicated and straightforward to use, we will not show a picture of it or describe its use further.

A companion command, REPLAY, displays a BMP, TGA, or TIF bitmap file in the AutoCAD graphics window. Unlike bitmap files inserted with the IMAGE command, these images are not attached to the drawing and you cannot access them or use them in any way. You can only view them. Any command that causes a screen redraw makes the image disappear. Nevertheless, REPLAY can be useful when you want to see what a certain bitmap file looks like (such as one that AutoCAD has for backgrounds or materials).

THE STATS COMMAND

Statistics about the last rendering are displayed by the STATS command. You can view this data within the Statistics dialog box, which is shown in Figure 9.19, or save it in a text file (a file type that can be read by the Windows Notepad accessory). Information about the parameters used for the rendering is included, along with data regarding times and the number surface faces involved.

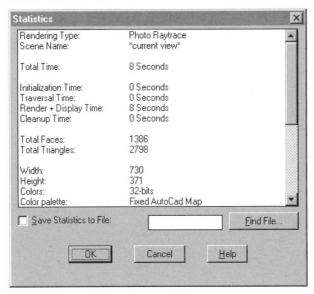

Figure 9.19

TRY IT! - RENDERING

Use file 3d_ch9_01.dwg on the CD-ROM that comes with this book to experiment with some of the parameters of the RENDER command. This is the file that was used for the three figures in the discussion of RENDER showing the differences between AutoCAD's three rendering types.

The checkerboard pattern in the rendering is from a bitmap file, named checkers.tga, that is attached to a 3D face. (We will describe how that is done later when we discuss materials.) If AutoCAD cannot find that file, the checkered pattern will not be included in your rendering, but that is not important for the objectives of this exercise, and you should not go to any trouble trying to find and include the image in your renderings. The worst consequence of not having the image will be that reflections during Photo Raytrace rendering will not show the checkered pattern.

Unless your computer's display is very small (in the 14- or 15-inch range) and has a relatively low screen resolution (800 by 600 pixels or less), you should divide AutoCAD's graphics area into four viewports—one in each quadrant of the screen. This will reduce rendering time and will allow you to compare the effects of different rendering settings.

You will not make any changes to the surface models or in the materials attached to them during this exercise. Instead, you will create a number of renderings to see how various parameters of the RENDER command work and how they affect renderings. The suggested steps for this exercise are:

1. Make renderings using each of the three rendering types AutoCAD sup ports: Render, Photo Real, and Photo Raytrace. After each rendering, use the STATS command to see how long the rendering took.

2. With the Photo Real rendering type active, deselect the Shadows checkbox and render the scene. Turn the Shadows checkbox back on.

3. Check the Crop Window checkbox. Draw a window that includes roughly one-fourth of the total area when you are prompted for a crop window.

4. Redraw the screen or move to a viewport showing the models as wire frames. Deselect the Crop Windows checkbox. Check the Query for Selection checkbox, and pick just one of the two glasses for a rendering object. A good technique for selecting it is to use a selection window.

5. Turn off the Query for Selection checkbox and render the entire scene (or move to a viewport containing a full rendering). Then turn the Query for Selection checkbox back on. Go to the Background dialog box, select the Merge radio button, and return to the main Render dia log box. Turn the Apply Materials checkbox off. Click the Render button, and select one of the glasses as a rendering object.

6. Now we will explore some background options:

- Zoom out until the objects are shown about half of their original size.

- Deselect the Query for Selection checkbox, and check the Apply Materials checkbox again.

- Go to the Background dialog box and turn on the Gradient radio button. Click the Top color button and set the top color to red. Click the Middle color button and set the middle color to green. Finally, click the Bottom color button and set it to blue.

- Return to the Render dialog box, select the current view as the scene to ren der (not the scene named EXERCISE), and click the Render button. Notice in the rendering that because a mirror-like material is attached to the two glass es and the pitcher, they reflect some of the background colors.

7. While this same viewport is current, initiate RENDER, go into the Background dialog box, and click the Image radio button.

- Click the Find File button and locate the file that is named CLOUD.TGA. The path to this file will most likely be C:\Program Files\Acad2002\Textures. See the tip in the discussion of bitmap files earlier in this chapter if you cannot file CLOUD.TGA.

- When you find CLOUD.TGA, click the Open button to attach the file to your background.

- Return to the Render dialog box and render the current view. The result will be a rather surrealistic image of the objects floating on a magic carpet, as shown in Figure 9.20.

Figure 9.20

If you have time, you can perform additional experiments with this file on your own. You could try a rotated gradient background, different sub sample settings, and directing the rendering to a file or to the render window.

LIGHT FUNDAMENTALS

AutoCAD makes renderings through the interaction of light and surfaces. You set up a 3D model for rendering by installing lights and adjusting the properties of the model's surfaces. We will first describe how to use lights.

We will explore the relationship between AutoCAD objects and lights, learn a little about the characteristics of real lights and how they relate to AutoCAD lights, and describe the different AutoCAD light types along with the command options for installing them. Then, we will use that knowledge to select and place lights in two renderings.

SURFACES, LIGHTS, AND SHADING

Lights only affect AutoCAD objects that have surfaces. All other objects disappear during rendering. If you want objects that normally do not have surfaces—text, for example—you must give them an extrusion thickness.

Although AutoCAD surface objects are affected by light, they do not block light. Rendering lights illuminate a surface and then pass completely through it to illuminate any surface that is behind it. Thus, as shown in Figure 9.21, the inside diameter of a tube is illuminated by a light outside the tube. Even objects on the far side of a wall from a light will be lighted as if the wall was not there. This characteristic can also be useful, but it can be bothersome when you want only certain objects to be affected by a light.

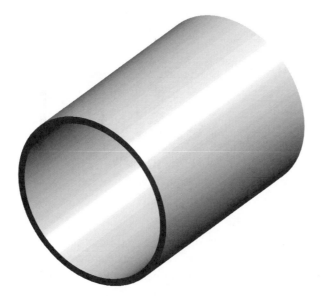

Figure 9.21

This light characteristic is not the same as transparency—the ability to see through objects. Transparency is a material property, rather than a light property, that will be discussed later.

Rendering creates a realistic 2D image of a 3D model by shading the model's surfaces. Although the model's basic color may be the same throughout a surface, the intensity—the shade—of that color is varied to give the illusion of form and depth. The tube in the previous figure, for example, is made to look round by giving it a relatively bright, light color on its side facing the light and gradually darkening the color around the tube's diameter to the shaded side.

Shading, in both the real world and in AutoCAD model space, results from the relative position of surfaces with light. AutoCAD shades a 3D model by adjusting the brightness of a surface according to the angle of that surface with incoming light. A surface is at its brightest when it directly faces incoming light, and as the surface's angle relative to incoming light increases, the shade of the surface becomes darker. This very important principle is illustrated in Figure 9.22.

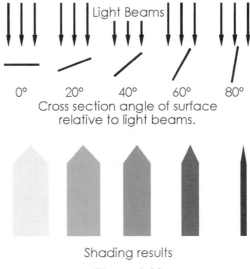

Figure 9.22

PROPERTIES OF LIGHTS

AutoCAD has four different types of lights. The differences between these light types are in how they handle light properties. Therefore, before we discuss AutoCAD's lights, we will take a quick look at some basic properties and characteristics of lights, and we will give special attention to how they relate to AutoCAD.

Intensity

Even though none of the AutoCAD render light types really emit light (as do, for instance, incandescent lightbulbs), their effect on surfaces is as if they do. A light causes surfaces that face the light to appear bright, with the degree of brightness depending on the light's intensity. In the real world, light intensity is measured in lumens (and indirectly in watts), but in AutoCAD intensity is simply a number—there are no units. The larger the number, the greater the effective brightness of the light.

Also, with some AutoCAD render light types, a light's intensity will diminish as distance from the light increases—objects close to the light are brighter than objects farther away. With other types of AutoCAD lights, the intensity remains constant regardless of distance.

Position

Every light used in a rendering is an object, and therefore must have position—a point in space where it is located. Position is important with lights that diminish in intensity with distance, but as we will see, the point position of some light types is of no consequence.

Position often implies how an object is oriented—the direction in which it is pointed—as well as its location in space. Direction is important for some types of lights, and these lights always have a target point as well as a light source point. Other types of light are not pointed in any specific direction and therefore, have only a source point.

Light Beams

Light travels in beams that are straight lines originating at the light's source (they are also called light rays). At one extreme are lights with beams that travel parallel to each other, as if they originated from an infinitely large light source. At the other extreme are lights with beams that travel in all directions, as if they originated from a point light source.

Light Color

Both real-world lights and AutoCAD lights can range from a single color to a mixture containing all of the basic colors. Lights with an even mixture of basic colors are white. Sunlight is white, as are most artificial lights. Although AutoCAD lights can have any color, their color only has an effect when it strikes a surface. You cannot see light beams from AutoCAD lights, regardless of whether they are colored or white. Consequently, a green AutoCAD light source does not look like a green lightbulb. Furthermore, light color mixes with the color of the surface to make a third color; and the results do not follow the standard rules for mixing colors.

AUTOCAD LIGHT TYPES

The light icon blocks shown in Figure 9.23 are used to mark the position of every AutoCAD light. These blocks allow you to see where a light is located and what type of light it is. Every light is given a unique name when it installed, and this name is shown in the block. The blocks also hold, in the form of attributes, information AutoCAD needs about the light. AutoCAD places these blocks in a layer named ASHADE. AutoCAD automatically creates this layer when a light is created.

Icon:	LT–NAME	LT–NAME	LT–NAME
Light Type:	Point	Distant	Spot
Block Name:	OVERHEAD	DIRECT	SH_SPOT

Figure 9.23

Ambient Lighting

Ambient lighting is the simplest type of light in AutoCAD render. In fact, it is not really a light because it has no particular source and it does not send out light beams. Ambient lighting is the relative brightness of surfaces that are not in the path of light beams. There is no light icon block for ambient lighting, and the only properties of ambient lighting are intensity and color. Intensity ranges from 0.0 to 1.00 and has a default setting of 0.30. A value of 0 turns ambient light off.

In most cases, you will want some ambient lighting in your renderings. Otherwise, surfaces not directly illuminated by another light may not show up. By itself, however, ambient light cannot produce realistic images since it shades all surfaces the same. Objects illuminated only with ambient light are evenly colored and flat in appearance, as shown in Figure 9.24.

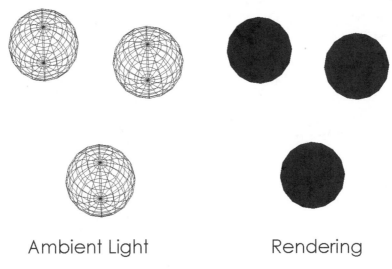

Ambient Light   Rendering

Figure 9.24

Distant Lights

Distant lights are comparable to sunlight as it strikes objects on earth. The light beams travel parallel to one another, in a specific direction, and their intensity remains constant regardless of distance. An object far from a distant light is just as bright as an object close to the light.

The intensity of a distant light can range from 0 to 1. If you want a distant light that is brighter than the maximum intensity allows, you can add additional lights pointed in the same direction. An intensity of 0, as you would expect, turns the light off.

As shown in Figure 9.25, direction is everything with a distant light and even objects that are behind the light are illuminated by it. AutoCAD places a block named

DIRECT, as shown in an earlier figure, at the light location point. The parallel lines of the block (representing parallel light beams) indicate the light's direction.

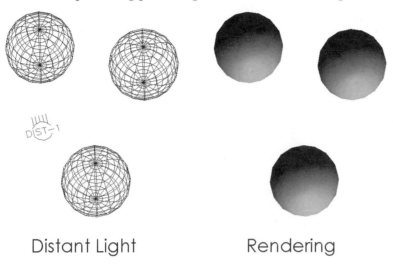

Distant Light   Rendering

Figure 9.25

Point Lights

A point light has light beams radiating in all directions from a single point, similar to a bare incandescent lightbulb. AutoCAD inserts a block named OVERHEAD, shown in Figure 9.26, in the location of a point light. This block has lines pointing in all directions just as a point light has light beams radiating in all directions.

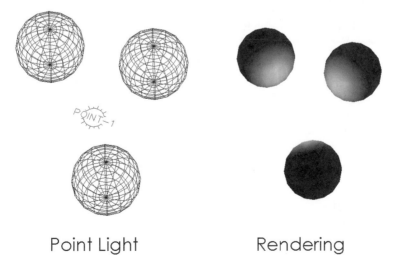

Point Light   Rendering

Figure 9.26

Intensity of point lights can be set to diminish as distance from the light increases. Then, as shown in Figure 9.27, objects close to the light will be brighter than objects further away. AutoCAD refers to this as attenuation. A consequence of attenuation is that two different intensity parameters are required for point lights. One is the source intensity and the other is the rate at which the intensity diminishes—the attenuation rate.

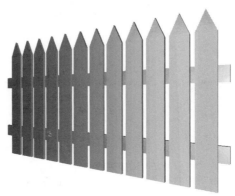

Figure 9.27

Source intensity is, as you would expect, the beginning intensity for the point light. The maximum source intensity depends on the rate of attenuation and on the area within the drawing extents. A point light for a large building may have a maximum source intensity in the neighborhood of 500,000, while one of a small part may have a maximum intensity of less than 2. If there is no attenuation, maximum intensity for point lights is 1.

Point light attenuation can be set to one of three rates—none, inverse linear, or inverse square. With no attenuation, objects far from the light are illuminated just as brightly as objects near the light. When inverse linear attenuation is used, brightness decreases in direct proportion to the distance from the light. As shown in the following table, an object twice as far from a point light as another object is only half as bright, while one that is four times as far away is one-fourth as bright. Inverse linear attenuation is the default setting for point lights.

Brightness decreases even faster when attenuation is set to the inverse of the square of the distance. An object twice as far from a point light as another object is one-fourth as bright, and one that is four times as far away is only one-sixteenth as bright. Brightness falls off so fast with inverse square attenuation that the light has little effect beyond 4 or 5 units.

Distance:	1	2	3	4	5	6	8
Inverse linear relative intensity	1	1/2	1/3	1/4	1/5	1/6	1/8
Inverse square relative intensity	1	1/4	1/9	1/16	1/25	1/36	1/64

Spotlights

Light beams from spotlights are concentrated in a cone-shaped pattern, as if the light had a reflector. This allows you to not only shine the light in a specific direction but also control the size of the illuminated area. This size is determined both by the taper angle of the light beam cone and by the spotlight's distance from the surface. The larger the cone angle, the larger the illuminated area; for a given cone angle, the further the surface is from the light, the larger the illuminated area. Consequently, the location of both the light and the light target point must be specified when a spotlight is installed.

Actually, spotlights have two cones of light. An interior cone, named the "hotspot cone," is centered within an outer cone of light, called the "falloff cone," These are shown in Figure 9.28.

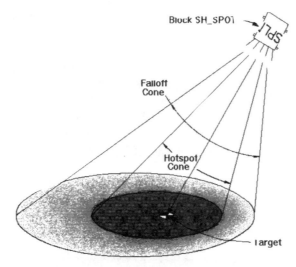

Block SH_SPOT

SPOT

Falloff Cone

Hotspot Cone

Target

Figure 9.28

The intensity of the hotspot cone remains constant from its center out to the edge of the cone, so that it makes a disk of light. The intensity of the falloff cone, on the other hand, gradually fades from the center of the cone to its edge. The taper angle of each of these cones can be set independently to any value between 0° and 160°, although the hotspot cone angle cannot be greater than the falloff cone angle. You must use Photo Real or Photo Raytrace rendering before these two cones are evident.

As with point lights, the attenuation of spotlight intensity can be set to none, inverse linear, or inverse square. The default setting is inverse linear falloff.

When attenuation is turned off, the spotlight will have a maximum intensity of 1.00. Otherwise, the maximum source light intensity depends on the size of the drawing extents and rate of attenuation.

AutoCAD places a block named SH_SPOT in the spotlight position. This block has five slightly diverging lines, representing light rays, pointing toward the target.

AUTOCAD SHADOWS

Shadows are controlled by the AutoCAD command that installs lights—the LIGHT command—and also by the RENDER and RPREF commands. When you install a light, you can specify that the light will cast shadows and the type of shadows that are to be cast. You can also adjust some parameters of those shadows. Then, you must use the RENDER or RPREF commands (discussed earlier in this chapter) to globally turn on shadow casting. The RENDER and RPREF commands also affect the type of shadows that are cast and control some shadow parameters.

AutoCAD's renderer can use any of three different methods in generating shadows. A description of those methods and of their characteristics follows.

Volumetric Shadows

The name for this type of rendering shadow comes from the technique that AutoCAD uses in computing shadow outlines. These shadows, as shown in Figure 9.29, always have sharply defined (hard) edges. Also, the color of shadows cast by semitransparent objects is affected by the color of the objects. An option within the LIGHT command determines whether or not volumetric shadows will be used. Moreover, rendering type, which is specified with the RENDER or RPREF commands must be set to Photo Real.

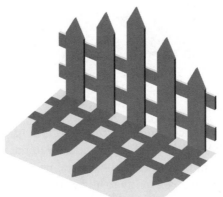

Figure 9.29

Ray-Traced Volumetric Shadows

AutoCAD uses this technique for generating shadows when volumetric shadows have been specified through the LIGHT command and Photo Raytrace rendering has been specified through the RENDER or RPREF commands. Similar to the volumetric shadows of Photo Real rendering, the shadows of ray-traced volumetric shadows have hard edges and are affected by the color of semitransparent objects. However, for complex geometries, the outlines of ray-traced volumetric shadows can be more accurate and the computation time can be faster than for photo real volumetric shadows.

Shadow Maps

In creating shadows of this type, AutoCAD makes a prerendering pass of the view and generates a bitmap of the shadows. Unlike volumetric shadows, these shadows have fuzzy (soft) edges, such as those in Figure 9.30, and the color of shadows cast by semitransparent objects is not affected by the color of the objects.

Figure 9.30

Shadow map shadows are turned on though an option in the LIGHT command, which also allows you to control the softness of the shadow edges, as well as the size of the shadow bitmap. The rendering type, as set by RENDER or RPREF, can be either Photo Real or Photo Raytrace.

You may have noticed in Figure 9.30 that the shadows are slightly detached from the objects. This is referred to as shadow map bias, and it can be controlled to some extent through the More Options button of the RENDER and RPREF commands. The More Options button of those two dialog boxes displays a dialog box labeled Depth

Map Shadow Controls, which is shown in Figure 9.31, that contains two edit boxes.

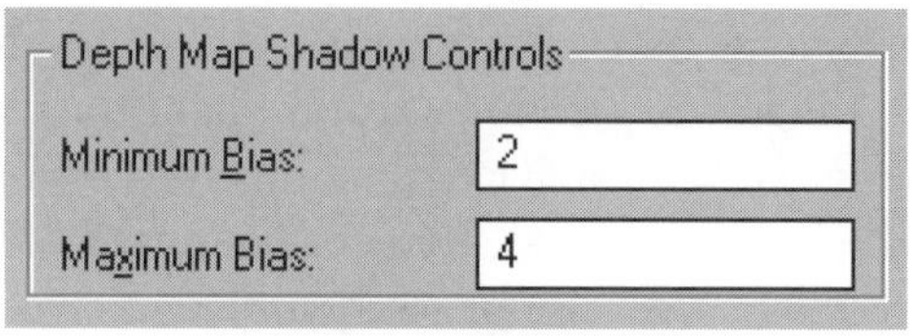

Depth Map Shadow Controls

Minimum Bias:	2
Maximum Bias:	4

Figure 9.31

Suitable values for shadow map bias depend on the scene being rendered; finding the proper values often requires numerous trial renderings. Usually, minimum bias should be set somewhere between 2 and 20, and maximum bias should be no more than 10 times the minimum bias value. The default values are 2 and 4. Assigning a low number, even 0, to Minimum Bias draws shadows closer to their objects, but can sometimes make the rest of the rendering appear grainy.

THE LIGHT COMMAND

All operations concerning lights are performed through the LIGHT command. A companion command is SCENE, which combines specific lights with specific views (created by AutoCAD's VIEW command) to be used by the RENDER command. The menu options and toolbar buttons for initiating these commands are shown in Figure 9.32.

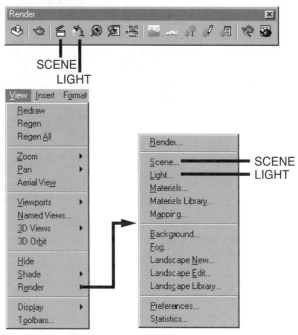

Figure 9.32

Whether you start the LIGHT command from the toolbar button, the pull-down menu, or the command line, the Lights dialog box shown in Figure 9.33 will be displayed.

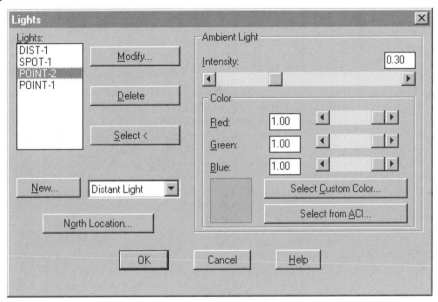

Figure 9.33

LIGHTS

The names of existing lights are shown in a list box on the left side of this dialog box. To modify or delete an existing light, highlight its name and click the Modify button or the Delete button. If you have forgotten which light is which in the drawing, click the Select button. AutoCAD will temporarily switch to the graphics screen so that you can select an installed light's block. The name of that light will then be highlighted in the list. When you press the Modify button, the Lights dialog box will be replaced by one appropriate for the type of light that is highlighted.

NORTH LOCATION

AutoCAD uses the positive direction of the WCS Y axis as the default north location for aiming distant lights. You can, however, use the North Location button in the Lights dialog box to specify another direction for north. A dialog box named North Location will be displayed so you can choose another coordinate system direction to represent north. This dialog box allows you to base the north direction on a named UCS rather than the WCS.

AMBIENT LIGHT

This cluster of options is for setting ambient light intensity and color. Increasing ambient light intensity from its default value of 0.30 will make unlighted areas of the 3D model brighter; reducing ambient light intensity will make them darker.

Although you can set the color of ambient light, you are not likely ever to need to do this. First of all, colored ambient light does not have a significant effect in renderings; second, the results depend on the properties of the surface, as well as those of the installed lights.

Ambient light color can be set by assigning a number between 0.00 and 1.00 to each of the three basic colors—red, green, and blue. These numbers may be either entered in an edit box or set with the slider bars. A value of 0 turns off that particular color. When each of the three colors equal 1.00, the resulting color is white (except when the graphics screen background color is white—then the light color is black). The button labeled Select Custom Color brings up a dialog box named Color that allows you to select a basic color or combine colors to create a custom color. The button labeled Select from ACI will bring up the same Select Color dialog box AutoCAD uses for setting layer color and for the DDCOLOR command.

The follow-up dialog boxes for installing and modifying distant lights, point lights, and spotlights all have a cluster of options for setting color that is identical to the one for setting ambient light color.

NEW

To install a new light, select a light type from the pull-down list, which is pictured in Figure 9.34, to the right of this button and then click the button. This will bring up a dialog box appropriate for the type of light.

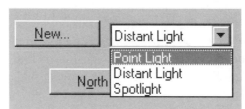

Figure 9.34

NEW/MODIFY DISTANT LIGHT DIALOG BOX

The dialog boxes for modifying an existing distant light and for installing a new distant light are identical except for their titles. An example of a Modify Distant Light dialog box is shown in Figure 9.35. The current settings will be shown in the dialog box when modifying an existing light, and the default settings will be shown for a new light.

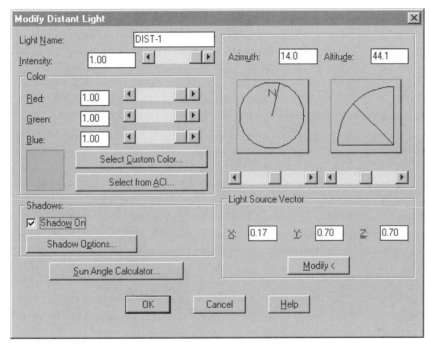

Figure 9.35

LIGHT NAME

The only mandatory entry for a new light is its name. Every light must have a single-word name, no more than eight characters in length. AutoCAD displays the name, using uppercase letters, in both the Light Name window and the block that will be inserted in the drawing at the light's location. When you are modifying a light, you can rename it by entering a new name in this edit box.

INTENSITY

The intensity of the distant light can be set with the slider bar or by typing in a value. Intensity can range from 0.00, which turns the light off, to 1.00.

COLOR

The color of the distant light is controlled by the cluster of options labeled Color. The methods for setting a color are the same as those in the main Lights dialog box for setting ambient light color.

AZIMUTH AND ALTITUDE

The direction of the distant light is controlled by the Azimuth and Altitude settings. Azimuth is the angle of the light's direction in the XY plane, and altitude is the angle of the light's direction from the XY plane. This is similar to the VPOINT rotation angles for setting a view direction, although azimuth uses the Y axis for a base rather than the X axis (unless you have used the option in the main Light dialog box to establish another direction for north). Both Azimuth and Altitude can be set by slider bars or by entering values in the edit boxes. When the edit box is used for setting altitude, a negative number can be entered to force the light to point up.

LIGHT SOURCE VECTOR

Light direction can also be set by this cluster of X, Y, and Z edit boxes. For example, if you wanted the light to point horizontally in the minus X direction, you could enter the values: X=1, Y=0, Z=0.

When the Modify button is clicked, AutoCAD will switch to the graphics screen and prompt for two points from the command line:

Enter light direction TO <current>: (*Specify a point or press ENTER.*)

Enter light direction FROM <current>: (*Specify a point or press ENTER.*)

These prompts use the word current without telling you what the current coordinates are. Pressing ENTER will leave the points in their current position. The light will point in the direction from the second point toward the first point.

AutoCAD initially sets the direction for a new distant light in the current view direction and inserts the light icon block in the center of the current viewport. A distant light can be moved without changing its direction by moving its icon block with the MOVE command and you can ROTATE the block to change the distant light's direction.

SUN ANGLE CALCULATOR

This button provides yet another way to set the direction of a distant light. Clicking it will bring up a dialog box titled Sun Angle Calculator in which you can specify a date, time, and a geographic location through world latitude and longitude coordinates. AutoCAD will then aim the direct light to match those parameters. To help you set a geographic location, this dialog box has supplementary dialog boxes in which you can indirectly set latitude and longitude by picking the name of a city in virtually any part of the world.

SHADOWS

If you want the light to cast shadows, select the Shadow On checkbox. Clicking the Shadow Options button displays the Shadow Options dialog box shown in Figure 9.36.

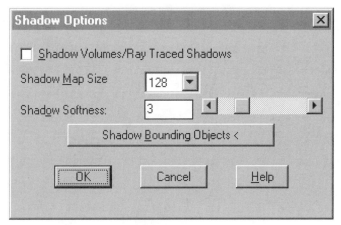

Figure 9.36

- Shadow Volumes/Ray Traced Shadows

- Selecting this checkbox will force AutoCAD to make volumetric shadows when Photo Real rendering is used and volumetric ray-traced shadows when Photo Raytrace rendering is used. When this checkbox is not selected, AutoCAD will use shadow maps in defining the outline of shadows.

- Shadow Map Size

- The size of the bitmap AutoCAD uses for shadows is set by the values in this drop-down list box. Values can range from 64 to 4096, with 128 as the default value. Larger values make for more accurate shadows, but they increase rendering time.

- Shadow Softness

- The value entered in this text box (or set by the slider) represents the number of pixels along the edges of shadows that are blended with the underlying images. Values can range from 1 to 10, with 3 being the default value.

- Shadow Bounding Objects

- This button is available only when shadow map shadows are in effect. When it is clicked, the dialog box will be temporarily dismissed so that you can select the objects that will cast shadows. Objects not selected will not cast shadows.

NEW/MODIFY POINT LIGHT DIALOG BOX

The dialog boxes for installing a new point light and for modifying an existing one are the same, except for their titles, as shown in Figure 9.37.

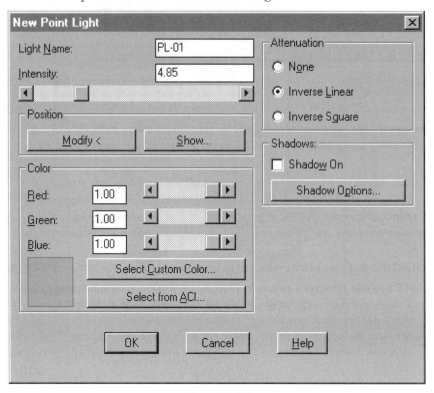

Figure 9.37

LIGHT NAME

Every light must be assigned a name having no more than eight characters. AutoCAD uses this name in the block it inserts at the point light location, and will display it in the main Lights dialog box.

INTENSITY

You can set the point light's source intensity by entering its value in the edit box. If you enter a number that is too large, AutoCAD will reduce it to the maximum. You can also use the slider bar below the text window to set intensity. The maximum value allowed depends on the type of attenuation and the drawing extents. If there is no attenuation, the maximum value is 1.00.

ATTENUATION

The rate of light intensity falloff (attenuation) is set by selecting one of these three radio buttons. When attenuation is set to none, objects far from the point light will be illuminated just as brightly as objects close to the light.

POSITION

By default, AutoCAD will place a new point light in the middle of the current viewport, but that will seldom be the location you want. Clicking the Show button will display the X, Y, and Z coordinates of the point light location. When the Modify button is clicked, AutoCAD will switch to the graphics screen and issue the prompt:

Enter light location <current>: (*Specify a point or press ENTER.*)

Although this prompt does not give the coordinates of the current light location, pressing ENTER will leave the light in its present location. You can specify a new point by typing in coordinates, pointing, or using object snap points.

AutoCAD places a point light icon block, which shows the name of the light, in the light's location. You can move this block to move the light.

COLOR

This cluster of options controls the color of the point light. The methods used in setting a color are the same as those in the main Lights dialog box for setting ambient light color.

SHADOWS

The options for point light shadows work the same as those in the New/Modify Distant Light dialog box.

NEW/MODIFY SPOTLIGHT DIALOG BOX

The dialog boxes for installing new spotlights and modifying existing spotlights are the same, as shown in Figure 9.38, except for their titles.

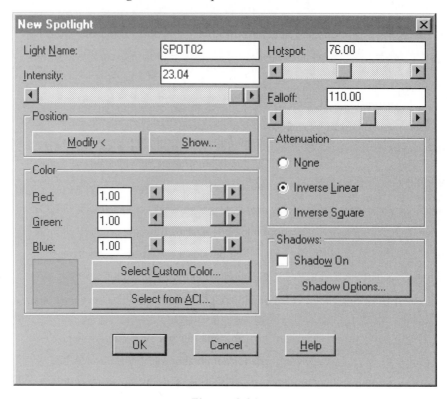

Figure 9.38

LIGHT NAME

Each spotlight must have a unique name that is no more than eight characters long. AutoCAD automatically converts lowercase letters to uppercase. This name is shown in the light's icon block and in the main Lights dialog box.

INTENSITY

Set the spotlight's source intensity by entering a value in the edit box or by using the slider bar below. If you enter an intensity value that is too large, AutoCAD will automatically reduce it to the maximum. If no attenuation is in effect, 1.00 is the maximum intensity. When attenuation is set to inverse linear or inverse square, maximum source intensity depends on the extents of the drawing.

ATTENUATION

The rate of light intensity falloff is set in this cluster of three radio buttons. When attenuation is set to none, objects far from the spotlight will be illuminated just as brightly as objects close to the light.

HOTSPOT/FALLOFF

The edge-to-edge angle of the hotspot cone and the falloff cone are set independently in the two edit boxes or in the slider bars below them. The angle of either cone can be set to any value between 0° and 160°; however, the hotspot cone angle cannot be larger than that of the falloff cone.

POSITION

AutoCAD initially places the target of a new spotlight in the center of the current viewport and positions the light so that it is approximately 1 unit away and points in the current view direction toward the target. The default positions will seldom be what you want.

- Show
- Clicking the Show button will display the current coordinates of both the light and the target, as shown in Figure 9.39.

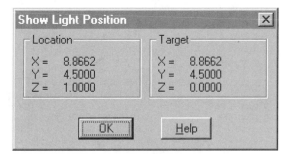

Figure 9.39

- Modify
- You can change the target and light locations by clicking this button. AutoCAD will switch to the graphics screen and prompt for two points from the command line:

Enter light target <current>: (*Specify a point or press ENTER.*)

Enter light location <current>: (*Specify a point or press ENTER.*)

In these prompts, AutoCAD does not tell you what the current coordinates are, although pressing ENTER will leave the points in their current position. You can use any of the standard methods for specifying points when setting new target and light locations.

AutoCAD places a block named SH_SPOT in the spotlight position. This block shows the name assigned to the spotlight and has five lines, representing light rays, pointing toward the target. You can use MOVE or ROTATE to re-aim a spotlight. The target point will automatically adjust to the new location or direction.

COLOR

This cluster of options controls the color of the spotlight. The methods used in setting a color are the same as those in the main Lights dialog box for setting ambient light color.

SHADOWS

The buttons for spotlight shadows work the same as those in the New/Modify Distant Light dialog box, except that you cannot select shadow bounding objects.

Tips: When selecting light locations, remember that the more inclined a surface is to incoming light beams, the darker the surface will be. Consequently, vary the light beam angle between surfaces to achieve maximum contrast between the surfaces. Figure 9–40 shows a hidden-line view of a cube on the left. In the center is the same cube rendered using a single direct light aimed with equal angles to all three visible sides of the cube. The result is a colored blob. In the rendering on the right, the direct light has been aimed so that the light beams are at different angles to each surface. Each of the three illuminated surfaces has a different shade, resulting in a 3D appearance.

Figure 9.40

It pays to think through and plan your lighting arrangement and to know the coordinates of key points on your 3D model. Often you will have one light that will set the overall tone of your rendering. You should install this light, then move it and change its intensity until the model looks right. Add a second light, if it is needed, to supplement the key light. This second light will usually be a different type than the first. Often a third light, and perhaps even a fourth or fifth one, is necessary.

Lights should be installed one at a time and the results tested before adding an additional light. When there are too many lights, they begin to cancel each other's effects. It also becomes difficult to tell which light is causing which effect.

You will seldom have any need to assign any color other than white to lights. You have more control and flexibility when you base colors on materials, rather than lights.

Spherical coordinates are a good way to establish spotlight locations because they can establish a point that is a specific distance and at any angle from another point. Use any method in locating the spotlight's target, then enter relative spherical coordinates to position the spotlight.

THE SCENE COMMAND

Scenes, which are a combination of lights and a view, are very useful when making a series of rendering from a single 3D model. SCENE, the command that creates and manages scenes, uses the Scenes dialog box, which is shown in Figure 9.41.

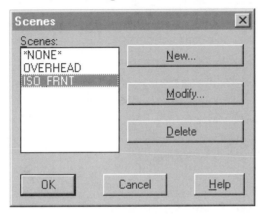

Figure 9.41

On the left side of the dialog box is a list of the names of existing scenes. There is always one named *NONE*, which represents the current view in the current viewport and all of the currently installed lights. If no lights are installed, AutoCAD uses a single distant light pointed in the current view direction. You can select a scene from this list to be modified or deleted.

On the right side of the Scenes dialog box are buttons to add a new scene and to modify or delete a selected existing scene. The New and the Modify buttons use dialog boxes that are identical except for their names (see Figure 9.42).

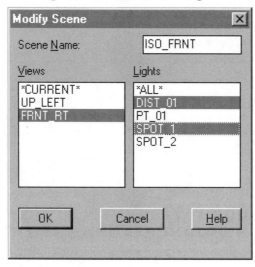

Figure 9.42

Each scene must be assigned a unique name having no more than eight characters. On the left side of the New/Modify Scenes dialog box, which is shown in Figure 9–42, is a list of all views that have been saved with the VIEW command. You may select just one view from this list. On the right side of the dialog box is a list of all lights currently installed in the drawing. You may select as many lights to be included in the scene as you want by holding down CTRL as you pick them. The selected view and the selected lights will be highlighted. Click the OK button to return to the Scenes dialog box.

TRY IT! - INSTALLING LIGHTS

Exercise One:

Retrieve and open the file 3d_ch9_02.dwg. It contains the solid model of a machine component that we will use for a rendering. The steps to install lights for the rendering are:

1. Before installing any lights you must decide on and set up a view for the rendering. For this model we will use a viewpoint that is rotated 8° degrees from the XY and -50° from the X axis in the XY plane, as shown in Figure 9.43. Use either the VPOINT or DVIEW command to set these viewing angles.

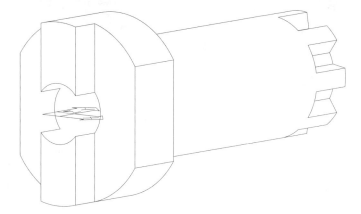

Figure 9.43

2. For general illumination of the model, we will use a single distant light that is aimed perpendicular to the model's center line and is pointed down at a fairly steep angle so that the shaded side is on the bottom. To install this light:

-   Initiate the LIGHT command. Make sure that ambient light intensity is set to its default value of 0.30, and all of its color values are 1.00 (so that its color is white). Then select Distant Light from the light type pull-down list, and click the New button.

-   In the New Distant Light dialog box, assign any name you want to this light. Make sure that the light's intensity is 1.00, all of the three color values are 1.00, and the Shadow On checkbox is off (there is no surface under the model, so shadows will have little effect).

-   Set the distant light's azimuth to 90°; set its altitude to 60°.

-   Click the OK button to return to the main Light dialog box, and click its OK button to complete the LIGHT command.

3. Start the RENDER command. In the Render dialog box, select Photo Real as the rendering type, then click the Render button.

4. The distant light did a good job with the cylindrical surfaces, as shown in Figure 9.44, but the front of the model is almost completely in the shade. We will add a spotlight to illuminate it.

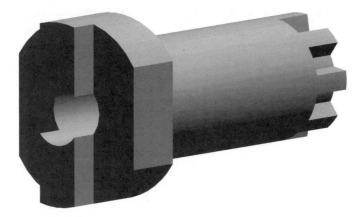

Figure 9.44

• Invoke the LIGHT command again. Select Spotlight as the light type and click the New button. In the New Spotlight dialog box, assign a name of your choice to the light. Make certain that the light color is white and that the Shadow On checkbox is cleared.

• To make the three separate but parallel surfaces stand out from each other we will use inverse square attenuation to make the more distant surfaces darker than closer surfaces. We also want the light to be relatively dim and to have wide cone angles. Therefore, set the following light parameters:

• Attenuation: Inverse square

• Intensity: 3.0

• Hotspot: 160.00

• Falloff: 160.00

• Click the Modify Position button. AutoCAD will temporarily dismiss the dialog box and issue command line prompts for the light's target and location points. Respond as follows:

Enter light target <current>: 0,0,0

Enter light location <current>: 1,-2,0

• Click OK to exit the New Spotlight dialog box, and then click OK to end the LIGHT command.

5.   Invoke the RENDER command again and click the Render button to render the current view. Your rendering should look similar to the one in Figure 9.45.

File 3d_ch9_03.dwg contains this model with the lights installed.

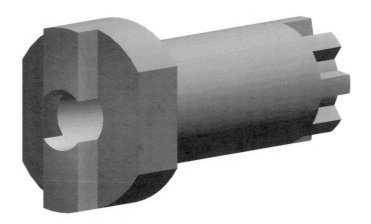

Figure 9.45

TRY IT! - INSTALLING LIGHTS

Exercise Two:

 Now open file 3d_ch9_04.dwg from the CD-ROM that comes with this book.

The model in this file, which is shown in Figure 9.46, is the 3D surface model room we con-structed in Chapter 3—with a ceiling, door, and wall mirror added. Also, the lamp, dresser, and chair have been rearranged slightly, and table has been shortened as well as moved. We will install a point light, a direct light, and two spotlights in the room.

Figure 9.46

Use AutoCAD's VIEW command to restore the view named PERSPECTIVE_01. This is a perspective view that has a front clipping plane that allows us to see inside the room.

You will install four lights in this room—one direct light, one point light, and two spotlights. Rather than go through the steps to install these lights, we will simply list the parameters you should use for them. Notice that feet and inch dimensions are given for point locations because this drawing file uses architectural units. Leave ambient light at its default setting of 0.30, and leave the color of each light at white.

You should install these lights in the listed order; and make a rendering after each light is installed so that you can see the results of the light. Also, using multiple viewports will help you compare results. Use the Photo Real rendering type and make certain that shadows are turned on.

First Light

Type: Point Light

Suggested Name: PT_01

Intensity: 120.0

Attenuation: Inverse Linear

Shadow: On

Shadow type: Volumetric

Light Location: X = 4' (48") Y = 5' (60") Z = 12' (144")

Comment: This light simulates a ceiling light.

Second Light

Type: Distant Light

Suggested Name: DIST_01

Intensity: 0.50

Azimuth: 150.0

Altitude: 45.0

Shadow: Off

Comment: This light provides general room illumination.

Third Light

Type: Spotlight

Suggested Name: SPOT_01

Intensity: 2.00

Attenuation: Inverse Linear

Hotspot: 90.00

Falloff: 160.00

Shadow: Off

Target Location: X = 4'4 (52") Y = 8'4 (100") Z = 8' (96")

Light Location: X = 4'4 (52") Y = 8'4 (100") Z = 5"6 (66")

Comment: This light, which is very dim, simulates light radiating from the top of the lampshade. It points directly up.

Fourth Light

Type: Spotlight

Suggested Name: SPOT_02

Intensity: 100.0

Attenuation: Inverse Linear

Hotspot: 0.00

Falloff: 160.00

Shadow: On

Shadow type: Volumetric

Target Location: X = 4'4 (52") Y = 8'4 (100") Z = 0' (0")

Light Location: X = 4'4 (52") Y = 8'4 (100") Z = 4' (48")

Comment: This light, which points directly down, simulates light radiating from the bottom of the lampshade.

Figure 9–47 shows, in grayscale, the rendered room when all four lights are in use.

File 3d_ch9_05.dwg contains the lights we have just installed.

Figure 9.47

Choosing and installing lights is a subjective process. There are many other light installations for this room that would be equally good, and you may want to use this surface model room for experiments of your own. Also, the colors used for objects in this file are extremely basic, and you may want to change them.

MATERIALS

So far, our discussions on rendering have centered on the effect that lights have on the appearance of surfaces. But the properties of those surfaces also have a major effect on appearance during rendering. Surfaces can be reflective, diffuse, transparent, smooth, rough, have one or more colors, and have patterns.

AutoCAD refers to these properties as attributes and manages them through the RMAT command. Before we describe how to use RMAT, we will briefly explore the relationships between light and material and point out the AutoCAD material attributes that control those relationships.

RELATIONSHIPS BETWEEN LIGHT AND MATERIAL

As Figure 9.48 shows, when a light beam strikes the surface of an object, part of the light beam is reflected and part of it penetrates the object. Incoming light is referred to as incident light, light penetrating the object is called refracted light, and light bouncing off the surface is called reflected light.

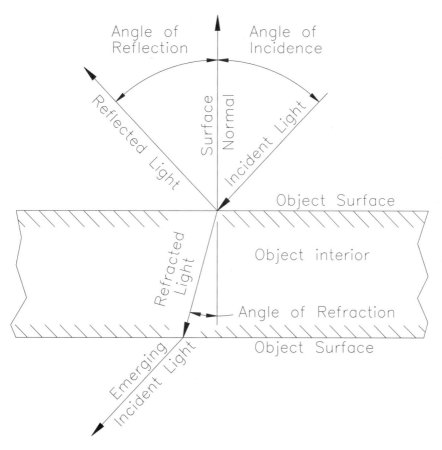

Figure 9.48

The percentage of incident light that is reflected depends on the object's material. Most materials reflect almost all incident light, whereas very little (often none) is refracted. These are opaque materials. However, there are some materials—glass and water for example—for which most of the light is refracted. These, of course, are transparent materials. AutoCAD allows you to control the relative transparency of a material through the RMAT command's Transparency attribute. Either Photo Real or Photo Raytrace rendering must be used for this attribute to have an effect.

As shown in Figure 9.48, the direction of a light beam changes as it passes through a transparent object, and it resumes its original angle of incidence when it emerges from the other side. This is why objects seen under water or through thick glass appear to be offset. The light beam's angle within an object is called the angle of refraction. When Photo Raytrace rendering is used, AutoCAD can handle refracted light through the material attribute named Refraction.

Figure 9–48 also points out that the angle at which light is reflected is the same as that of the incoming light beams. Or, as physics textbooks say: The angle of reflection equals the angle of incidence. Although this is true even if the surface is rough and uneven, a rough surface reflects incident light in many directions because its individual sub-surfaces are tilted in many directions. This is illustrated in a highly magnified view of a surface cross-section in Figure 9.49.

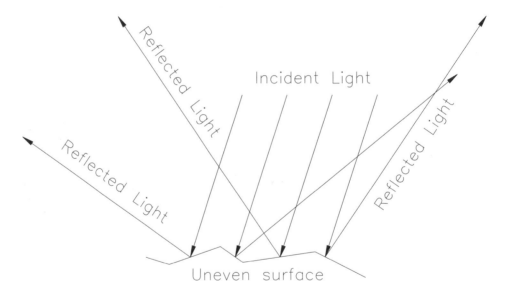

Figure 9.49

Scattered reflected light is called diffuse reflection, and materials that have diffuse reflection are often said to have a dull, or matte finish. Since there is no predominant reflection direction, diffuse surfaces appear to be uniformly illuminated from all viewpoints. Brightness of a diffuse material is still affected by the angle of incidence, as we explained in our discussion of lights, but it is not affected by the viewing angle.

On the other hand, when a surface is smooth, all of the reflected light beams have the same angle. This condition is called specular reflection, and materials with specular reflection are referred to as being shiny or glossy. If you view a shiny surface from the angle of reflection you will be looking directly into the reflected light beams, which makes the surface appear to have the same color as the incident light. The surface will be highlighted. Because most lights—such as sunlight—are white, highlights usually are also white.

In AutoCAD's renderer, the relative amount of diffuse reflection is controlled by the material attribute named Color, whereas the relative amount of specular reflection is controlled by the Reflection material attribute.

With perfect reflectors the angle of incidence and the angle of specular reflection are the same, and you must view the surface from this angle in order to see the highlight. Not many surfaces, however, are perfect reflectors. Most have a range of viewing angles at which highlights can be seen. The material on the left in Figure 9.50 has a narrow range of specular reflection, while the material on the right has a wide range of specular reflection. The viewer on the left will see a small but intense highlight; the viewer on the right will see a larger, but less bright, highlight. AutoCAD refers to the range of the specular reflection angle as roughness, and allows you to control it through the RMAT command.

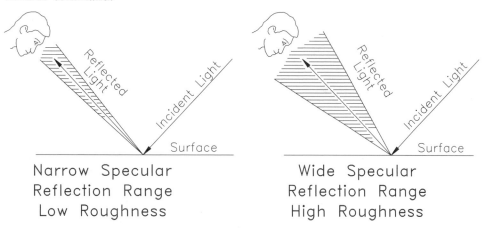

Figure 9.50

MATERIAL MAPPING

In addition to the properties, or attributes, we have just described—color, reflection, roughness, transparency, and refraction—you can also use bitmap images when defining a material. This is often called mapping, and the bitmap images are sometimes referred to as maps.

AutoCAD 2002 includes about 150 bitmap files that can be used for defining materials. The default AutoCAD installation places these files in the AutoCAD 2002\Textures folder. All of these Autodesk-supplied bitmap files are in the TGA file format, but other formats, such as GIF, BMP, TIF, JPG, and PCX, can be used as well.

The process to define a standard material involves specifying its attributes and, if needed, supplementing those attributes with maps. You can supplement the attributes with up to four different forms of mapping: texture maps, bump maps, reflection maps, and opacity maps.

Texture Maps

Texture maps (AutoCAD also calls them pattern maps) are bitmap images of floor tiles, wood, rusted metal, granite, bricks, wallpaper, or just about anything else. See Figure 9.51 for an example. You must, of course, have a suitable bitmap file available. About 50 of the bitmap files that come with AutoCAD are suitable for materials. A texture map is specified through the RMAT command's Color attribute.

Figure 9.51

By default, the bitmap image will expand, compress, or stretch as necessary to fit the size and shape of a surface, but you can adjust and scale the image to meet your needs. These adjustments are done through the RMAT and SETUV commands. Moreover, you must make some adjustments when a nonplanar surface is mapped and when a planar surface that is not parallel to the WCS XY plane is mapped. We will describe these adjustments when we discuss the SETUV command.

If the scale of the image is such that it does not completely fill the surface, you can either have the image repeated, which is called tiling, or have it appear just once, which is called cropping. See Figure 9.52 for examples.

Tiled Bitmap   Cropped Bitmap

Figure 9.52

Texture maps that have patterns do not work well with 3D solid models. AutoCAD treats the entire surface of a 3D solid model as a single object, and the bitmap image is severely distorted when it is applied to faces having various orientations. However, AutoCAD has three special materials, called template materials, that do work well with 3D solids. We will describe template materials shortly.

Bump Maps

A bump map is a bitmap image that is laid over a texture map to create an embossed, or bumpy, effect. The bump map image has a pattern similar to the texture map, but it has different colors (usually lighter) and it may be offset slightly to enhance the raised appearance (see Figure 9.53). Bump maps are specified through a special RMAT attribute called Bump Map. To be effective, a corresponding texture map must also be specified.

Figure 9.53

Reflection Maps

Reflection maps, sometimes referred to as environment maps, simulate an image reflected on a shiny surface. They are not the same as the mirrored reflections of Photo Raytrace renderings. Reflection maps are specified through the Reflection material attribute. They work best on surfaces that are rounded, and they do not work well at all on planar surfaces.

Opacity Maps

Opacity maps, which are specified through the Transparency material attribute, control the degree of transparency (and conversely, the degree of opaqueness) for a surface. In an opacity map, pure white areas are completely opaque, while pure black areas are completely transparent. Transparency from shades of gray is proportional to the relative darkness of the gray, and transparency from colors depends on the equivalent grayscale values of the colors. As shown in Figure 9.54, opacity maps are a convenient way to make a hole, or a window, through a surface.

This bitmap image
on a 3D Face...   makes a window.

Figure 9.54

TEMPLATE MATERIALS

The standard AutoCAD rendering materials we have just described define materials through attributes, and the attributes can be supplemented with mapping. In addition to these standard materials, AutoCAD has three special materials that simulate wood, granite, and marble. They are called template materials because they start with the basic features of the material they simulate. You can then modify them in ways appropriate for the material. Template materials are shown only when Photo Real or Photo Raytrace rendering is used.

Although these template materials, which AutoCAD sometimes calls solid materials, share some similarities with texture maps, they are different in two significant ways. First, AutoCAD generates their patterns at rendering time—they are not based on a bitmap file. Second, their patterns are 3D.

While template materials can be attached to planar objects, such as 3D faces, their 3D characteristics work especially well with 3D solids. As shown in Figure 9.55, when a material made from the wood template is attached to a 3D solid box primitive, the grain of the wood appropriately matches the sides of the box. If this box were made from 3D faces, you would have to add separate bitmaps to each surface.

Figure 9.55

WORKING WITH MATERIALS

When you work with materials, you will spend the bulk of you time within the RMAT command. With this one command, you will specify the properties that a material is to have and attach the material to objects. Although RMAT gives you some control over the scale and positioning of bitmaps, you will have to use the SETUV command to control their orientation and their scale on individual objects.

Establishing the properties of a material can be time consuming. Therefore, AutoCAD helps you organize the materials you have created into libraries, so that they can be efficiently reused. It does this through the MATLIB command.

The menus and toolbar buttons for initiating these commands are shown in Figure 9.56. Yet another command for working with materials is SHOWMAT. This command, which simply tells you what material is attached to a selected object, is unique among the rendering commands in that it does not use a dialog box. You must initiate SHOWMAT from the command line.

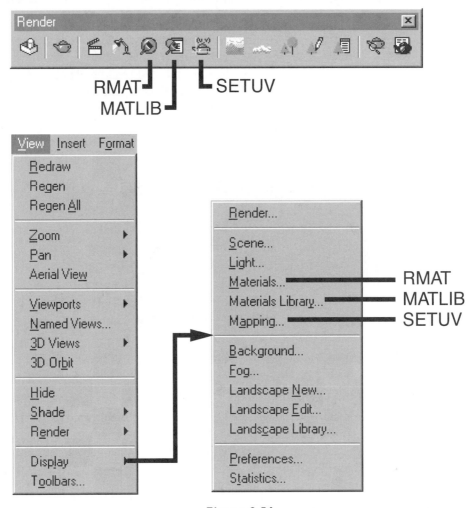

Figure 9.56

THE RMAT COMMAND

Two steps are required for an object to have material properties. The first step is to define a set of material characteristics, or attributes as AutoCAD calls them. This set of attributes, which is referred to as a material, will have a unique name. You can use a predefined material from a materials library or define your own material.

The second step is to assign the named material to the object. The object may be explicitly selected to receive the material, or it may receive the material due to its color or its layer. Both of these steps are done through the RMAT command.

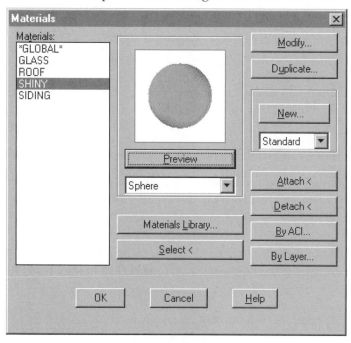

Figure 9.57

The command starts with the Materials dialog box, which shows a list of existing materials in a text box; see Figure 9.57. There will always be a material named *GLOBAL* in this list. Every object will be rendered using the material properties of *GLOBAL* unless another material is specifically attached to it. Most operations in the Materials dialog box involve selecting a material name in the list, and then clicking a button to initiate an action with that material.

PREVIEW

A sphere or cube having the attributes of the selected material will be displayed when this button is pressed. A drop-down list allows you to specify whether a sphere or a cube is be used in the preview.

MATERIALS LIBRARY

This button brings up the Materials Library dialog box for importing predefined materials into the Materials dialog box. The Materials Library dialog box will be discussed when the MATLIB command is covered.

SELECT

Use this button to find out what material is attached to a specific object. The Materials dialog box will be dismissed to allow you to select one AutoCAD object. Then the dialog box will reappear with the name of the material attached to that object displayed in the lower margin. The SHOWMAT command performs the same function as this button.

MODIFY

This button brings up a dialog box to be used for changing the attributes of the specified material.

DUPLICATE

This button allows you to define a new material based on an existing material. A dialog box containing the settings of the highlighted material will be displayed.

NEW

This button is for defining a new material. The drop-down list below this button, as shown in Figure 9–58, allows you to specify whether the new material is be a standard material or one of the template materials—granite, marble, or wood. A dialog box appropriate for the material type will be displayed.

Figure 9.58

ATTACH

This button is explicitly for attaching the highlighted material to one or more AutoCAD objects. The dialog box will temporarily disappear for you to select the objects. Explicitly attached materials take precedence over materials attached by other methods.

DETACH

Allows you to remove the highlighted material from an AutoCAD object. The dialog box is dismissed until you finish selecting objects.

BY ACI

This button is for assigning a material to objects having a specific color. The Attach by AutoCAD Color Index dialog box, which is shown in Figure 9.59, will be displayed. Within this dialog box you specify a material and a color. Then you click the Attach button to attach the specified material to all objects having that color. You can also detach the material from the selected color. Materials attached by ACI take precedence over materials attached by layer, but not over materials attached explicitly.

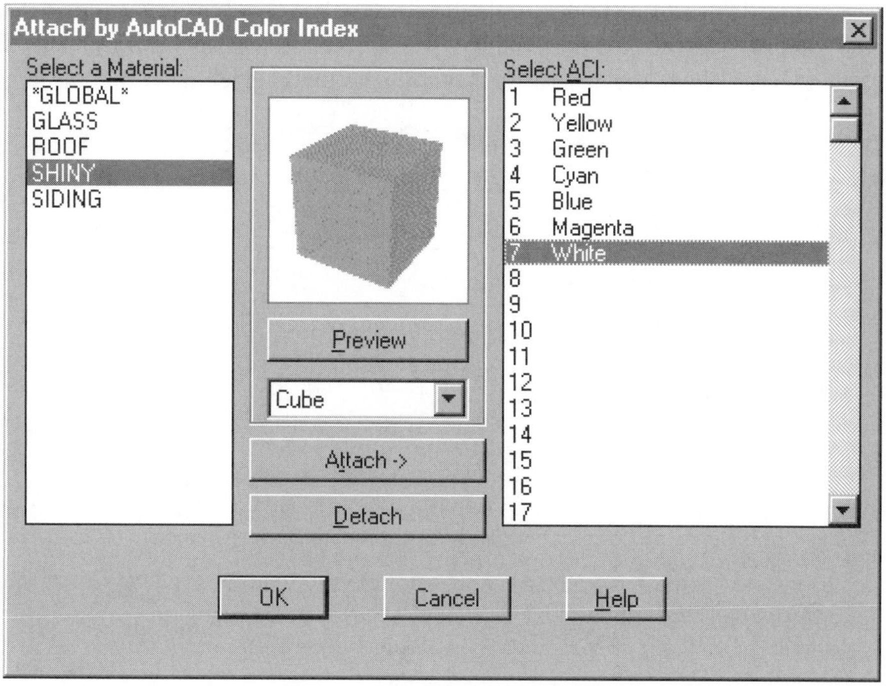

Figure 9.59

BY LAYER

This button is for assigning materials to all objects within a specific layer. The Attach by Layer dialog box will be displayed for you to use in specifying a material and a layer and to attach the material to or detach it from objects within that layer. This dialog box is the same as the one used by the By ACI button, except that a list of layers is shown rather than colors.

NEW/MODIFY STANDARD MATERIAL DIALOG BOXES

These two dialog boxes, which are identical except for their names, are for defining the properties of a standard material. See Figure 9.60. Because the buttons in these two dialog boxes are interrelated, we will not discuss them on a button-by-button basis.

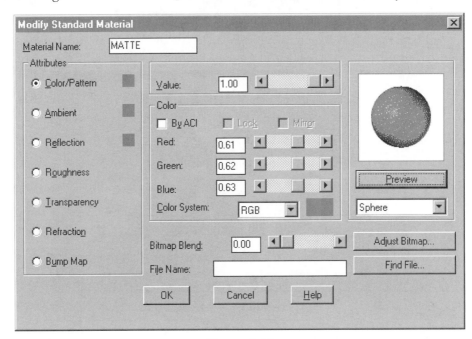

Figure 9.60

Near the top of the dialog box is a space for the material's name. If you are modifying an existing material, its name will be shown. If you are creating a new material, you must give it a name having no more than 16 characters. Spaces within the name are allowed. If you enter a new name for an existing material, the material will be renamed.

On the left side of the dialog box are seven radio buttons for setting the material's attributes—its characteristics—one at a time. You select an attribute, and then assign it a numerical value using the edit box labeled Value or the slider bar. Three of the attributes—Color, Reflection, and Transparency—also permit you to specify a bitmap for the material, and the sole purpose of one attribute—Bump Map—is for specifying a bitmap. The attributes, with their high and low values and their default settings, along with the property they control, and the type of mapping associated with them, are shown in Table 9–2.

Table 9–2 Attributes

Attribute Name	Attribute Value Range	Default Value	Property Controlled by Attribute	Associated Map Type
Color/Pattern	0–1.0	0.70	Color and relative amount of diffuse (matte) reflection	Texture map
Ambient	0–1.0	0.10	Color and brightness in shaded areas	None
Reflection	0–1.0	0.20	Color and relative amount of specular (shiny) reflection	Reflection map
Roughness	0–1.0	0.10	Relative size of highlights. Roughness has an effect only when Reflection has a non-zero value	None
Transparency	0–1.0	0.00	Fraction of light that passes through the material. Also enables mirror reflections	Opacity map
Refraction	0–100	1.00	The relative degree at which light beams bend as they pass through transparent material	None
Bump Map	N/A	N/A	Specifies the file name for a bump map and sets its parameters	Bump map

The Refraction attribute will have an effect only when Transparency has a nonzero value and Photo Raytrace rendering is used. Although it accepts values up to 100, you will probably keep the refraction setting somewhere between 1 and 3. The value you use will depend on the shape of the transparent object, its background, the view, and the desired effect. When Refraction is set to its default value of 1.00, no light refraction occurs. In Figure 9.61, refraction for the wedge on the left is 1.00, and for the wedge on the right it is slightly more than 1.00.

Figure 9.61

The Color/Pattern, Ambient, and Reflection attributes can have separate colors. Color is controlled by a cluster of options, labeled Color, located in the center of the dialog box. By default, all of the buttons in this cluster will be grayed out, except for the one labeled By ACI. This button is a checkbox that is selected by default. When it is on, attribute color is the same as that of the object. When By ACI is not checked, you can set the attribute's color by assigning values from 0 to 1.00 to each of the three basic color components.

The drop-down list box below the slider bars allow you to choose between the RGB and HLS color system. The RGB color system uses red, green, and blue for basic colors, and HLS controls hue, lightness, and saturation.

The checkbox labeled Lock, located above the slider bars, works only when the By ACI box is deselected, and even then it only affects the Ambient and Reflection attributes. When the By ACI checkbox is deselected (which means you are assigning colors to the attribute), the Lock checkbox controls whether or not you can assign separate colors to the Ambient or Reflection attribute. When Lock is selected, its default setting, Ambient and Reflection take on the same color that is assigned to the Color attribute. If you want either of these two attributes to have their own color, turn the Lock option off.

The remaining checkbox, labeled Mirror and located above the color slider bars is enabled only for the Reflection attribute. When it is checked, materials will show mirror-like reflections of other objects, provided Photo Raytrace rendering is used.

Clicking the Preview button will give you an idea of the effects of the current settings on a sphere or cube.

When the attribute selected is Color/Pattern, Reflection, Transparency, or Bump Map, you can assign a bitmap file to the attribute (see the previous discussion on material mapping for an explanation of how the bitmap files relate to the attributes). You can enter the bitmap file name, along with its path, directly in the edit box or click the Find File button to browse for the file.

The effect of Bitmap Blend, which can be set with the slider bar or by entering a value from 0 to 1.00 in the edit box, depends on the bitmap's attribute. For Color/Pattern, it represents the degree to which the bitmap image shows through the assigned color. When the blend is 0, only the attribute's assigned color will be shown—the bitmap will not even appear. When the blend is 1.0, none of the attribute's assigned color will show through the bitmap. For bump maps, bitmap blend represents the degree of bumpiness. Usually, blend values in the neighborhood of 0.10 work well for bump maps. For the Reflection attribute, bitmap blend represents the amount of reflection; for the Transparency attribute it represents the amount of transparency.

The Adjust Bitmap button brings up a dialog box titled Adjust Material Bitmap Placement for you to adjust the relative size of the bitmap, its origin, and whether it is to be tiled (repeated) or cropped (not repeated). This button will be grayed out for the Reflection attribute bitmap. As a dialog box almost identical to this one is used by the SETUV command, we will not describe the contents of the Adjust Material Bitmap Placement dialog box here.

NEW/MODIFY MARBLE MATERIAL

These two similar dialog boxes are for creating or modifying one of the three template materials. These materials have patterns and colors that simulate marble, wood, and granite, as shown in Figure 9.62. Unlike texture maps their patterns are generated during a rendering—they are not based on bitmap files.

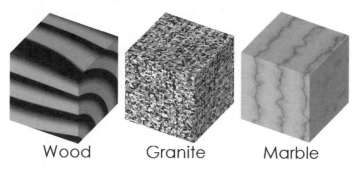

Wood   Granite   Marble

Figure 9.62

AutoCAD sometimes refers to these three materials as solid materials. Because the template materials are not based on bitmap files, they cannot be used in advanced rendering and animation programs, such as 3D Studio.

As shown in Figure 9.63, on the left side of the New/Modify Marble Material dialog box are radio buttons for the material's attributes. These work just as they do in the New/Modify Standard Material dialog box—you pick an attribute and set its value.

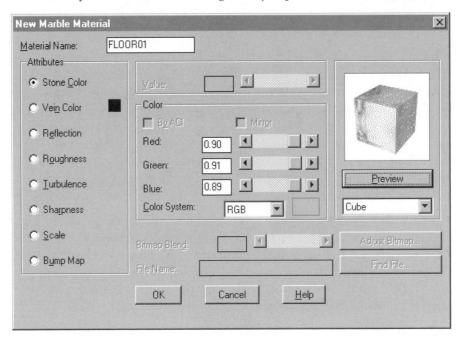

Figure 9.63

Stone Color is the general color of the marble and Vein Color is the color of its veins. Although you can assign each of these attributes a color, the color value edit box and the By ACI button are not available. The Reflection and Roughness attributes are the same as those of standard materials. A bitmap file can be assigned to the Reflection attribute.

Turbulence controls the extent to which the veins oscillate and swirl, while sharpness determines the extent to which the vein colors blend with the stone color. Scale sets the overall size of the veins. Bump Map works the same as it does with standard materials.

NEW/MODIFY WOOD MATERIAL

This template material simulates the growth rings of wood. As shown in Figure 9.64, radio buttons for setting the material's attributes are in a cluster on the left side of its dialog box.

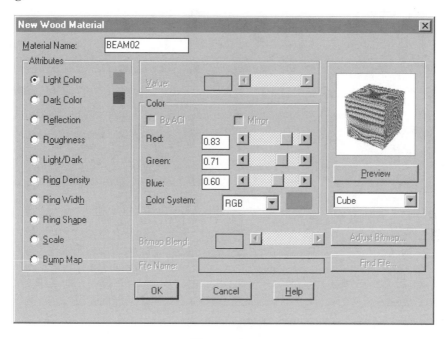

Figure 9.64

The Dark Color attribute sets the color of the rings, while Light Color sets the wood color between the rings. Assigning colors to these attributes works the same as it does for standard materials, except that the color value edit box and the By ACI button are inactive. Reflection and Roughness work the same as they do for standard materials.

The Light/Dark attribute controls the overall intensity of the rings. The wood is almost completely dark when a value of 0 is used, and is almost completely light when value is set to 1.0. Ring Density controls the number of rings within a given cross-section area, with higher values resulting in more rings.

Ring Width determines the width of the light-colored rings relative to the dark rings. Higher values result in relatively wider light rings. When the Ring Shape attribute is set to 0, the growth rings are perfectly round; as its value increases the rings become more irregular. Scale sets the overall size of the wood pattern. The Bump Map attribute works the same as it does for standard materials.

NEW/MODIFY GRANITE MATERIAL

This template material simulates granite stone. Radio buttons for controlling the material's attributes are in a column on the left side of its dialog box, as shown in Figure 9.65.

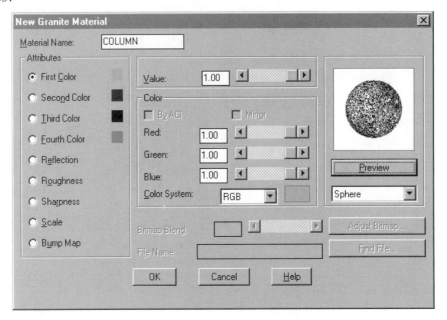

Figure 9.65

The material consists of random patterns in four different colors. Setting these four colors works the same as it does for standard materials, except that the By ACI button is inactive. Unlike the colors for the other template materials, you can set a value for each of the four granite colors.

Reflection and Roughness work the same as they do in standard materials. When Sharpness is set to 1.0, the colored patterns are distinct and have sharp boundaries; when sharpness is set to its other extreme of 0, the colors are blended completely together. Scale controls the overall size of the color patterns. Bump Map works as it does for standard materials.

Tips: Assigning a color to an attribute is often useful because it allows you to render objects in colors unrelated to the color in which they were drawn. Also, the color white for the Reflection attribute (which results in white highlights) tends to make objects look shinier than other colors.

You do not need to be precise in setting color values because differences between value settings are hardly noticeable. You can generally set the values of the Color, Ambient, and Reflection attributes to 1.0; and then assign a low value to Roughness to give the material a shiny finish, or a high value to give it a matte finish. For extremely shiny, realistic finishes, turn the Mirror toggle on and use Photo Raytrace rendering.

Texture maps are an excellent way to enhance renderings, especially those of architectural models. They enable you to add tile floors, wallpaper, roof shingles, and wood patterns to your models.

Materials based on the wood template can be useful, especially if you are working with solid models. The other two template materials, however, are not so useful. AutoCAD comes with some bitmap files that give more realistic images of marble and granite. The names of these files are GRANITE.TGA, GRAYMARB.TGA, IMARBLE1.TGA, MARB-PALE.TBS, MARBTEAL.TGA, and PINKMARB.TGA.

THE SHOWMAT COMMAND

SHOWMAT performs the same function as the Select button in the Materials dialog box of the RMAT command. It reports the name of the material attached to an object, and the method used to attach it. The command is called from the command line:

Command: SHOWMAT

Select object: (*Select a single AutoCAD object.*)

AutoCAD will respond with one of the following messages on the command line:

Material <Material Name> is explicitly attached to the object.

Material <Material Name> is attached by ACI to ACI <color number>.

Material <Material Name> is attached by layer to layer <layer name>.

Material *GLOBAL* is attached by default or by block.

THE MATLIB COMMAND

AutoCAD comes with about 150 predefined materials in a file named REN-DER.MLI ready for you to use, and you can make additional material library files yourself. The command that manages these material files is MATLIB, which uses the dialog box named Materials Library (see Figure 9.66).

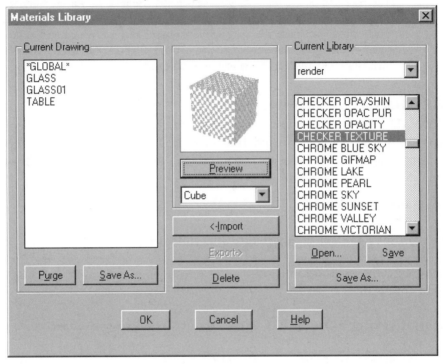

Figure 9.66

On the left side of this dialog box is the Materials List, a list of materials currently in the drawing—and on the right is the Library List, a list of materials in the library file. You can select a material in either list for further action.

Two buttons operate only on the Materials List:

- Purge
- Removes all unattached materials from the Materials List.
- Save
- Saves all of the materials in the Materials List in a material library file, having the .MLI extension. AutoCAD will display a dialog box named Library File for you to use in naming the file.

The following two buttons operate only on the Library List:

- Open

- Opens a new material library file. AutoCAD will display the Library File dialog box for you to select a saved library file.

- Save

- Saves the current library list to a material library file. It also uses the Library File dialog box.

The following buttons operate on materials selected in either list:

- Import

- Adds a material selected in the Library List to the Materials List.

- Export

- Adds a material selected in the Materials List to the Library List.

- Preview

- Displays a sphere or cube with the attributes of the selected material.

- Delete

- Deletes the selected material. If the material is currently attached to an object, AutoCAD will display a warning message.

 Tips: The Materials Library dialog box can also be accessed through the Materials Library button in the Materials dialog box used by the RMAT command. In fact, if you intend to use library materials, it is better to access the Materials Library dialog box through the Materials dialog box than directly with the MATLIB command. Then you can import materials from the library list as well as attach them to objects.

The Delete button in the Materials Library dialog box is the only way you can delete a material from a drawing. There is no provision in the RMAT command for deleting materials.

THE SETUV COMMAND

By their very nature, the bitmap images used for material mapping are 2D. But often they need to be attached to 3D objects. Furthermore, even if they have a similar geometric shape, the proportions of the bitmap may not conform to the proportions of the object to which it is attached. For example, a bitmap image that is square may be attached to a planar surface that is long and slender. In some cases these mismatches in geometry and proportion do not cause problems, but they often cause patterns and images to be unacceptably distorted.

To solve these problems, AutoCAD has a command that fits images to object geometry and allows you to adjust the size, shape, origin, and orientation of the images. This command is SETUV. The letters U and V refer to the mapping coordinates AutoCAD uses in establishing directions and origins of bitmap images. Although sometimes these directions correspond to the current X and Y directions, there is no direct relationship between them. The UV coordinate system is completely independent from the UCS. When template materials are involved, a W direction is used as well.

When SETUV is initiated, AutoCAD will ask you to select the object or objects that are to receive the UV mapping coordinates. The objects you select do not have to have a material attached to them. You can establish mapping coordinates on an object first, and then attach a material that uses bitmaps to that object. When you have finished the object selection set, SETUV will display the Mapping dialog box, shown in Figure 9.67.

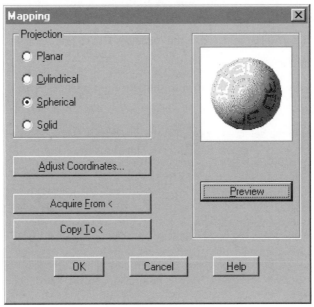

Figure 9.67

Typically, you will first click on a radio button from the cluster labeled Projection to select the projection type that comes closest to the shape of the object or objects you have selected. Examples of cylindrical, spherical, and planar projections are shown in Figure 9.68. The Solid radio button is for assigning mapping coordinates to objects that have one of the three template materials (wood, granite, or marble) attached.

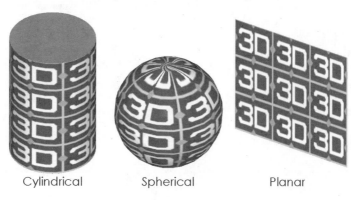

Cylindrical   Spherical   Planar

Figure 9.68

Next, you will click the Adjust Coordinates button to bring up a dialog box appropriate for setting mapping coordinates on the projection geometry you have selected.

The Acquire From button allows you to assign the mapping coordinates that exist on another object to the current selection set, and the Copy To button allows you to copy the mapping coordinates of the current selection set to other objects.

ADJUST PLANAR COORDINATES

When the Planar projection radio button is active in the Mapping dialog box, clicking that dialog box's Adjust Coordinates button will bring up the Adjust Planar Coordinates dialog box (see Figure 9.69).

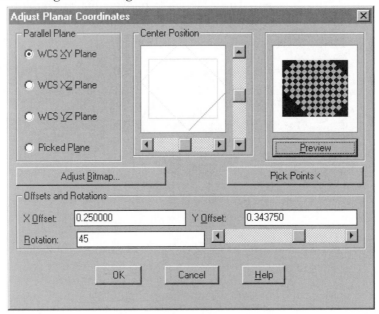

Figure 9.69

The cluster of radio buttons labeled Parallel Plane are for specifying the orientation of the mapping plane. You must specify one of the three principal WCS planes or click the Picked Plane button if your surface is askew to the principal planes. Notice that the principal planes are relative to the WCS, not the UCS. The object does not have to lie in one of these planes; it only has to be parallel to the specified plane.

When the Picked Plane radio button is selected, AutoCAD will temporarily dismiss the dialog box and issue prompts for you to pick three points on the plane to be used for orienting it. You can change an existing planar orientation by clicking the Pick Points button.

The Center Position image box shows the current location and orientation of the bitmap. You can use the slider bars to the side of the image box to adjust the center location. You can also use the edit boxes near the bottom of the dialog box to change the image's X and Y offsets; you can also rotate the image through the Rotation edit box or slider bar. If your selected object has a bitmap attached, the Preview pane will show you what it looks like.

Clicking the Adjust Bitmap button leads to the Adjust Object Bitmap Placement dialog box, which will be described shortly.

ADJUST CYLINDRICAL COORDINATES

Clicking the Adjust Coordinates button in the Mapping dialog box when the Cylindrical projection radio button is active brings up the Adjust Cylindrical Coordinates dialog box, which is shown in Figure 9.70.

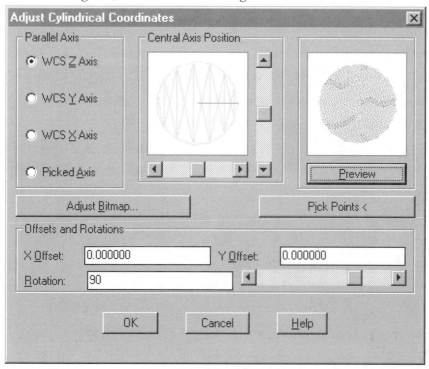

Figure 9.70

Use the radio buttons in the upper left corner of the dialog box to specify the direction of the cylinder's axis relative to the WCS. If you pick points for the axis, AutoCAD will prompt for three points—two to establish the axis, and one to establish the wrap line (the start of the bitmap image).

The Central Axis Position image box shows the selected object's mesh lines projected onto a plane perpendicular to current axis, as well as the location of the axis (by way of a circle), and the wrap line. By default, the wrap line is in the minus Y direction when the axis is in the Z direction, and in the minus Z direction when the axis is in the X or Y directions. Clicking the Preview button will show you the results of the current settings, provided a bitmap is attached to the selected object.

The X and Y Offsets can be changed by entering new values in their edit boxes. The

Rotation edit box and slider rotates the wrap line about the cylinder axis by the specified angle.

Clicking the Adjust Bitmap button leads to the Adjust Object Bitmap Placement dialog box, which is described later.

ADJUST SPHERICAL COORDINATES

When the Spherical projection radio button is active in the Mapping dialog box, clicking the Adjust Coordinates button will bring up the Adjust Spherical Coordinates dialog box. See Figure 9.71.

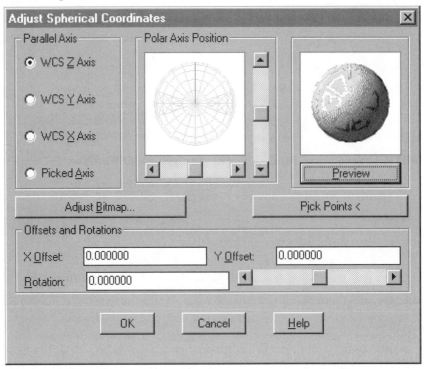

Figure 9.71

The controls in this dialog box work basically in the same way as those in the Adjust Cylindrical Coordinates dialog box.

ADJUST UVW COORDINATES

Clicking the Adjust Coordinates button in the Mapping dialog box when the Solid projection radio button is active brings up the Adjust UVW Coordinates dialog box. This dialog box, which is shown in Figure 9.72, pertains only to the three template materials—wood, granite, and marble.

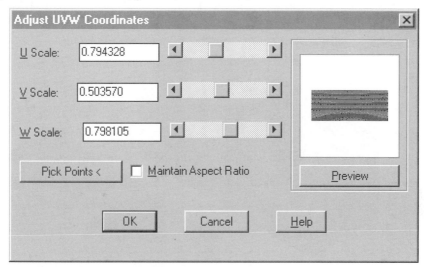

Figure 9.72

You can change the scale of the image in each direction with the three edit boxes or slider bars. When the Maintain Aspect Ratio checkbox is selected, changing the value in one direction changes the values in the other two directions as well.

You can change the U, V, and W directions by clicking the Pick Points button. AutoCAD will temporarily dismiss the dialog box and prompt for four points—the origin and the U, V, and W axis directions. This allows you to not only reorient the image but also fix it on the object, so that if you rotate the object the image will rotate also.

An example of reorienting a template material is shown in Figure 9.73. The same wood material has been attached to both of the 3D solids. The 3D solid on the left has the default wood material orientation, which orients the end view of the wood's annual growth rings on a plane parallel with the WCS YZ plane. The U, V, and W directions have been reoriented on the 3D solid on the right, so that the growth rings are on the smallest face of the 3D solid. Moreover, they will remain on that face even when the 3D solid is rotated.

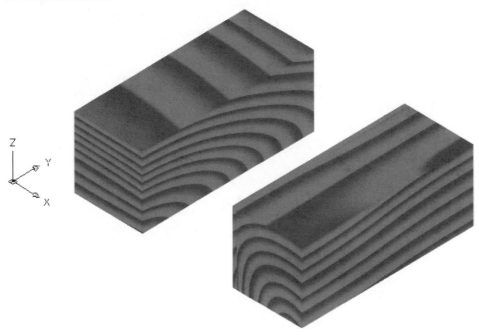

Figure 9.73

ADJUST OBJECT BITMAP PLACEMENT

When you click the Adjust Bitmap button in the Adjust Planar Coordinates, Adjust Cylindrical Coordinates, or Adjust Spherical Coordinates dialog boxes, a dialog box titled Adjust Object Bitmap Placement will appear. You will use this dialog box, which is shown in Figure 9.74, often as you change the scale of bitmap images, adjust their position on the object to which they are attached, and specify whether the image is to be tiled or cropped. An almost identical dialog box is displayed by the RMAT command's Adjust Bitmap button. Furthermore, the dialog box for the BACK-GROUND command has an Adjust Bitmap button that brings up a dialog box that is similar to this one.

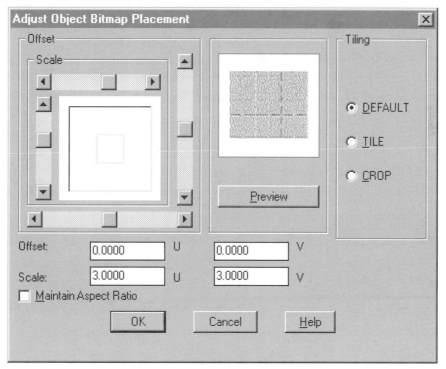

Figure 9.74

The scale of a bitmap represents the number of times that the bitmap is repeated. You can have a scale in the U direction that is different than the scale in the V direction. Scale is set by the slider bars to the top and left of the image tile, or by entering a value in the Scale edit boxes.

Offset represents a shift of the center point of the images. It can be given a value ranging from -1.0 to 1.0, and like scale, it can be different in the two directions. You can change the bitmap offset with the slider bars below and to the right of the image tile, or by changing the contents of the Offset edit boxes.

An idiosyncrasy with both the Scale and Offset edit boxes is that they are limited to seven digits, and you cannot type over existing digits—you must delete some digits before you can change them. When the Maintain Aspect Ratio checkbox is selected, the U and V directions are locked together when either scale or offset values are changed.

When the TILE radio button is selected, the bitmap image will be repeated as many times as necessary to completely fill the selected surface. Conversely, when the CROP radio button is selected, the bitmap image will be appear just once on the selected surface. The remaining areas will be rendered in the object's material color. Scale must be greater than one before either of these two radio buttons have any effect.

Bitmap placements adjusted through the RMAT command apply to the material as a whole, whereas bitmap placements adjusted through the SETUV command apply to specific objects. Consequently, the material attached to an object can have bitmap scales and offsets that are different than those of the object. When this is the case, offsets are added and scales are multiplied.

Tiling can also differ between the object and the material. As a result, the Adjust Object Bitmap Placement dialog box for the SETUV command has a radio button labeled DEFAULT. When this button is active, the bitmap is either tiled or cropped depending on the setting for the material as a whole. Clicking the TILE or CROP radio button will override the overall material setting.

 Tip: You must use a one-line AutoLISP function to remove UV mapping from an object. Enter the following on the command line:

(c:setuv "d" (ssget))

You will be prompted to select the objects whose UV mapping you want removed.

We will lead you through three exercises in creating and applying materials. They will all be for material on the exterior of the 3D surface model house shown in Figure 9.75. You last saw this house in Chapter 4 when you used it in an exercise in setting up perspective views.

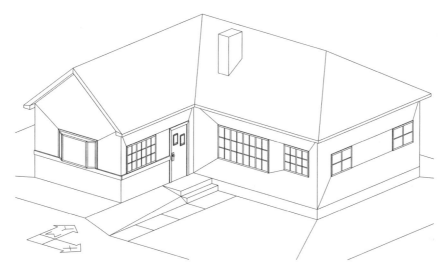

Figure 9.

Retrieve the file 3d_ch9_06.dwg from the accompanying CD-ROM and open it. You will work with this file in all three exercises.

TRY IT! - CREATING AND APPLYING MATERIALS

Exercise One

In this first exercise, you will create a material that will serve as the siding of the house. This material will not use any bitmaps; you will simply create a color for the material.

Initiate the RMAT command, make certain that the material type is Standard, and click the New button. In the New Standard Material dialog box, name the new material SIDING, and set the value of the Color/Pattern, Ambient, and Reflection attributes to 1.00. Activate the Color/Pattern attribute, click the By ACI checkbox off, set the Red and Green color values to 0.80, and the Blue color value to 0.50. Last, click the OK button to return to the Materials dialog box.

In the Materials dialog box, click the By Layer button to bring up the Attach by Layer dialog box. Highlight SIDING in the left-hand list box, and highlight the layer named Wall_sf in the right-hand list box. Then, click the Attach button. Click the OK button in this dialog box, as well as in the Materials dialog box to exit the RMAT command.

Finish this exercise by rendering the scene. It does not make any difference what type of rendering you use. The walls of your house should have a light, brownish-yellow color.

TRY IT! - CREATING AND APPLYING MATERIALS

Exercise Two

In this exercise, you will create a material for the roof of the house that uses a bitmap file of cedar shingles. After you attach that material to the roof surfaces you will use the SETUV command to adjust the bitmap on those surfaces.

Start by making the layer named Roof_sf the current layer and freezing all of the other layers except for Ashade (AutoCAD uses the layer Ashade while making renderings). This will allow you to concentrate on and have easy access to the five 3D faces that make the roof's surfaces.

Rotate the view point 320° in the XY plane from the X axis, and 40° from the XY plane. The resulting view should be similar the one in Figure 9.76.

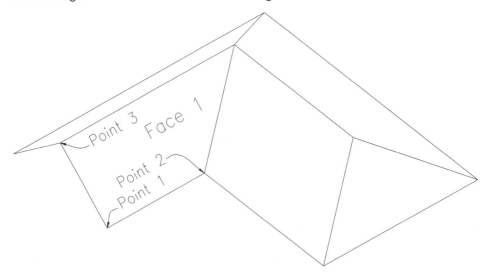

Figure 9.76

Start the RMAT command, make sure that the material type is Standard, and click the New button. Name the new material ROOF. Click the Color/Pattern attribute radio button, and then click the Find File button. Locate a file named ishingl2.tga and bring it into your material by double-clicking its file name or by clicking the Open button (with the default AutoCAD installation, this file will be in a folder named Textures). Set the Bitmap blend to 1.00. That is all there is to creating the material. If you want to get an idea of what the image looks like, preview the image in its cube form.

Click the OK button to return to the main Materials dialog box. Then, with the material

ROOF highlighted in the Materials list box, click the Attach button. AutoCAD will dismiss the dialog box. Individually pick each of the five 3D faces, and then press ENTER to bring back the dialog box. Click the OK button to end the RMAT command.

You should now render the scene to ensure that the material is attached and displays its image. For this, and for all of the renderings that follow, you must use Photo Real rendering. The scale and orientation of the bitmap images in this rendering will not be correct. They must be adjusted with the SETUV command.

Start the SETUV command. When you are prompted to select an object, select the 3D face labeled Face 1 in Figure 9–76. In the Mapping dialog box, make certain that the projection type is Planar, and click the Adjust Coordinates button.

Because this 3D face is not parallel with any of the three principal WCS planes, select the Picked Plane radio button. The dialog box will disappear, and AutoCAD will issue the following three command line prompts. Respond to these prompts by picking the points shown in Figure 9.76.

Place the lower left corner of the mapping plane: (*Pick point 1.*)

Place the lower right corner of the mapping plane: (*Pick point 2.*)

Place the upper left corner of the mapping plane: (*Pick point 3.*)

The Adjust Planar Coordinates dialog box will reappear. The U direction for the bitmap will be the direction from point 1 to point 2, and V is the direction from point 1 to point 3. Press the Adjust Bitmap button to bring up the Adjust Object Bitmap Placement dialog box. In this dialog box, set the offset to:

   U = 0.000000

   V = 0.025000

and the scale to:

   U = 10.00000

   V = 8.000000

Lastly, click three OK buttons to exit the SETUV command.

When you render the scene, the pattern on this 3D face should be similar to that in Figure 9–77. You may need to use Render's Query for Selections option for this material to render correctly.

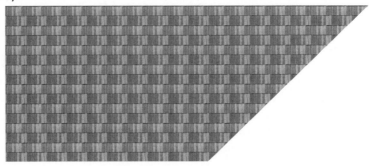

Figure 9.77

Use the SETUV command four more times to set the coordinates for the other four 3D faces. Use Figure 9.78 (which shows the plan view of the house) to establish the corners of each mapping plane. The lower left corner of each mapping plane is at point 1, the lower right corner is at point 2, and the upper left corner is at point 3.

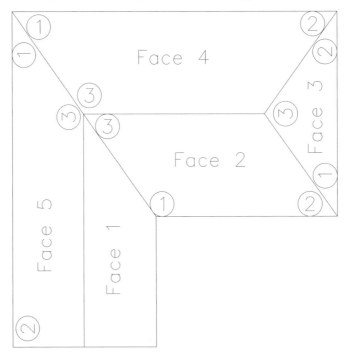

Figure 9.78

Set the UV scale for each mapping plane as shown in the following table. These scales, which are based on the relative dimensions of the 3D faces, will make the shingles on each 3D face approximately the same size. You do not need to make any offset adjustments for these faces.

Direction	Face 2	Face 3	Face 4	Face 5
U	14.280000	15.700000	20.570000	20.570000
V	11.440000	11.440000	11.440000	11.440000

When you have set the mapping directions and scales, render all five faces of the roof. The shingle rows should be parallel with the edges of the roof, and the shingles should be approximately the same size on each 3D face. A close-up of a portion of the rendered roof is shown in Figure 9.79. We would like to have more variety and randomness in the shingles, but ishingl2.tga is the only bitmap file of shingles that comes with AutoCAD.

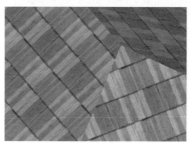

Figure 9.79

TRY IT! - CREATING AND APPLYING MATERIALS

Exercise Three

In this last exercise, you will create a material that has a texture map and a bump map of bricks, and you will attach that material to the house. You will also have to use SETUV to adjust the scale and orientation of the mapping planes, but it will not be as involved as in the previous exercise.

Thaw and switch to the layer named Brick02, and freeze all the other layers. Then, rotate the viewpoint 330° in the XY plane from the X axis, and 20° from the XY plane. Only two objects, both of them 3D faces, are in this layer.

Invoke the RMAT command and start a new standard material. Name the material BRICK. With the Color/Pattern attribute active, click the Find File button to locate and bring into the material a bitmap file named BMBRICK.TGA (This material type is located on the CD supplied in this book). Since the image in this file is very dark, set the bitmap blend for it to 0.50. Then, turn off the By ACI checkbox and set the RGB color values to: Red = 1.0, Green = 0.70, and Blue = 0.50. Set the value of Color/Pattern to 1.00. This color will show through the brick images.

Click the Bump Map attribute radio button, and then click the Find File button to locate and bring a bitmap file named BMBRICB.TGA into the material (BMBRICB.TGA can be found on the CD supplied with this book). Set its bitmap blend to 0.10, and click the OK button to return to the main Materials dialog box. With BRICK highlighted in the Materials list box, click the Attach button, and pick the two 3D faces.

A rendering at this time will not be satisfactory because the two 3D faces are perpendicular to the WCS XY plane. By default, AutoCAD's renderer assumes that planar surfaces are parallel with the WCS XY plane.

Initiate the SETUV command and pick 3D face 1 shown in Figure 9.80. With the Planar projection radio button active, click the Adjust Coordinates button in the Mapping dialog box. In the Adjust Planar Coordinates dialog box, click the WCS XZ Plane radio button, and then click the Adjust Bitmap button. In the Adjust Object Bitmap Placement dialog box, set the Scale to 6.0 in the U direction and 2.0 in the V direction. You do not need to offset the bitmap. Then exit the SETUV command dialog boxes.

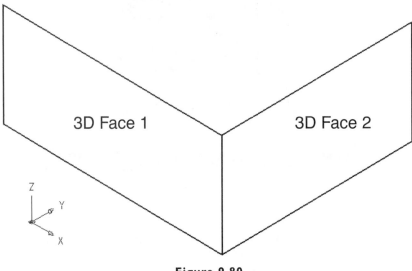

Figure 9.80

Use these same steps to adjust the mapping coordinates for 3D face 2. The projection plane for this face is parallel with the WCS YZ Plane, and the bitmap scale is 4.0 in the U direction and 2.0 in the V direction. No offset is needed.

Now render these two 3D faces. The rendering should be similar to the close-up of a rendering shown in Figure 9.81.

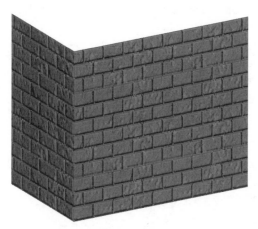

Figure 9.81

That finishes our exercises with rendering materials. You should thaw all of the layers and render the surface model.

 The 3D house, with all of the materials created through these exercises is in file 3d_ch9_07.dwg on the accompanying CD-ROM.

The model in this file also has the BRICK material applied to the chimney; the drawing file contains two distant lights, plus a background image from the bitmap file CLOUD.TGA. The path for the bitmap files in this drawing is C:\Program Files\Acad2002\Textures. If your AutoCAD files are arranged differently, you may have to change this path before the bitmap files will be incorporated into the materials.

A portion of a rendering of the house is shown in Figure 9.82. Objects other than those that have attached materials are rendered in their object color. If you have time, you may want to experiment on your own and add more materials to the model. You now have all of the knowledge you need to do this.

Figure 9.82

You will probably agree that the siding on the house is a bit too "perfect." A bitmap-based texture would help, but AutoCAD does not include any suitable bitmap files. The grass would also look more realistic if a texture map was applied, but again, AutoCAD does not include a suitable bitmap file. If you have suitable bitmap files, you can certainly use them in materials for this model.

LANDSCAPE OBJECTS

You have seen how bitmap files can serve as backgrounds for renderings and how they can add textures and patterns to materials to create realistic objects. Now you will see how bitmaps can be used as landscape objects to add images of trees, (such as the one in Figure 9.83) shrubs, people, and so on to renderings. Landscape objects are different than materials in that they are attached to their own objects; you do not have to use them with the attribute of a material to be attached to a surface object.

Figure 9.83

Three commands are used with landscape objects. LSNEW is for inserting landscape objects in a drawing, LESEDIT is for editing existing landscape objects, and LSLIB helps you manage libraries of landscape objects. A fourth command, FOG, can add to the realism of a rendering by fading specific areas of it to emphasize distance. Fog can be set to affect all objects in a rendering, including landscape objects and the background. The menus and toolbar buttons to initiate these commands are shown in Figure 9.84.

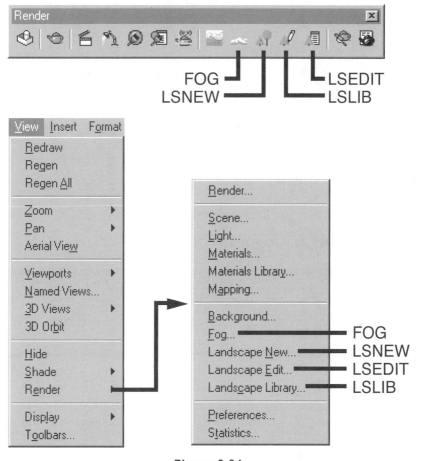

Figure 9.84

THE LSNEW COMMAND

This command inserts landscape objects into a drawing. It uses the Landscape New dialog box shown in Figure 9.85.

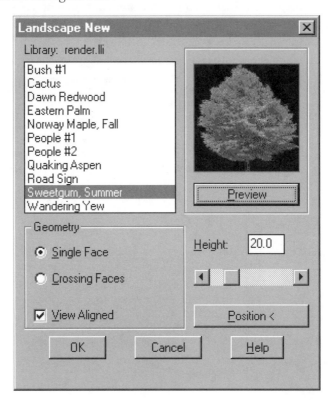

Figure 9.85

In the upper left corner of the dialog box is a list box showing the landscape objects in the current landscape library. See the description of the LISLIB command for information about landscape libraries. Select the object you want inserted in the drawing by clicking on its name. You can select only one object. You can see what the selected object looks like by clicking the Preview button.

The inserted landscape object will look like a rectangular or triangular 3D face with its landscape object name written on it. However, during a rendering it will assume the image of the bitmap file it is based on. Render type must be set to either Photo Real or Photo Raytrace for the landscape objects to render properly. Landscape objects will even cast shadows when the light and render settings activate shadows and Photo Raytrace rendering is used.

Landscape objects are special AutoCAD objects that will report to be object type Plant by the LIST command. Nevertheless, they have grips and respond to most AutoCAD editing commands. They can be moved, copied, scaled, and erased.

Below the list box of landscape objects is a cluster of options that control the geometry of the selected object. The radio buttons labeled Single Face and Crossing Faces determine whether the landscape object will be projected onto one face or onto two mutually perpendicular faces. Figure 9.86 shows an example of a single face landscape object, and Figure 9.87 shows an example of a crossing faces landscape object.

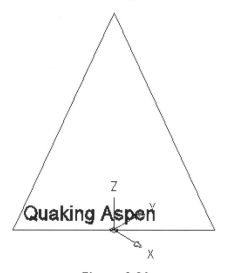

Figure 9.86

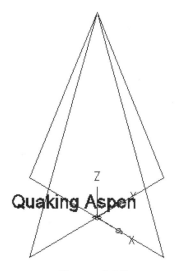

Figure 9.87

Figure 9.88 shows rendered versions of these landscape objects. The crossing faces object is on the left, and the single face object is on the right.

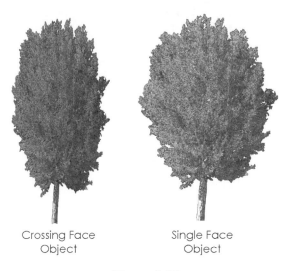

Crossing Face
Object

Single Face
Object

Figure 9.88

When the View Aligned checkbox is selected, the landscape object will always face the current view direction. Whenever the viewpoint is changed, the object will automatically rotate to match the new view direction. The landscape object will be triangular, and the AutoCAD ROTATE command will have no effect on it.

When the View Aligned checkbox is deselected, the landscape object will be aligned with the current UCS XZ plane and its orientation will not change when the view direction changes. The landscape object will have a rectangular shape, rather than triangular, and it can be rotated. Figure 9.89 shows a typical view aligned landscape object in both wireframe and rendered views.

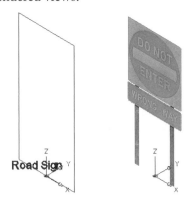

Figure 9.89

The height of a landscape object, in current drawing units, can be specified by entering a value in the edit box labeled Height or with the slider bar below the edit box. You must use the edit box to specify heights greater than 100. The height of landscape objects is always in the positive Z direction of the UCS.

By default, landscape objects are always placed at the origin of the UCS. But, when the Position button is clicked AutoCAD will temporarily dismiss the dialog box so you can specify another location.

 Tips: Landscape objects of nonplanar objects, such as trees and people, are best inserted in the View Aligned mode to ensure that they always face the view direction. Signs and other planar objects, on the other hand, are generally best inserted in the Non-View Aligned mode.

You can tell which way a non-view-aligned landscape object is facing by the appearance of its name. If the letters in the name are backward, then the front of the object is away from the camera.

THE LSEDIT COMMAND

When this command is initiated, AutoCAD issues a command line prompt for you to select one landscape object. Then it will display the Landscape Edit dialog box, which has the same features as the Landscape New dialog box. However, except for the selected object, the names of all landscape objects in the list box will be grayed out. You can edit only the selected object. See the description of the LSNEW command for an explanation of the features of the Landscape Edit dialog box.

THE LSLIB COMMAND

This command determines which landscape objects will be shown in the Landscape New list box of the LSNEW command. It uses the Landscape Library dialog box, shown in Figure 9.90.

Figure 9.90

A list box shows the objects in the current landscape library, and the name of the library is shown above the list box. To change libraries, click the Open button and select a new file from the Open Landscape Library dialog box. Landscape libraries always have a file extension of .LLI.

If you want to remove a landscape object from the library, highlight its name and click the Delete button. Clicking the Save button will cause the Save Landscape Library dialog box to be displayed. This is a standard file management dialog box that enables you to save the library in a different folder or with a different file name.

If you want to add another landscape object to the library, click the New button; if you want to modify an existing landscape object, click the Modify button. Both of these buttons bring up the same dialog box, shown in Figure 9.91. Only their names are different. The Modify button displays a dialog box titled Landscape Library Edit, and the New button displays a dialog box titled Landscape Library New.

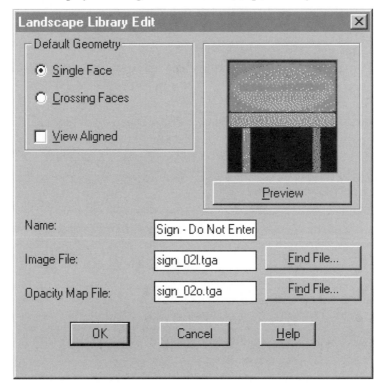

Figure 9.91

Enter or change the name of the object in the edit box. This name appears on the landscape object when it is inserted into a drawing. Names can be as long as 19 characters, including spaces.

You must specify two files for a landscape object. One of them, the Image File, will show the landscape object image on a black background. The other file, the Opacity Map File, shows the area of the image in white, and it also has a black background. In Figure 9.92, the content of an image file is shown on the left and the content of its corresponding opacity file is shown on the right.

Figure 9.92

This combination of black and white ensures that the image will be displayed in a rendering. You will recall from our discussion on material mapping that the color white is opaque and black is transparent. You can enter these file names directly in the edit boxes, or use the Find File button to browse for them. The Preview button will cause the image to be displayed.

The Default Geometry cluster of options sets the default geometry conditions in the Landscape New dialog box. You can, of course, override the default settings in that dialog box.

Tip: Although landscape objects are very useful for adding realism to rendering, AutoCAD comes with only 11 landscape objects. Furthermore, if you selected Typical during the installation of AutoCAD, the default landscape library file, render.lli, will have only one object. If this is the case on your computer, copy the render.lli file from the Acad\Support folder on your AutoCAD 2002 installation CD-ROM to your computer's Acad2002\Support folder. You must also copy all of the files in the Acad\textures folder of the AutoCAD CD-ROM to the Acad2002\Textures folder of your computer, if you have not already done so.

TRY IT! - LANDSCAPING A 3D HOUSE

We will add some landscape objects to the 3D house that you used in the exercise on materials. Retrieve drawing file 3d_ch9_08.dwg from the accompanying CD-ROM. The version of the model in this file has the roof, brick, and house siding materials attached, along with some extra 3D faces to prevent the background from showing under the house in relatively distant views. We will add some shrubs and a tree to the model. When you open this drawing file, it should show the house in a perspective view. If it doesn't, use the VIEW command to restore the view named 50_FOOT_PERSPECTIVE.

Invoke the LSLIB command. If the Landscape Library dialog box shows that the current landscape library is render.lli and there are 11 items in the library, you do not need to make any changes. You can exit LISLIB. If render.lli is not the current landscape library, click the Open button to find that file (it will probably be in a folder named Support). When you find it, highlight it and click on the Open button to bring it into the Landscape Library dialog box.

If render.lli does not have 11 landscape objects, click the New button of the dialog box, and add the files for two landscape objects—one for a bush and one for a tree. These files are located in the Acad2002\Textures folder. The file names for the landscape objects you need in the library are:

> Landscape Object 1:   Image file—8bush02l.tga
>
> Opacity map file—8bush02o.tga
>
> Landscape Object 2:   Image file—maple.tga
>
> Opacity map file—maple_op.tga

The last character in the opacity file for the first object is the letter o; not the number zero. Be certain to save the modified render.lli file when you exit it.

Now, start the LSNEW command. Highlight the object named Bush #1 (or whatever you named it). Set the geometry to Crossing Faces and activate the View Aligned checkbox. Set the Height of the bush to 50 (this will make it slightly over 4 feet tall), and click the Position button. AutoCAD will prompt for an insert position. Because you are in a perspective view, you will have to type in the coordinates of the insert point. Therefore, on the command line enter: 36',15',-1'4. If you are uncomfortable with feet and inches input, you can enter 432,180,-16 instead. Be certain that you include the minus sign with the Z coordinate. Click the OK button to end the LSNEW command, and the bush will appear in front of the house in the form of two intersecting triangles.

Next, you will make two copies of the bush. Use the AutoCAD COPY command to place one copy 5 feet (60 units) in the X direction, and a second copy 10 feet (120 units) in the X direction. Because of the Perspective mode, the best way to make these copies is to enter relative displacements on the command line.

Rendering with landscape objects and shadows is often slow, so we will place the tree before making a rendering. Start the LSNEW command again, and highlight the object named Norway Maple, Fall (or its equivalent name if your landscape library is different). Set the geometry to Crossing Faces and make certain that the View Aligned checkbox is active. Set the tree's Height to 360 (30 feet). You will have to enter this value directly into the edit box because the slider is limited to a maximum height of 100. Click the position button, and enter the coordinates of 53',12',-1'4 (or 636,144,-16).

If you look at the landscape objects from an isometric-type viewpoint, such as in Figure 9.93, they will appear to be simply crude triangles containing the object's name and having the height specified by the LSNEW command. Nevertheless, they contain the data needed for them to look like three bushes and one tree when they are rendered.

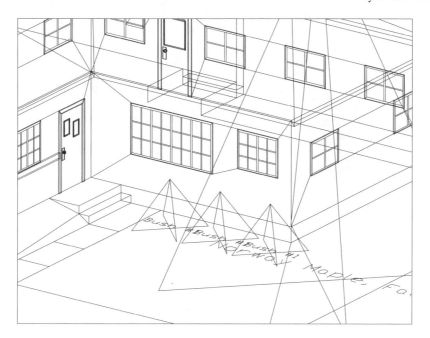

Figure 9.93

Make certain that the view 50_FOOT_PERSPECTIVE is current and start the RENDER command. Click the Background button to bring up the Background dialog box. Click the Image radio button, and find and load the drawing file cloud.tga for the background image. Return to the Render dialog box. Either Photo Real or Photo Raytrace rendering can be used, but Photo Raytrace will probably be faster because volumetric shadows are being used. Furthermore, only Photo Raytrace rendering will enable shadow casting from landscape objects. Click the Render button in the dialog box.

Your rendered scene should look similar to Figure 9.94. Notice that the landscape objects cast shadows as if they were true 3D objects.

 The 3D house with the four landscape objects inserted is in file 3d_ch9_09.dwg on the accompanying CD-ROM.

Figure 9.94

THE FOG COMMAND

This command adds an extra color to objects in a rendering; the extent to which an object is colored depends on its distance from the camera. The purpose of this extra color is to give an illusion of distance or depth. If the color is light, such as white, the effect is similar to fog or haze; if it is dark, objects fade and become dimmer as their distance from the camera increases.

Fog settings are controlled through the Fog/Depth Cue dialog box shown in Figure 9.95. This dialog box is displayed when the FOG command is invoked, but it can also be accessed from the button labeled Fog/Depth Cue in the dialog boxes used by the RENDER and RPREF commands.

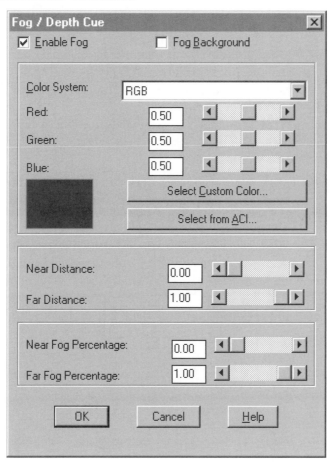

Figure 9.95

Fog is turned on and off by the checkbox labeled Enable Fog located near the top of the dialog box. The current settings remain in effect even when it is turned off, so that you do not have to reset them.

When the Fog Background checkbox is selected, the fog settings will also apply to the rendering background. This is useful if you are using an image as a background.

The cluster of options for color control the color of the fog. You can set each of the three color values independently, using either the RGB or the HLS color system. Clicking the Select Custom Color button will bring up a dialog box showing a wide range of colors you can use in selecting a color, and pressing the Select from ACI button will bring up the same Select Color dialog box used by the LAYER command. Pure white is a good choice for fog color, while pure black is good for having objects fade with distance.

The values established by the Near Distance and Far Distance edit boxes or their slider bars determine where the fog starts and ends. They represent the percentage of the distance from the camera to the back clipping plane. You will recall that clipping planes are established by the DVIEW command, which we discussed in Chapter 4. Although you can use the CAMERA and 3DCLIP commands to establish clipping planes, you will probably prefer to use DVIEW, because it allows you to more precisely set target and clipping plane locations.

The Near Fog Percentage and Far Fog Percentage values determine the percentage of fog at the near and far distances. You can set them with the slider bars or by entering a number ranging from 0 to 1 in the edit boxes.

TRY IT! - USING FOG

The 3D house we have been working with does not have enough relative distance to be suitable for demonstrating fog, so we will use an artificial example instead. Retrieve and open the file 3d_ch9_10.dwg. The objects in this file consist of 11 3D faces that are perpendicular to the line of sight, as shown in Figure 9.96. The 3D faces are evenly spaced in the Y direction, and each face is numbered so that you can identify its location.

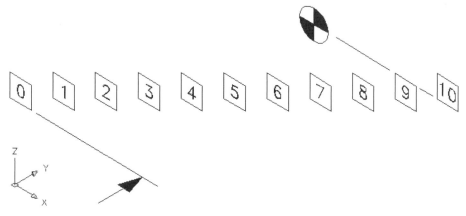

Figure 9.96

The camera point is even with the first 3D face, the target point is even with the last 3D face, and a back clipping plane has been established at the target. Figure 9–96 shows this set-up, but when you open the drawing you will simply see a row of 11 numbered squares. Also, the AutoCAD graphics area will be divided into four horizontal viewports, as shown in Figure 9.97. The view direction coordinates in each viewport are 0,-1,0. You should leave the screen divided into four viewports, and you should not change their view directions.

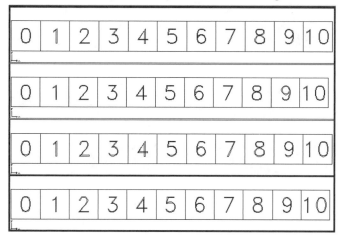

Figure 9.97

The exercise will consist simply of making renderings within each viewport using various fog settings and visually comparing the results. You should use the Fog/Depth Cue button in the Render dialog box to set the fog parameters, rather than repeatedly invoking the FOG command. You should experiment on your own using a variety of fog color, distance, and percentage combinations. We suggest, that you start with a black fog color and set the fog parameters in each of the viewports as shown below.

First Viewport:

Near Distance: 0.00   Far Distance: 1.00

Near Fog Percentage: 0.00   Far Fog Percentage: 1.00

Second Viewport:

Near Distance: 0.00   Far Distance: 1.00

Near Fog Percentage: 0.50   Far Fog Percentage: 1.00

Third Viewport:

Near Distance: 0.50   Far Distance: 1.00

Near Fog Percentage: 0.00   Far Fog Percentage: 1.00

Fourth Viewport:

Near Distance: 0.00   Far Distance: 0.50

Near Fog Percentage: 0.00   Far Fog Percentage: 1.00

The results from these settings will be similar to those in Figure 9.98.

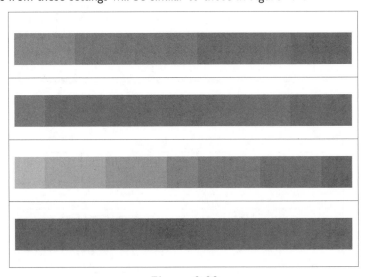

Figure 9.98

COMMAND REVIEW

DVIEW

The camera and back clipping plane locations, which are used by the **FOG** command, are established and controlled by **DVIEW**.

BACKGROUND

The Background dialog box can also be accessed through the **BACKGROUND** command.

FOG

The dialog box used by the Fog/Depth Cue button in the Render dialog box can be reached directly by invoking the **FOG** command.

LSEDIT

Existing landscape objects can be edited with this command.

LSNEW

The landscape objects of the current library will be displayed in the dialog box used by this command, and from there they can be inserted in a drawing.

LSLIB

The current landscape library, and consequently the landscape objects that appear in **LSNEW**'s list box, are controlled by the **LSLIB** command.

MATLIB

This command displays the Materials Library dialog box and a list of ready-made materials that can be imported into the current drawing.

RENDER

This command creates shaded images of 3D models and controls whether shadows will or will not be shown.

RMAT

This command creates, modifies, and attaches materials.

RPREF

The RPREF command uses a dialog box identical to the one used by **RENDER**, except it does not have a Render button.

SHOWMAT

This command lists the material attached to an object.

SETUV

The mapping of bitmap images to specific geometries and objects is accomplished by the **SETUV** command.

CHAPTER REVIEW

Directions: Fill in the blanks as indicated in each of the following.

1. Indicate which of the three rendering types—Render, Photo Real, or Photo Raytrace—supports each feature in the following list (some features in this list are supported by more than one rendering type).

    _____ a. Bitmap images

    _____ b. Light beam refraction

    _____ c. Reflected images

    _____ d. Shading

    _____ e. Shadows

    _____ f. Transparent materials

Directions: Answer the following questions.

2. What is the difference between a vector graphic file and a bitmap graphic file?

3. What is a scene?

4. What do the Query for Selections and the Crop Window buttons in the Render dialog box do?

5. List the methods AutoCAD provides for capturing a rendering to a file.

6. What constitutes the background of a rendering?

7. What is the relationship between surface orientation and light beams in a ren dering?

8. What is ambient light in a rendering?

9. Attenuation is the falloff in intensity of a light. List the light types that support attenuation. Which attenuation type falls off faster, inverse linear or inverse square?

10. How do you make rendered shadows that have sharp edges?

11. When aiming distant lights, what is the normal direction for north?

12. What is the difference between the hotspot and falloff cones of a spotlight?

13. What is refracted light? Which AutoCAD rendering types can handle refracted light? What is the angle of refraction, and which AutoCAD rendering type can account for it?

14. Which AutoCAD material attribute controls the relative size of highlights on reflective surfaces?

15. List the four types of material maps and the material attributes that are used with them.

16.   What are the results from setting the scale of a bitmap image in a material map to 2.0 in both the U and V directions?

17.   What is the difference between tiled and cropped bitmap images? What condition is necessary for tiling and cropping to have an effect?

18.   Name the three template materials. How do they differ from material maps?

19.   List the three different ways that a material can be attached to objects.

20.   Which material attribute and which checkbox in the New/Modify Standard Material dialog will cause a surface to reflect surrounding objects? Which of the three rendering types must be used for the surface to be reflective?

21.   What is the purpose of a material library? How do you transfer a material from a library into the current drawing file so that it can be attached to an object?

22.   When working with materials that incorporate bitmaps, why is the SETUV command needed?

23.   How do landscape objects differ from materials?

24.   What are the two types of bitmap files needed to make a landscape object? How do the two bitmaps differ?

25.   What do the fog near and far distances control, and what do their values represent?

Directions: Match the following items as indicated.

26.   Match the background type in the list on the left with it characteristic in the list on the right.

      _____ a. Gradient   1.   Blends two or three colors across the back ground.

      _____ b. Image   2.   Colors the background with a single color.

      _____ c. Merge   3.   Displays a bitmap file in the background.

      _____ d. Solid   4.   Does not clear the screen prior to render ing.

27.   Match the light type in the list on the left with the light beam properties in the list on the right.

      _____ a. Distant light   1.   Light beams are parallel to one another and travel in one specific direction. No target is used.

      _____ b. Point light   2.   Light beams radiate in all directions from a single source.

      _____ c. Spotlight   3.   Light beams travel in a cone-shaped pattern toward a target.

Directions: Circle the letter corresponding to the correct response in each of the following.

28.   In renderings, surfaces reflect light but do not block it.

   a. true

   b. false

29.   If a color is assigned to a light, the light will be rendered in that color. A red point light, for example, will show up in a rendering as a red spot of light.

   a. true

   b. false

30.   Distant lights will illuminate objects that are behind the light source.

   a. true

   b. false

31.   Landscape objects can be inserted so that they always face the current view direction.

   a. true

   b. false

32.   Landscape objects have the ability to cast shadows.

   a. true

   b. false